AF597967

NEUROMETHODS

Series Editor
Wolfgang Walz
University of Saskatchewan
Saskatoon, SK, Canada

For further volumes:
http://www.springer.com/series/7657

Expression Profiling in Neuroscience

Edited by

Yannis Karamanos

Laboratoire de Physiopathologie de la Barrière Hémato-encéphalique, Université d'Artois, Lens, France

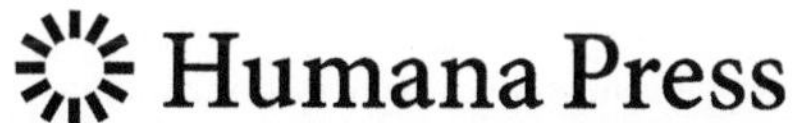

Editor
Yannis Karamanos
Laboratoire de Physiopathologie de la
Barrière Hémato-encéphalique
Université d'Artois - Faculté des Sciences
rue Jean Souvraz SP18, 62307, Lens, France
yannis.karamanos@univ-artois.fr

ISSN 0893-2336 e-ISSN 1940-6045
ISBN 978-1-61779-447-6 e-ISBN 978-1-61779-448-3
DOI 10.1007/978-1-61779-448-3
Springer New York Dordrecht Heidelberg London

Library of Congress Control Number: 2011944335

Printed on acid-free paper

Humana Press is part of Springer Science+Business Media (www.springer.com)

Preface to the Series

Under the guidance of its founders Alan Boulton and Glen Baker, the Neuromethods series by Humana Press has been very successful since the first volume appeared in 1985. In about 17 years, 37 volumes have been published. In 2006, Springer Science + Business Media made a renewed commitment to this series. The new program will focus on methods that are either unique to the nervous system and excitable cells or which need special consideration to be applied to the neurosciences. The program will strike a balance between recent and exciting developments like those concerning new animal models of disease, imaging, in vivo methods, and more established techniques. These include immunocytochemistry and electrophysiological technologies. New trainees in neurosciences still need a sound footing in these older methods in order to apply a critical approach to their results. The careful application of methods is probably the most important step in the process of scientific inquiry. In the past, new methodologies led the way in developing new disciplines in the biological and medical sciences. For example, Physiology emerged out of Anatomy in the nineteenth century by harnessing new methods based on the newly discovered phenomenon of electricity. Nowadays, the relationships between disciplines and methods are more complex. Methods are now widely shared between disciplines and research areas. New developments in electronic publishing also make it possible for scientists to download chapters or protocols selectively within a very short time of encountering them. This new approach has been taken into account in the design of individual volumes and chapters in this series.

Wolfgang Walz

Preface

Transcriptomics and proteomics, studying the profile of the expression of nucleic acids and proteins respectively, are increasingly applied to gain a mechanistic insight into a wide spectrum of investigation. The use of expression profiling studies for the central nervous system and brain function helps researchers to understand neurodegenerative disorders and tumor development mechanisms. Those approaches also have diagnostic potential, revealing biomarkers, especially in the field of primary brain tumors. In the past, the analysis of the transcriptome and/or the proteome was restricted to dedicated scientists with specific skills. The technology and the instruments evolved and are now accessible to every scientist. The challenge became the sample preparation, which is the checkpoint between biological conditions and relevant results. This volume, dedicated to gene and protein expression profiling, provides a survey of the most commonly used approaches in the field that include corresponding background information, tested laboratory protocols, and step-by-step methods for reproducible laboratory experiments. The information enclosed scans the different ways of studying the central nervous system/brain environment through expression profiling. The first part of this book addresses the gene expression profiling of the brain at a large scale or to a specific cell type such as blood–brain barrier endothelium. Then, the second part describes the protein expression studies and the different technologies applied. Two reviews, one on the past and the perspectives of MALDI imaging technology and one on the use of new nanotechnology applications to perform single cell analysis, complete the presentation and suggest new directions for future investigations.

Lens, France *Yannis Karamanos*

Preface

Contents

Preface to the Series . . . . . *v*
Preface . . . . . *vii*
Contributors . . . . . *xi*

1 DNA Microarrays for Gene Expression Analysis in Brain Tissue and Cell Lines 1
Gustavo A. Barisone and Elva Díaz

2 Gene Expression Profiling Using the Terminal Continuation RNA Amplification Method for Small Input Samples in Neuroscience 21
Stephen D. Ginsberg, Melissa J. Alldred, and Shaoli Che

3 Expression Profiling in Brain Disorders 35
Peter J. Gebicke-Haerter

4 Endothelial Cell Heterogeneity of Blood–Brain Barrier Gene Expression: Analysis by LCM/qRT-PCR 63
Tyler Demarest, Nivetha Murugesan, Jennifer A. Macdonald, and Joel S. Pachter

5 Gene Expression Profiling Using 3′ Tag Digital Approach 77
Yan W. Asmann, E. Aubrey Thompson, and Jean-Pierre A. Kocher

6 Sharing Expression Profiling Data with Gemma 89
Anton Zoubarev and Paul Pavlidis

7 Two-Dimensional Protein Analysis of Neural Stem Cells 101
Martin H. Maurer

8 iTRAQ Proteomics Profiling of Regulatory Proteins During Oligodendrocyte Differentiation 119
Mohit Raja Jain, Tong Liu, Teresa L. Wood, and Hong Li

9 Protein Profiling of the Brain: Proteomics of Isolated Tissues and Cells 139
Nicole Haverland and Pawel Ciborowski

10 The Proteome of Brain Capillary Endothelial Cells: Toward a Molecular Characterization of an In Vitro Blood–Brain Barrier Model 161
Sophie Duban-Deweer, Christophe Flahaut, and Yannis Karamanos

11 MALDI Imaging Technology Application in Neurosciences: From History to Perspectives 181
Michel Salzet, Céline Mériaux, Julien Franck, Maxence Wistorski, and Isabelle Fournier

12 Profiling of HIV Proteins in Cerebrospinal Fluid 225
Melinda Wojtkiewicz and Pawel Ciborowski

13 Proteomic Profiling of Cerebrospinal Fluid 245
Gwenael Pottiez and Pawel Ciborowski

14 New Nanotechnology Applications in Single Cell Analysis: Why and How? 271
Gradimir N. Misevic, Gerard BenAssayag, Bernard Rasser, Philippe Sales, Jovana Simic-Krstic, Nikola Misevic, and Octavian Popescu

Index *283*

Contributors

MELISSA J. ALLDRED • *Center for Dementia Research, Nathan Kline Institute, Orangeburg, NY, USA; Department of Psychiatry, New York University Langone Medical Center, Orangeburg, NY, USA*
YAN W. ASMANN • *Division of Biomedical Statistics and Informatics, Department of Health Sciences Research, Mayo Clinic Comprehensive Cancer Center, Rochester, MN, USA*
GUSTAVO A. BARISONE • *Department of Pharmacology, UC Davis School of Medicine, Davis, CA, USA*
GERARD BENASSAYAG • *Centre d'Elaboration de Matériaux et d'Etudes Structurales, CNRS, Toulouse, France*
SHAOLI CHE • *Center for Dementia Research, Nathan Kline Institute, Orangeburg, NY, USA; Department of Psychiatry, New York University Langone Medical Center, Orangeburg, NY, USA*
PAWEL CIBOROWSKI • *Mass Spectrometry and Proteomics Core Facility, Durham Research Center, University of Nebraska Medical Center, Omaha, NE, USA*
TYLER DEMAREST • *Blood–Brain Barrier Laboratory, Department of Cell Biology, University of Connecticut Health Center, Farmington, CT, USA*
ELVA DÍAZ • *Department of Pharmacology, UC Davis School of Medicine, Davis, CA, USA*
SOPHIE DUBAN-DEWEER • *Laboratoire de physiopathologie de la barrière hémato-encéphalique, Faculté des Sciences Jean Perrin, Université d'Artois, Lens, France*
CHRISTOPHE FLAHAUT • *Laboratoire de physiopathologie de la barrière hémato-encéphalique, Faculté des Sciences Jean Perrin, Université d'Artois, Lens, France*
ISABELLE FOURNIER • *MALDI Imaging Team, Laboratoire de Neuroimmunologie et Neurochimie Evolutives, CNRS FRE 3249, University of Lille1, Villeneuve d'Ascq, France*
JULIEN FRANCK • *MALDI Imaging Team, Laboratoire de Neuroimmunologie et Neurochimie Evolutives, CNRS FRE 3249, University of Lille1, Villeneuve d'Ascq, France*
PETER J. GEBICKE-HAERTER • *Department of Psychopharmacology, Central Institute for Mental Health, Mannheim, Germany*
STEPHEN D. GINSBERG • *Center for Dementia Research, Nathan Kline Institute, Orangeburg, NY, USA; Departments of Psychiatry and Physiology & Neuroscience, New York University Langone Medical Center, Orangeburg, NY, USA*

NICOLE HAVERLAND • *University of Nebraska Medical Center, Omaha, NE, USA*
MOHIT RAJA JAIN • *Center for Advanced Proteomics Research and Department of Biochemistry and Molecular Biology, New Jersey Medical School Cancer Center, University of Medicine and Dentistry of New Jersey, Newark, NJ, USA*
YANNIS KARAMANOS • *Laboratoire de Physiopathologie de la Barrière Hémato-encéphalique, Université d'Artois - Faculté des Sciences, Lens, France*
JEAN-PIERRE A. KOCHER • *Division of Biomedical Statistics and Informatics, Department of Health Sciences Research, Mayo Clinic Comprehensive Cancer Center, Rochester, MN, USA*
HONG LI • *Center for Advanced Proteomics Research and Department of Biochemistry and Molecular Biology, New Jersey Medical School Cancer Center, University of Medicine and Dentistry of New Jersey, Newark, NJ, USA*
TONG LIU • *Center for Advanced Proteomics Research and Department of Biochemistry and Molecular Biology, New Jersey Medical School Cancer Center, University of Medicine and Dentistry of New Jersey, Newark, NJ, USA*
JENNIFER A. MACDONALD • *Blood–Brain Barrier Laboratory, Department of Cell Biology, University of Connecticut Health Center, Farmington, CT, USA*
MARTIN H. MAURER • *Department of Physiology, University of Heidelberg, Heidelberg, Germany*
CÉLINE MÉRIAUX • *MALDI Imaging Team, Laboratoire de Neuroimmunologie et Neurochimie Evolutives, CNRS FRE 3249, University of Lille1, Villeneuve d'Ascq, France*
GRADIMIR N. MISEVIC • *Gimmune GmbH, Zug, Switzerland*
NIKOLA MISEVIC • *Institute of Brain-Research, University of Bremen, Bremen, Germany*
NIVETHA MURUGESAN • *Blood–Brain Barrier Laboratory, Department of Cell Biology, University of Connecticut Health Center, Farmington, CT, USA*
JOEL S. PACHTER • *Blood–Brain Barrier Laboratory, Department of Cell Biology, University of Connecticut Health Center, Farmington, CT, USA*
PAUL PAVLIDIS • *Department of Psychiatry, Centre for High-Throughput Biology, University of British Columbia, Vancouver, BC, Canada*
OCTAVIAN POPESCU • *Molecular Biology Center and Institute for Interdisciplinary Experimental Research, Babes-Bolyai University, Cluj-Napoca, Romania; Institute of Biology, Romanian Academy, Bucharest, Romania*
GWENAEL POTTIEZ • *University of Nebraska Medical Center, Omaha, NE, USA*
BERNARD RASSER • *Orsay Physics, Fuveau, France*
PHILIPPE SALES • *Centre d'Elaboration de Matériaux et d'Etudes Structurales, CNRS, Toulouse, France*
MICHEL SALZET • *MALDI Imaging Team, Laboratoire de Neuroimmunologie et Neurochimie Evolutives, CNRS FRE 3249, University of Lille1, Villeneuve d'Ascq, France*
JOVANA SIMIC-KRSTIC • *Department of Biomedical Engineering, Faculty of Mechanical Engineering, University of Belgrade, Belgrade, Serbia*
E. AUBREY THOMPSON • *Department of Cancer Biology, Mayo Clinic Comprehensive Cancer Center, Rochester, MN, USA*

MAXENCE WISTORSKI • *MALDI Imaging Team, Laboratoire de Neuroimmunologie et Neurochimie Evolutives, CNRS FRE 3249, University of Lille1, Villeneuve d'Ascq, France*
MELINDA WOJTKIEWICZ • *University of Nebraska Medical Center, Omaha, NE, USA*
TERESA L. WOOD • *Department of Neurology and Neuroscience, New Jersey Medical School Cancer Center, University of Medicine and Dentistry of New Jersey, Newark, NJ, USA*
ANTON ZOUBAREV • *Department of Psychiatry, Centre for High-Throughput Biology, University of British Columbia, Vancouver, BC, Canada*

Chapter 1

DNA Microarrays for Gene Expression Analysis in Brain Tissue and Cell Lines

Gustavo A. Barisone and Elva Díaz

Abstract

Microarray expression profiling of the nervous system provides a powerful approach to ascribe activities to genes involved in distinct phases of neural development and function. Expression profiling of neural tissues and cell lines requires isolation of high-quality RNA, amplification of the isolated RNA, and hybridization to DNA microarrays. In this chapter, theoretical background for expression profiling as well as protocols for reproducible microarray experiments from brain tissue and cell lines derived from brain tissue will be presented in the context of neural proliferation and tumorigenesis in the cerebellum.

Key words: Cerebellum, Expression profiling, Medulloblastoma, DNA microarrays, Neural proliferation, Tumorigenesis, Sonic hedgehog

1. Introduction

1.1. Sonic Hedgehog Signaling, Cerebellar Proliferation, and Medulloblastoma

In the cerebellum, Sonic hedgehog (Shh) is secreted by Purkinje neurons to regulate granule neuron precursor (GNP) proliferation during postnatal development (1) via Nmyc (2–4). Shh binding to its receptor Patched (Ptc) relieves Ptc-mediated inhibition of Smoothened (Smo) signaling (5). Medulloblastoma, the most common pediatric brain tumor (6), is thought to arise from cerebellar GNPs in which the Shh pathway is over-activated (7). Although advances in treatment of medulloblastoma have resulted in a significantly better outcome for many patients (8), about one-third of the patients with medulloblastoma present progression or recurrence of the tumor within 5 years of the initial diagnosis, even after complete surgical resection. Thus, a better understanding of the molecular mechanisms that underlie growth and recurrence of this tumor will lead to improved therapies and novel targets to treat this disease.

Yannis Karamanos (ed.), *Expression Profiling in Neuroscience*, Neuromethods, vol. 64,
DOI 10.1007/978-1-61779-448-3_1,

In previous studies, we have shown that the Mad family member Mxd3 is a critical regulator of cerebellar GNP proliferation and Nmyc expression (9). Mxd3 is a member of the Myc/Max/Mad family of basic helix–loop–helix leucine zipper (bHLHZ) transcriptional regulators (10) to which Nmyc also belongs. Intriguingly, Mxd3 is upregulated in preneoplastic cells and tumors derived from *ptc* heterozygous mice (9), a widely used mouse model of medulloblastoma (11), as well as human tumor samples (12), suggesting that inappropriate activation of Mxd3 might play a role in tumor progression. We hypothesize that misregulation of Mxd3 in cerebellar GNPs causes alterations in gene expression that ultimately lead to medulloblastoma development. However, the transcriptional targets of Mxd3 in cerebellar GNPs or tumors are at present unknown.

Expression profiling with DNA microarray technology provides a straightforward approach to identify differentially expressed genes during tumor progression that might be regulated by Mxd3. To illustrate this approach, in this chapter we provide details for the analysis of gene expression in primary brain tumors and cell lines derived from primary brain tumors.

1.2. Principle of DNA Microarray Expression Profiling

A DNA microarray is a multiplex technology that consists of a miniature arrayed series of thousands of unique sequences of nucleic acids called features. Features are probed by hybridization to fluorescently labeled target sequences derived from a biological sample of interest. Probe–target hybridization is detected by quantification of labeled targets using a microarray scanner. A typical microarray consists of tens of thousands of unique features allowing high-throughput genome-wide analysis of probe–target hybridization.

The core principle behind microarrays is hybridization between two strands of complementary nucleic acid sequences by the formation of hydrogen bonds between complementary base pairs. A high number of complementary base pairs in a nucleotide sequence results in higher noncovalent bonding between the two strands. Subsequent washing of microarrays leads to removal of nonspecific hybridization such that only strongly paired sequences will remain hybridized. Thus, fluorescently labeled target sequences that bind strongly to a probe sequence will generate a signal that depends on the strength of the hybridization determined by the number of paired bases, the hybridization conditions (such as temperature), and washing after hybridization. The total strength of the signal, from a feature, depends upon the amount of target sample binding to the probes present on that spot. Microarrays use relative quantification in which the intensity of a feature is compared to the intensity of the same feature under a different condition, and the identity of the feature is known by its position.

In this chapter, commercially available DNA microarrays from Agilent are used to measure gene expression differences in primary

Step		Protocol	Approx. time	Stop point?	See additional
Isolate RNA	from tissue	**3.1.1**	4 h	yes	*notes* **1-3, 5-8**
	from cells	**3.1.2**	1 h	yes	*notes* **1-3, 5, 10**
⬇					
Perform QC of RNA sample		**3.1.3**	1.5 h	yes	*notes* **9, 11**; *fig.* **2**
⬇					
Amplify RNA		**3.2.1** (steps 1-12)	3 h	no	*note* **4**
⬇					
Label RNA		**3.2.1** (steps 13-21)	3.5 h	yes*	*note* **12**
⬇					
Prepare hybridization mixture		**3.2.2** (steps 1-6)	40 min	no	*note* **13**
⬇					
Load sample		**3.2.2** (steps 7-10)	10-20 min	no	*notes* **14, 15, 18**
⬇					
Hybridize microarray		**3.2.2** (steps 11-13)	17 h	no	*notes* **4, 16, 19**
⬇					
Wash microarray		**3.2.2** (steps 14-29)	5-10 min	no	*notes* **17, 18**
⬇					
Scan microarray		-	0.5-1 h	yes	*notes* **20, 21**

Fig. 1. Flow chart of steps in expression profile analysis, from sample preparation to microarray scanning. Time is approximate, and estimated for 2–4 samples. Optional stop points are indicated as follows: (yes) stop and store samples at –80°C; (yes*) stop and store samples at –80°C for no more than 2 days; (no) do *not* stop.

brain tumors compared with brain tumor cell lines as an example of this approach. A typical gene expression profiling experiment entails the isolation of high-quality RNA samples from the tissues and/or cell lines of interest, amplification and labeling of the samples, and hybridization to DNA microarrays (Fig. 1).

2. Materials

2.1. RNA Isolation

The products used are listed below. All solutions and reagents must be RNase free (see Note 1). Products marked (*) are used for isolation of RNA from cell lines in Sect. 3.1.2.

1. Brain tissue samples dissected by standard procedures (see Note 2). Upon dissection, tissues (50–100 mg per sample) should be transferred to 1.5-ml microfuge tubes, snap-frozen in liquid nitrogen, and stored at –80°C until ready for purification.
2. 1.5-ml microfuge tubes.
3. Trizol reagent (Invitrogen, store at 4°C).
4. Barrier pipette tips.
5. Needles (22-1/2 gauge) and 3-ml syringes.
6. Refrigerated microfuge set to 4°C.
7. Nutator mixer (BD Diagnostics).
8. Chloroform.
9. Linear acrylamide (Ambion, store at –20°C).
10. Isopropanol.
11. Ice-cold 70% ethanol (store at –20°C).
12. 10 mM Tris buffered at pH 7.5.
13. 100 mM $MgCl_2$.
14. RNase-free DNase I.
15. Phase-lock gel tubes (5 PRIME).
16. Microfuge.
17. RNase-free water, see Note 3.
18. Acid phenol:chloroform (Fisher Bioreagents or equivalent commercial source).
19. 7.5 M ammonium acetate.
20. *Phosphate-buffered saline (PBS)
21. *0.05% Trypsin-EDTA
22. *RNeasy kit (Qiagen), including:
    (a) Mini spin columns
    (b) Buffer RLT
    (c) Buffer RW1
    (d) Buffer RPE.
23. *QIAshredder spin columns (Qiagen).
24. NanoDrop2000 (ThermoScientific).

2.2. Quality Control Analysis of RNA Isolation with Bioanalyzer

1. Bioanalyzer (Agilent), with nanochip priming station and nanochip vortexer.
2. RNA 6000 Nano Kit (Agilent cat. #5067-1511); includes size ladder, marker, gel matrix, dye, electrode cleaners, spin filters, and nanochips.
3. Control Total RNA (optional, used to assure assay accuracy).
4. Heat block set to 65°C.

5. RNaseZap (Ambion).
6. 1-ml syringes.

2.3. Amplification, Labeling, and Microarray Hybridization

1. 50–5,000 ng of total RNA or poly(A)$^+$ RNA in a volume no larger than 8.3 μl.
2. Agilent's Quick Amp Labeling Kit, two-color (cat. #5190-0444), Illumina TotalPrep RNA Amplification kit (Ambion cat. #IL1791) or equivalent.
3. Thermocycler, or water baths at 37, 40, 65, and 80°C.
4. 0.2-ml thin wall tubes.
5. Barrier pipette tips.
6. Vortex mixer.
7. RNase-free water, see Note 3.
8. Microfuge.
9. RNeasy kit (Qiagen).
10. NanoDrop2000 (ThermoScientific).
11. Agilent's Gene Expression Hybridization kit, including:
    (a) Blocking agent
    (b) Fragmentation buffer
    (c) Hybridization buffer
    (d) Wash buffers 1 and 2
    (e) Triton X-102
    (f) Slide rack and staining dishes.
12. Roche NimbleGen Hybridization System, including:
    (a) 4-bay or 12-bay hybridization unit with humid cover (cat. #05223652001 or #05223679001)
    (b) A4 mixer hybridization chambers
    (c) A4 assembly/disassembly tool
    (d) Gasket brayer
    (e) Port plugs and port plug insertion tool, or port stickers.
13. Agilent's Whole Human Genome (4x44) Oligo Microarrays (cat. # G4112F), see Note 4.
14. M-100 positive displacement pipette and tips.
15. One glass or plastic dish ("disassembly dish") of approximately 10×15 cm.
16. Two beakers big enough to hold the slide rack hanging at a safe distance from a magnetic bar.
17. Magnetic stirrer and bars.
18. Microarray dryer (such as NimbleGen's cat. # 05223636001) or centrifuge equipped with slide adaptor.

19. 50-ml conical tubes.
20. Argon gas or ozone-free workspace.
21. Microarray fluorescent scanner (such as Molecular Devices' GenePix 4000B or Agilent's G2565BA) connected to a computer with feature-extraction software (such as GenePix Pro 6.0).

3. Methods

3.1. RNA Isolation, see Note 5

3.1.1. RNA Extraction from Brain Tissue with Trizol Reagent

Brain tissues are rich in lipids, which can complicate the RNA extraction process and make it difficult to obtain pure RNA. This protocol was developed to generate high-quality RNA free of brain contaminants. We have found that this method, while fairly involved, provides consistent and reliable results for downstream microarray applications. A modification to make this protocol more streamlined is to combine the Trizol-based RNA isolation approach described below with the column-based purification protocol outlined in Sect. 3.1.2. In this case, at step 20 one would skip to Sect. 3.1.2 (see Note 6).

1. Homogenize 50–100 mg of brain tissue in 1 ml Trizol reagent by pippeting up and down at least 5 times with a 1-ml pipette tip.
2. Further homogenize the brain tissue in Trizol reagent by passing the sample slowly 5 times through a 22-1/2 gauge needle attached to a 3-ml syringe.
3. Incubate the homogenized sample for 5 min at room temperature on a Nutator mixer. Samples can be incubated longer (~15 min) if tissue chunks still remain after 5 min.
4. Spin sample at 11,000× *g* for 5 min in a refrigerated microfuge set to 4°C to pellet tissue that does not dissolve, and transfer supernatant to a new 1.5-ml microfuge tube.
5. Add 200 μl of chloroform per 1 ml Trizol reagent and shake by hand for 15 s to mix.
6. Incubate the sample for 5 min at room temperature and then spin the sample at 11,000× *g* in a refrigerated microfuge set to 4°C for 15 min to separate phases.
7. Transfer the upper (aqueous) phase to a new tube (~60% of original Trizol volume).
8. Add 0.5 volume of Trizol reagent (300 μl) and shake by hand to mix.
9. Repeat steps 5–7.
10. Add an equal volume of chloroform (~600 μl) and shake by hand to mix.

11. Spin the sample at 11,000× *g* for 15 min in a refrigerated microfuge set to 4°C and transfer the upper (aqueous) phase to a new tube.
12. Add 2 μl (10 μg) of linear acrylamide and vortex gently to mix. See Note 7.
13. Add 0.7 volume of isopropanol (~500 μl) and vortex gently to mix.
14. Incubate the sample at −20°C for 30 min to precipitate RNA.
15. Spin the sample at 11,000× *g* for 25 min in a refrigerated microfuge set to 4°C to pellet RNA precipitate.
16. Wash the pellet with 1 ml ice-cold 70% ethanol and spin at 7,500× *g* for 5 min in a refrigerated microfuge set to 4°C.
17. Remove wash by aspiration (see Note 8) and allow the pellet to air-dry briefly.
18. Resuspend the pellet in 44 μl of 10 mM Tris, pH 7.5.
19. Add 5 μl of 100 mM $MgCl_2$.
20. Add 1 μl of DNase I (RNase free) and incubate at 37°C for 15 min.
21. During the last minute of the above spin, prepare 1.5 ml phase-lock gel (PLG) tubes by spinning briefly at room temperature to pellet gel.
22. Add 100 μl of acid phenol:chloroform solution to PLG tubes.
23. After incubation at 37°C (step 20), add 50 μl of RNase-free water to tubes with RNA samples and transfer the entire volume to the prepared PLG tubes.
24. Mix the PLG tubes by hand and spin for 5 min at 14,000× *g* in a microfuge at room temperature.
25. Add 100 μl chloroform and repeat spin.
26. Remove the aqueous phase (~100 μl) to a new tube.
27. Add 2 μl (10 μg) of linear acrylamide (see Note 7) and 50 μl of 7.5 M ammonium acetate and vortex gently to mix.
28. Add 150 μl of isopropanol and vortex gently to precipitate RNA.
29. Incubate the sample at −20°C for 30 min to precipitate RNA.
30. Spin the sample at 11,000× *g* for 25 min in a refrigerated microfuge set to 4°C to pellet RNA precipitate.
31. Wash the pellet with 1 ml ice-cold 70% ethanol and spin at 7,500× *g* for 5 min in a refrigerated microfuge set to 4°C.
32. Remove wash by aspiration (see Note 8) and allow the pellet to air-dry briefly.
33. Resuspend the pellet in 11 μl of 10 mM Tris, pH 7.5, and incubate the sample at 37°C to aid resuspension.

34. Use 1 μl of sample for RNA quantification by NanoDrop2000. See Note 9 for interpretation of results.
35. Store RNA samples at −80°C.

3.1.2. RNA Extraction from Cell Lines with RNeasy Columns

Extracting RNA from cell lines has the technical advantage of a much easier homogenization step, which results in a faster processing time and usually higher, more reproducible RNA quality. In addition, cells can usually be grown in large quantities, making it possible to obtain very high RNA yields. In this section, we present a protocol modified from the RNeasy kit manufacturer's instructions, with the conditions that have consistently worked best in our hands.

1. Prepare fresh lysis reagent by adding 10 μl β-mercaptoethanol per 1 ml of buffer RLT. Do not store longer than 1 month.
2. Harvest cells. It is recommended to use no more than 1×10^7 cells per extraction. See Note 10.
   (a) For cells grown in suspension, pellet cells and remove the culture medium completely. Add 1 ml lysis reagent and resuspend cells by pipetting up and down 3–4 times using a P-1000 micropipettor with barrier tips. Transfer to a 1.5-ml microfuge tube and incubate at RT for 5 min.
   (b) For cells grown as a monolayer in tissue culture flasks, wash cells with PBS, add 0.05% Trypsin-EDTA, and incubate at 37°C until the cells detach (5–10 min). Add serum-containing medium to inactivate the trypsin, and pellet the cells by centrifugation at 200× *g* for 5 min at room temperature. Completely remove medium. Add 1 ml lysis reagent and resuspend cells by pipetting up and down 3–4 times using a P-1000 micropipettor with barrier tips. Transfer to a 1.5-ml microfuge tube and incubate at room temperature for 5 min.
   (c) For cells grown as a monolayer in tissue culture dishes, remove culture medium completely. Add lysis reagent (see suggested volumes below). Swirl the plate to make sure that the entire surface is covered. Collect cells with a cell-scraper. Transfer to a 1.5-ml microfuge tube and pipette up and down 3–4 times using a P-1000 micropipettor with barrier tips. Incubate at room temperature for 5 min.

Plate	Lysis reagent (ml)
6-Well plate, per well	0.6
100-mm dish	1
140-mm dish	2

3. To homogenize the sample, put a QIAshredder column (Qiagen) in a 2-ml collection tube. Pipette the sample into the column and centrifuge for 2 min at RT at maximum speed.
4. Pipette the sample into an RNeasy mini column. Centrifuge for 20 s at room temperature at maximum speed. Discard the flow-through.
5. Add 700 μl Buffer RW1 to the column. Centrifuge for 20 s at room temperature at maximum speed. Discard the flow-through and collection tube.
6. Transfer the RNeasy column into a fresh 2-ml collection tube.
7. Add 500 μl Buffer RPE. Centrifuge for 20 s at room temperature at maximum speed. Discard the flow-through.
8. Add another 500 μl Buffer RPE. Centrifuge for 2 min at room temperature at maximum speed. Discard the flow-through.
9. Transfer the RNeasy column to a fresh 2-ml collection tube. Centrifuge for 1 min at room temperature at maximum speed.
10. Transfer the RNeasy column to a 1.5-ml tube. Pipette 30–50 μl of elution solution directly onto the column membrane. Centrifuge for 1 min at room temperature at maximum speed to elute RNA.
11. Measure RNA concentration with NanoDrop2000 spectrophotometer. See Note 9 for interpretation of results.
12. Store RNA samples at −80°C.

3.1.3. Quality Control Analysis of RNA Isolation with Bioanalyzer

1. Pipette 550 μl of the gel matrix into a spin filter. Centrifuge at 1,500 × *g* for 10 min at room temperature. Aliquot the filtered gel (65 μl) and store at 4°C for up to 1 month.
2. Vortex the dye concentrate (previously equilibrated to RT) for 10 s and add 1 μl of dye to a 65-μl aliquot of filtered gel. Vortex well and centrifuge at 13,000 × *g* for 10 min at room temperature. Use within 1 day.
3. Position a chip on the priming station and follow the chip manufacturer's instructions to load the gel–dye mix.
4. Pipette 5 μl of the marker in each of the 12 sample wells and in the ladder well.
5. Pipette 1 μl of the prepared ladder in the ladder well.
6. Pipette 1 μl of each sample into sample wells.
7. Pipette 1 μl of marker in each unused sample well.
8. Vortex the chip for 1 min at 2,400 rpm.
9. Position the chip in the bioanalyzer and run within 5 min. See Note 11 for interpretation of results.

3.2. Amplification, Labeling, and Microarray Hybridization

3.2.1. Amplification and Labeling

Labeling of RNA samples for microarray analysis typically incorporates an amplification step based on T7 RNA polymerase (13). The method consists of generating double-stranded cDNA from the RNA sample by reverse transcription and then using that cDNA to generate cRNA by in vitro transcription in the presence of labeled ribonucleotides (typically biotinylated or fluorescent), therefore producing microgram quantities of labeled RNA ready for array hybridization. The choice of amplification/labeling methods depends on the subsequent microarray platform to be used. In this section, we describe the generation of fluorescently labeled RNA using Agilent's Quick Amp labeling kit.

1. Transfer 50–5,000 ng of RNA to a 1.5-ml tube, in a total volume of no more than 8.3 µl.
2. Add 1.2 µl of T7 primer.
3. Add RNase-free water to a final volume of 11.5 µl.
4. Incubate at 65°C for 10 min to denature. It is recommended to use a thermocycler with a hot lid to prevent condensation in the top of the tube.
5. Prewarm 5× first strand buffer to 80°C for 5 min. Vortex and spin. Keep at room temperature until ready to use in step 8.
6. After the 10 min incubation at 65°C, immediately transfer the sample tubes to ice for 5 min.
7. Prepare cDNA master mix. Mix components by flipping the tube 4 times (do not vortex) and briefly spin down. Per reaction, add:
   (a) 4 µl 5× first strand buffer
   (b) 2 µl 0.1 M DTT
   (c) 1 µl 10 mM dNTPs
   (d) 1 µl MMLV-RT enzyme
   (e) 0.5 µl RNase Out.
8. Briefly spin down the samples and store on ice.
9. Add 8.5 µl master mix per tube. Mix by pipetting up and down.
10. Incubate at 40°C for 2 h.
11. Incubate at 65°C for 10 min.
12. Incubate on ice for 5 min.
13. Prepare transcription master mix. Per reaction, add:
    (a) 15.3 µl nuclease-free water
    (b) 20 µl 4× transcription buffer
    (c) 6 µl 0.1 M DTT
    (d) 8 µl NTPs
    (e) 6.4 µl 50% PEG (prewarmed to 40°C and vortexed to ensure resuspension)

   (f) 0.5 μl RNase Out
   (g) 0.6 μl inorganic pyrophosphatase
   (h) 0.8 μl T7 RNA polymerase
   (i) 2.4 μl Cy3 or Cy5 CTP.
14. Briefly spin samples to collect contents at the bottom of the tube.
15. Add 60 μl Transcription Master Mix per sample.
16. Incubate at 40°C for 2 h.
17. Add 20 μl RNase-free water.
18. Purify labeled cRNA using Qiagen's RNeasy mini columns:
   (a) Add 350 μl buffer RLT
   (b) Add 250 μl 100% ethanol. Mix by pipetting
   (c) Proceed with purification as outlined in Sect. 3.1.2, step 4.
19. Quantify cRNA using NanoDrop2000 spectrophotometer in Microarray Measurement mode. Select RNA-40 as sample type.
20. Initialize the instrument and blank with water
21. Measure samples and print or record the following values:
   (a) Cy3 or Cy5 dye concentration (pmol/μl)
   (b) 260/280 nm ratio
   (c) cRNA concentration (ng/μl)
   (d) Determine yield (= concentration x volume) and specific activity (=1,000 × dye concentration/cRNA concentration). See Note 12 for interpretation of these parameters.

3.2.2. Microarray Hybridization

1. Prepare hybridization sample by mixing the two Cy-labeled cRNAs in a 1.5-ml tube as follows:
   (a) 825 ng Cy3-cRNA (usually the "treated" sample)
   (b) 825 ng Cy5-cRNA (usually the "reference" sample)
   (c) 11 μl 10× blocking agent
   (d) Bring volume to 52.8 μl with RNase-free water
   (e) 2.2 μl fragmentation buffer.
2. Mix gently but thoroughly by vortexing at low speed.
3. Incubate at 60°C for exactly 30 min. See Note 13.
4. Add 55 μl 2× hybridization buffer.
5. Mix by pipetting, taking care not to introduce bubbles. Do not vortex.
6. Centrifuge for 1 min at maximum speed to collect the content at the bottom of the tube. Place on ice. Immediately load sample on the array.

7. Prepare the hybridization chamber (see Note 14):
   (a) Place a microarray slide in the assembly/disassembly tool. The barcode side must be inserted first.
   (b) Open an A4 mixer and expose the adhesive. Hold the array and assembly/disassembly tool with your thumb and index finger so that the slide will not move. Place the A4 mixer on the slide, far end first, making sure it is correctly aligned to the tool. It is safe to slightly bend the mixer during this process.
   (c) Remove the slide/mixer assembly from the assembly/disassembly tool by pulling from the mixer. If the adhesion is correct, the slide/mixer should easily come as one piece.
   (d) Using the braying tool, remove all air bubbles trapped between the adhesive gasket and the slide. Take extra care to make sure the gasket is tightly glued to the slide all around the hybridization chamber. Small air bubbles can be seen more easily if the assembly is placed on a black surface during the process.
8. Using a positive displacement pipette, load 100 μl of the hybridization sample prepared in step 6. Position the tip on the sample port hole and push down to make sure that it sits tightly inside the port hole. Slowly dispense the sample. Watch the solution as it covers the array area, making sure you are dispensing smoothly and slowly enough so that air is not trapped in the array area. Never try to suck the sample back up. Once the entire sample has been dispensed, keep the tip on the sample port hole. Do not release the plunger. The sample will continue to flow and eventually reach the vent port on the opposite side of the array. Once liquid comes out of the vent port, quickly remove the pipette. Do not release the plunger until the tip is away from the slide.
9. Remove excess liquid at either port by briefly touching with absorbent paper. Avoid prolonged contact as this will draw sample out of the array area itself.
10. Cover the port holes with stickers or port plugs. See Note 15.
11. Place the array slide on the hybridization station (see Note 16), making sure that the mixer bladder holes are correctly positioned on the corresponding O-rings in the station. Click-lock the slide cover.
12. Set the station to mixing mode B.
13. Hybridize at 65°C for 17 h.
14. Add Triton X-102 to Wash Buffer 1 to a final concentration of 0.005%. Keep at room temperature. See Note 17.
15. Add Triton X-102 to Wash Buffer 2 to a final concentration of 0.005%. Keep at 37°C overnight.

16. At the end of the hybridization, fill the disassembly dish (i.e., the cover of a 1,000-μl tip box) with Wash Buffer 1. See Note 18.
17. Place a slide rack and a stir bar in slide staining dish #1. Fill with Wash buffer 1 to cover the slide rack completely. Place on a stir plate at room temperature.
18. Place a stir bar on slide staining dish #2 and place on a stir plate. Do not fill with buffer yet.
19. Remove the array slide from the hybridization station. Place it in the assembly/disassembly tool and submerge the whole assembly in the disassembly dish. See Note 19.
20. Use your thumb and index finger to hold one end of the assembly/disassembly tool and the slide on the opposite side so that it will not move during removal of the A4 mixer.
21. Using your other hand, carefully but steadily "peel" the A4 mixer off the slide. Be aware that the adhesive is quite strong and significant effort will be required, but the array slide will not break as long as it is kept fully inserted in the assembly/disassembly tool. Take special care not to scratch the array area. Discard the A4 mixer.
22. Quickly place the array slide in the rack in staining dish #1. Minimize exposure to the air.
23. If processing more than 1 slide, repeat steps 19–22 for up to 7 more slides. Do not wash more than 8 arrays at a time.
24. Stir the buffer in staining dish #1 for 1 min at medium speed.
25. Meanwhile, fill staining dish #2 with prewarmed Wash Buffer 2.
26. Transfer the slides to staining dish #2. Stir for 1 min.
27. Remove the slide rack from staining dish #2. Do this as if in slow motion, to minimize droplets on the slide. It should take about 10 s.
28. Dry the arrays by spinning for 2 min in an array dryer or a centrifuge. Alternatively, dry each of them individually by blowing argon gas over them.
29. If necessary, place the slides in an ozone-free environment for transportation. See Note 20.
30. Scan immediately to avoid signal loss due to exposure to environmental oxidants. See Note 21.

4. Notes

1. We do not find it necessary to purchase RNase-free certified disposable plasticware if the plasticware is from unopened containers or from containers in which care was taken to avoid

contamination. With regard to RNase-free solutions, we typically purchase RNase-free 1 M stock solutions from a commercial vendor such as Ambion and make the appropriate dilution with RNase-free water (see Note 2). Reagents such as isopropanol and chloroform should be from unopened containers designated exclusively for RNA work.

2. In our hands, the largest source of variability occurs during the dissection process. For example, we routinely identify genes encoding hemoglobin subunits as "differentially expressed" in brain samples due to variable amounts of blood that contaminate each tissue dissection. To combat this issue, it is imperative that at least three biological replicates are analyzed in any given experiment. Technical replicates (i.e., analysis of the same RNA sample on different microarrays) do not provide any usefulness with regard to the detection of false positives.
3. We have found that water purified by the Milli-Q ultrafiltration system (Millipore) is sufficient for these procedures. However, care must be taken to ensure that the BioPak point-of-use filter connected to the water purification system remains protected from RNases.
4. The choice of microarray will depend on the experiment being performed and the source of RNA material. We have found Agilent's 4×44K format to be the most cost-effective array platform with the greatest coverage, quality, and flexibility.
5. Most researchers are acutely aware of the risk of RNase contamination. The risk of contamination is typically from performing procedures that use RNases (e.g., plasmid purifications) in close proximity to the area in which RNA will be used. To avoid RNase contamination, before any procedure involving RNA, all workspaces and micropipettors are cleaned with Ambion's RNaseZap. Fresh, clean gloves must be worn throughout the procedure. We do not routinely find it necessary to treat the microcentrifuge tubes used for RNA if they are from unopened bags or from bags in which care was taken to avoid contaminating the tubes. We do recommend that gloves be worn when handling any reagents or reaction vessels.
6. One alternative is to skip to Sect. 3.1.2 and proceed with the RNA isolation with RNeasy columns. We favor the method using phase-lock gel tubes because the costs are lower compared to that with RNeasy spin columns, especially if a large number of samples are being analyzed.
7. We routinely include linear acrylamide as carrier to facilitate alcohol precipitation of nucleic acids even when the quantity of RNA is not limiting. The inclusion of linear acrylamide also helps to visualize the nucleic acid pellet.

8. To remove the supernatant without disruption of the pellet, we first use gentle suction to remove all but the last ~20 μl of solution. The tube is then spun briefly to bring any remaining drops of solution to the bottom of the tube. Then a fresh unattached pipette tip is used to collect the solution at the bottom of the tube by capillary action.
9. The Thermo Scientific NanoDrop2000 is a micro-volume spectrophotometer that requires sample volumes as small as 0.5 μl and provides a full spectral output of each sample analyzed. The ratio of the absorbance at 260 nm and 280 nm ($A_{260/280}$) is used to assess the purity of nucleic acids. For pure RNA, the $A_{260/280}$ is ~2. If the $A_{260/280}$ is less than 2, then this result indicates the presence of protein contaminants since proteins (in particular, the aromatic amino acid residues) absorb light at 280 nm. Contamination by phenol can lead to overestimation of RNA concentration. Phenol absorbs with a peak at 270 nm, and the $A_{260/270}$ for samples uncontaminated by phenol should be around 1.2. Absorption at 230 nm can be caused by other organic contaminants such as thiocyanates present in Trizol reagent. For a pure RNA sample, the $A_{260/230}$ should be around 2.
10. RNA collected directly from cultured cells following the protocol in this section is usually of high-enough quality for microarray-oriented downstream applications. In some cases, however (such as when the number of samples or complexity of the experiments makes it impossible to immediately process all the samples at once), it may be desirable to collect the sample in Trizol reagent for added stability. Proceed as follows: use Trizol instead of Lysis reagent, proceeding exactly as outlined in Sect. 3.1.2, step 2 (a, b, or c). Add 0.2 ml chloroform per 1 ml Trizol. Shake or vortex vigorously for at least 15 s. Incubate at room temperature for <5 min. Centrifuge at 12,000× g for 15 min at 4°C. Transfer the aqueous (upper) phase to a new 1.5-ml tube. Measure the volume. Add 1 vol of 70% ethanol. Mix by flipping the tube or pipetting (do not vortex). Pipette the sample into an RNeasy mini column and proceed as outlined in Sect. 3.1.2, step 4.
11. We recommend the use of Agilent's bioanalyzer instrument to assess RNA quality. This microfluidics-based system consists of an instrument (often available through campus core facilities) and a disposable "chip" for analysis of the RNA samples. The assay is inexpensive, reproducible, and fast: 12 samples can be run on 1 chip in approximately 40 min. RNA quality is reported as a quantitative parameter, the RNA Integrity Number (RIN), which allows quality comparison between samples. Moreover, the RIN can be used to validate sample quality thresholds for a particular downstream application. For example, we have found that RNA samples with RIN ≥ 5.2 provide good

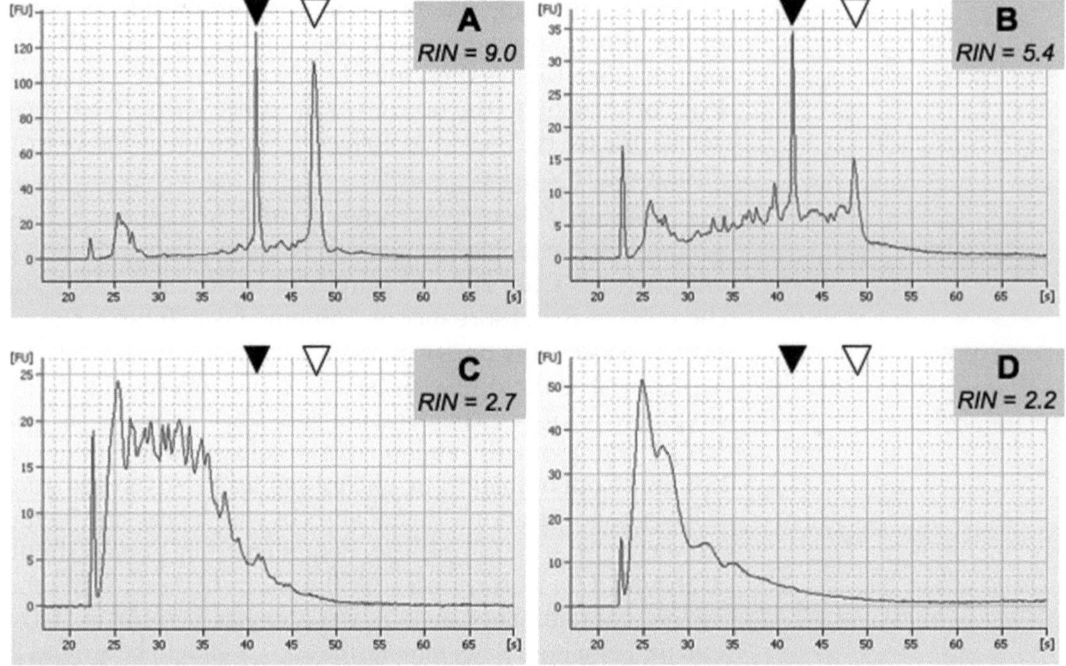

Fig. 2. Example of four RNA samples isolated from human brain tumor tissue. RNA was isolated using the protocol presented in Sect. 3.1.1. Sample quality was assessed using the bioanalyzer on an RNA 6000 chip as described in Sect. 3.1.3. Samples are representative of 4 distinct RNA qualities: (**a**) very good RNA quality (RIN > 9); (**b**) partially degraded RNA (RIN 5–6, note the significantly lower peak for 28S rRNA); (**c**) highly degraded RNA (RIN < 3); (**d**) almost completely degraded RNA (RIN = 2). We have successfully used samples with qualities similar to or better than **b** (RIN > 5.2) for downstream microarray applications. *Solid* and *open arrowheads* indicate the position of the 18S and 28S rRNA species, respectively.

quality, reliable microarray data (Fig. 2). Agilent maintains a "RIN database" (http://www.chem.agilent.com/rin/_rinsearch.aspx) which can be searched to validate one's results for a particular downstream application.

12. To proceed to hybridization, it is necessary to achieve a yield of ≥825 ng of RNA *and* a specific activity of ≥8 pmol/μg. It is recommended to repeat the cRNA preparation unless both conditions are met.

13. This is the cRNA fragmentation step, which ensures that labeled probes of the correct (average) size are generated from the amplified cRNA population for hybridization to the array. It is very important to incubate exactly as described. We recommend the use of a thermocycler to set the temperature accurately and a heated lid to prevent condensation. Longer incubation times, higher temperature, or increased salt concentration (resulting from condensation) will result in excessive fragmentation; too small cRNA fragments will be washed off the array, resulting in underestimation of the corresponding transcripts. Shorter incubation times or lower temperature will result in

insufficient fragmentation; long cRNA fragments may span more than one feature, resulting in nonspecific signal or false positives.

14. This protocol describes the use of Agilent's 4×44K arrays with the Roche NimbleGen hybridization system and A4 mixer chambers (formerly MAUI hybridization system). We have found that this provides the best quality results and highest reproducibility, probably due to the fact that the systems provide active mixing of a small volume of concentrated sample over the area of the microarray. When using other types of arrays, however, it is essential to check compatibility with the mixer chamber. After the recent corporate acquisition by Roche, the A4 mixers have been discontinued. As of the moment of this writing, they can still be found through other sellers such as Kreatech Diagnostics (http://www.kreatech.com; this website also offers a convenient array compatibility tool), but stock availability is uncertain. Other hybridization systems are available (for example, from Agilent); however, provided compatible mixers can be found for the array being used, we recommend Roche NimbleGen's system for its quality and reproducibility.

15. When using port plugs, avoid making excessive pressure with the insertion tool, as this will force sample out of the chamber through the opposite port. If using stickers, handle them with fine tweezers to minimize contact of the glued side with any surface before applying to the slide; this is essential to ensure that no leakage occurs during hybridization.

16. The hybridization station should be left to equilibrate at 65°C for at least 2 h before hybridization begins. Use a conventional thermometer to make sure that the temperature is accurate.

17. The use of Triton in the washing steps is optional but recommended since it reduces background and artifacts. Triton X-102 can be added to the whole bottle of buffer the first time it is opened. It is important to prewarm Wash Buffer 2 overnight; only prewarm the volume needed for the experiment in a sterile, RNase-free bottle.

18. Special care should be taken to ensure that all equipment used in the hybridization and wash are clean. It is recommended to use dedicated dishes. Wash all dishes, racks, and stir bars by placing the latter in the dishes and completely filling with Milli-Q water. Empty and fill again at least 5 times. Store dry equipment in a covered container or cabinet away from dust.

19. Before removing the array from the hybridization station, observe it carefully. Formation of air bubbles during hybridization is not uncommon. Note if the air bubbles are moving freely. If they are, they will not have any significant impact on the data quality. If they are not, the area of the array will probably not have been hybridized properly, and data from the features on that area will not be reliable.

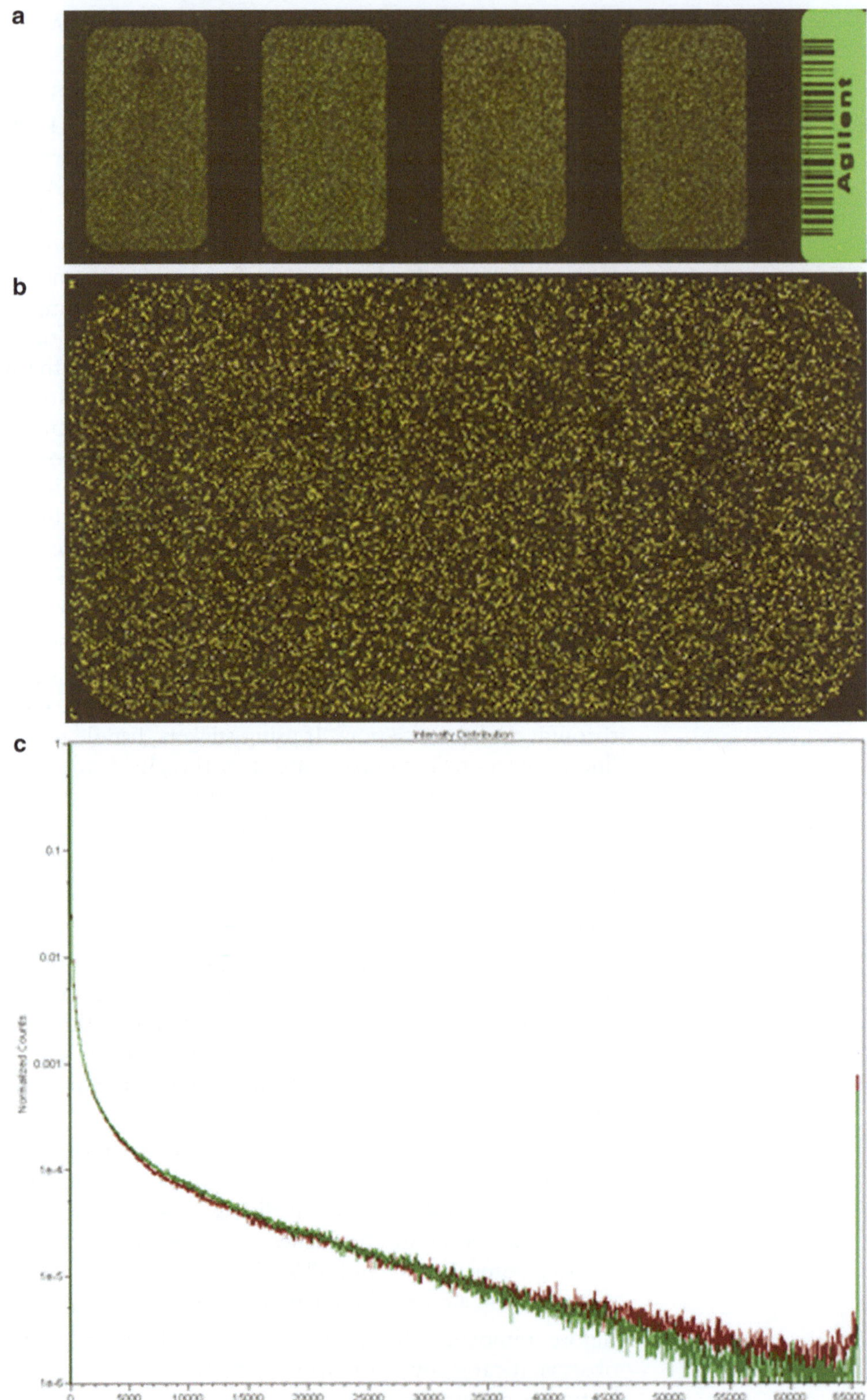

Fig. 3. Examples of array images and scatter plots. (**a**) Whole slide preview, showing the four arrays in the 4 × 44K Agilent format. (**b**) High-resolution scan of one of the array areas. (**c**) Scatter plot of the image in **b**, note that the signals fully overlap and no more than 1×10^{-6} features are at saturating intensity.

20. Cy5 dye is sensitive to ozone, especially in its excited state during scanning. Depending on the geographical location and time of the year, ozone levels can have a significant impact on the signal quality. To minimize exposure, it is recommended (although not essential) to (a) scan microarrays immediately after washing and drying; (b) transport microarrays to the scanner location in a 50-ml conical tube filled with argon gas, tightly closed and protected from light; and (c) house the scanner in an ozone-free enclosure, which can easily be built as described in http://cmgm.stanford.edu/pbrown/protocols/Ozone_Prevention.pdf.
21. Make sure that the scanner is on to avoid delays, as some models require warm-up time. Be familiar with the scanner and software. Scanner models differ significantly. If more than one option is available, evaluate the following parameters: (a) resolution and speed: 10 μm/pixel is acceptable for most arrays. 5 μm/pixel scanners will generate higher quality images at the cost of more than double the scanning time and file size; (b) laser/filter choices and compatible dyes; (c) scanning mechanism: simultaneous scanning usually results in shorter time than sequential channel scanning; (d) slide capacity: single-slide scanners require the user to stand by the instrument when more than one slide is to be scanned. Agilent's G2565BA and Molecular Devices GenePix 4200AL, for example, can load 48 and 36 slides, respectively, requiring no human attention during the lengthy (5–7 h) scan process; (e) real-time display of the image being acquired, which allows the user to pause–adjust–restart the instrument easily to find the optimal conditions (area, gain, and dynamic range) for each particular experiment; (f) maximum scan area and slide type; (g) additional features, such as laser output settings, focus adjust, calibration, and bar code reading capabilities; (h) feature extraction/analysis software: some applications require the reference sample to be labeled with a specific dye; for example, the software will assume by default that the reference is labeled with Cy5 and, therefore, calculate Cy3/Cy5 ratios. Although this can later be changed with any spreadsheet software, it may result in unnecessary complications and confusion if the wrong choice is made when labeling your samples. Examples of high-quality images and scatter plots are shown in Fig. 3.

Acknowledgments

This work was supported by grants to E. D. from the Alfred P. Sloan Research Foundation, the James S. McDonnell Foundation Twenty-first Century Award Program, the UC Davis Health System Research Award Program, an individual allocation of the

UC Davis Institutional Research Grant from the American Cancer Society, a Career Development Award from the UCSF Brain Tumor SPORE Program, and a NIH Director's New Innovator Award. G. A. B. was supported in part by a postdoctoral fellowship from the California Institute of Regenerative Medicine.

References

1. Wechsler-Reya, R. J., and Scott, M. P. (1999) Control of neuronal precursor proliferation in the cerebellum by Sonic Hedgehog, *Neuron* 22, 103–114.
2. Kenney, A. M., Cole, M. D., and Rowitch, D. H. (2003) Nmyc upregulation by sonic hedgehog signaling promotes proliferation in developing cerebellar granule neuron precursors, *Development (Cambridge, England)* 130, 15–28.
3. Knoepfler, P. S., Cheng, P. F., and Eisenman, R. N. (2002) N-myc is essential during neurogenesis for the rapid expansion of progenitor cell populations and the inhibition of neuronal differentiation, *Genes & development* 16, 2699–2712.
4. Oliver, T. G., Grasfeder, L. L., Carroll, A. L., Kaiser, C., Gillingham, C. L., Lin, S. M., Wickramasinghe, R., Scott, M. P., and Wechsler-Reya, R. J. (2003) Transcriptional profiling of the Sonic hedgehog response: a critical role for N-myc in proliferation of neuronal precursors, *Proceedings of the National Academy of Sciences of the United States of America* 100, 7331–7336.
5. Ingham, P. W., and McMahon, A. P. (2001) Hedgehog signaling in animal development: paradigms and principles, *Genes & development* 15, 3059–3087.
6. McNeil, D. E., Coté, T. R., Clegg, L., and Rorke, L. B. (2002) Incidence and trends in pediatric malignancies medulloblastoma/primitive neuroectodermal tumor: A SEER update, 39, 190–194.
7. Wechsler-Reya, R., and Scott, M. P. (2001) The developmental biology of brain tumors, *Annual review of neuroscience* 24, 385–428.
8. Brian, R. R., Tobey, J. M., and Roger, J. P. (2004) Current treatment of medulloblastoma: Recent advances and future challenges, *Seminars in oncology* 31, 666–675.
9. Yun, J. S., Rust, J. M., Ishimaru, T., and Diaz, E. (2007) A novel role of the Mad family member Mad3 in cerebellar granule neuron precursor proliferation, *Mol Cell Biol.* 27, 8178–89
10. Hurlin, P. J., Queva, C., Koskinen, P. J., Steingrimsson, E., Ayer, D. E., Copeland, N. G., Jenkins, N. A., and Eisenman, R. N. (1995) Mad3 and Mad4: novel Max-interacting transcriptional repressors that suppress c-myc dependent transformation and are expressed during neural and epidermal differentiation, *The EMBO journal* 14, 5646–5659.
11. Goodrich, L. V., Milenkovic, L., Higgins, K. M., and Scott, M. P. (1997) Altered neural cell fates and medulloblastoma in mouse patched mutants, *Science (New York, N.Y* 277, 1109–1113.
12. Barisone, G. A., Yun, J. S., and Diaz, E. (2008) From cerebellar proliferation to tumorigenesis: new insights into the role of Mad3, *Cell cycle (Georgetown, Tex* 7, 423–427.
13. Van Gelder, R. N., von Zastrow, M. E., Yool, A., Dement, W. C., Barchas, J. D., and Eberwine, J. H. (1990) Amplified RNA synthesized from limited quantities of heterogeneous cDNA, *Proceedings of the National Academy of Sciences of the United States of America* 87, 1663–1667.

Chapter 2

Gene Expression Profiling Using the Terminal Continuation RNA Amplification Method for Small Input Samples in Neuroscience

Stephen D. Ginsberg, Melissa J. Alldred, and Shaoli Che

Abstract

The process of RNA amplification is a stepwise series of molecular manipulations intended to amplify transcriptomic signals from small quantities of starting materials, including single cells and homogeneous populations of individual cell types for microarray analysis and other high-throughput downstream genetic approaches. An RNA amplification methodology named terminal continuation (TC) RNA amplification has been developed by our group to amplify RNA from small starting material inputs. In brief, an RNA synthesis promoter is attached to the 3′ and/or 5′ region of the isolated population of cDNAs utilizing the TC approach. Amplified RNAs are in either an antisense or a sense orientation depending on the placement of the T7 RNA polymerase promoter sequence. TC RNA amplification is utilized for many downstream applications including gene expression profiling and cDNA library/subtraction library construction, among others. Input sources of RNA can originate from a myriad of in vivo and in vitro tissue sources. Notably, fresh, frozen, and fixed tissues can be employed for TC RNA amplification, enabling precise and reproducible cell type and tissue specificity of downstream transcriptome-based assessments.

Key words: Functional genomics, In vitro transcription, Microarray, Postmortem human brain, RNA amplification, Transcriptome

1. Introduction

Standard molecular biology methods enable gene expression analysis using a plethora of experimental paradigms and model systems. Most of these methods evaluate the abundance of individual elements one at a time (or a few at a time), including in situ hybridization, Northern analysis, and ribonuclease (RNase) protection assay, among others. In contrast, high-throughput genomic methodologies allow the assessment of dozens to hundreds to thousands

Yannis Karamanos (ed.), *Expression Profiling in Neuroscience*, Neuromethods, vol. 64,
DOI 10.1007/978-1-61779-448-3_2,

of genes simultaneously in a coordinated manner (1, 2), but often require abundant, high-quality RNA for effective experimentation. Our research has focused upon microarray and related expression profiling of individual populations of neurons following experimental injury and age-related neurodegeneration (3–10), and has required the development of RNA amplification strategies that enable the use of small amounts of input RNA.

Nucleic acids for transcriptomic analyses can originate from a variety of in vivo and in vitro sources including plant and animal materials. Fresh, flash-frozen, and fixed tissues are potentially useful, depending on tissue quality and the experimental design of the specific study. Isolation of genomic DNA and total RNA is not only routinely performed using fresh and frozen tissues (11–13), but can also originate from paraffin-embedded fixed tissues (14–18). Use of paraffin-embedded tissue is a viable resource, as often fresh and/or frozen tissues are not accessible, whereas archived fixed human and animal model tissues exist in a myriad of laboratories and tissue repositories.

Although most emphasis lies in the downstream genomic output of microarrays and related high-throughput technologies, it is extremely important to note that RNA species are highly sensitive to RNase degradation. RNases are found in virtually every cell type (19), and RNase inhibition is a key parameter to consider for the success of RNA-based experimentation. RNases are quite stable and retain their activity over a broad pH range (19, 20). Therefore, RNase-free precautions must be employed during RNA extraction procedures and throughout RNA amplification methods.

To date, an optimal method to prepare brain tissues for downstream genetic analyses has yet to be agreed upon. A consensus protocol has not yet been promulgated. RNAs are routinely harvested from fixed tissues where RNA Integrity Number (RIN) values have been ascertained via bioanalysis (21, 22). RNA from tissue samples using cross-linking fixatives including 10% neutral buffered formalin and 4% paraformaldehyde as well as precipitating fixatives such as 70% ethanol buffered with 150 mM sodium chloride is another resource, although RNA quantity and quality are typically less than that of fresh and frozen samples (14, 23–27). Another method for assessing RNA quality in tissue sections prior to performing expression profiling studies is acridine orange (AO) histofluorescence. AO is a fluorescent dye that intercalates selectively into nucleic acids and has been used to detect RNA and DNA in brain tissues (10, 28–30). Importantly, individual RNA species (e.g., mRNA, rRNA, and tRNA) cannot be delineated by AO histofluorescence. A more thorough examination of RNA quality can be obtained via bioanalysis (e.g., 2100 Bioanalyzer, Agilent Technologies), which utilizes capillary gel electrophoresis to detect RNA quality and quantitate relative abundance (such as a RIN number) (5, 31, 32).

The advent of high-throughput microarray analysis has enabled significant progress in expression profiling studies in many disciplines, and microarray analysis has emerged as a useful tool to assess transcript levels in a myriad of systems and paradigms. A disadvantage of conventional microarray technology is a requirement for significant amounts of high-quality input sources of RNA for sensitivity and reproducibility. The quantity of RNA harvested from a single cell, estimated to be approximately 0.1–1.0 picograms, is not enough to employ standard RNA extraction procedures (11, 33). Therefore, molecular biological methods have been implemented to increase the amount of input RNA for downstream genetic analyses, including exponential PCR-based amplification and linear RNA amplification approaches. Unfortunately, PCR-based protocols are not optimal, as exponential amplification often can skew quantitative relationships between genes from the initial population (34, 35). In contrast, linear RNA amplification methods allow for the analysis of relative gene expression levels. A linear RNA amplification procedure typically entails generating quantities of RNA species through in vitro transcription (IVT) (31, 35–37). Hybrid PCR/linear RNA amplification methods (38, 39) as well as isothermal RNA amplification (40, 41) procedures have also been developed that generate faithful representation of original input RNA.

A variety of strategies have been developed to improve RNA amplification efficiency (38, 39, 42–44). An obstacle that hinders RNA amplification protocols is the difficulty with second-strand synthesis efficiency and specificity (35, 38, 39, 45). Technical improvements include obviating second-strand cDNA synthesis and enabling flexibility in placement of bacteriophage transcriptional promoter sequences for both antisense and sense amplification.

A procedure has been developed in our laboratory termed terminal continuation (TC) RNA amplification (31, 32, 35, 46–48) which satisfies these objectives (Fig. 1). TC RNA transcription can be performed using a promoter sequence (e.g., T7, T3, or SP6) attached to either the 3′ or the 5′ oligonucleotide primers. Therefore, transcript orientation can be in an antisense orientation when the bacteriophage promoter sequence is placed on the first-strand poly d(T) primer, or in a sense orientation when the promoter sequence is attached to the TC primer, depending upon the design of the experimental paradigm (Fig. 1) (32, 47–49). TC RNA amplification of genetic signals includes synthesizing first-strand cDNA complementary to the mRNA template, subsequently generating second-strand cDNA complementary to the first-strand cDNA (when antisense RNA is being generated; if sense RNA is being synthesized, the second-strand step is obviated) (47, 48), and finally performing IVT using the double-stranded cDNA as template (31, 32, 35, 46). First-strand cDNA synthesis complementary to the template mRNA entails the use of two oligonucleotide primers: a first-strand poly d(T) primer and a TC primer. The poly d(T) primer is similar to conventional

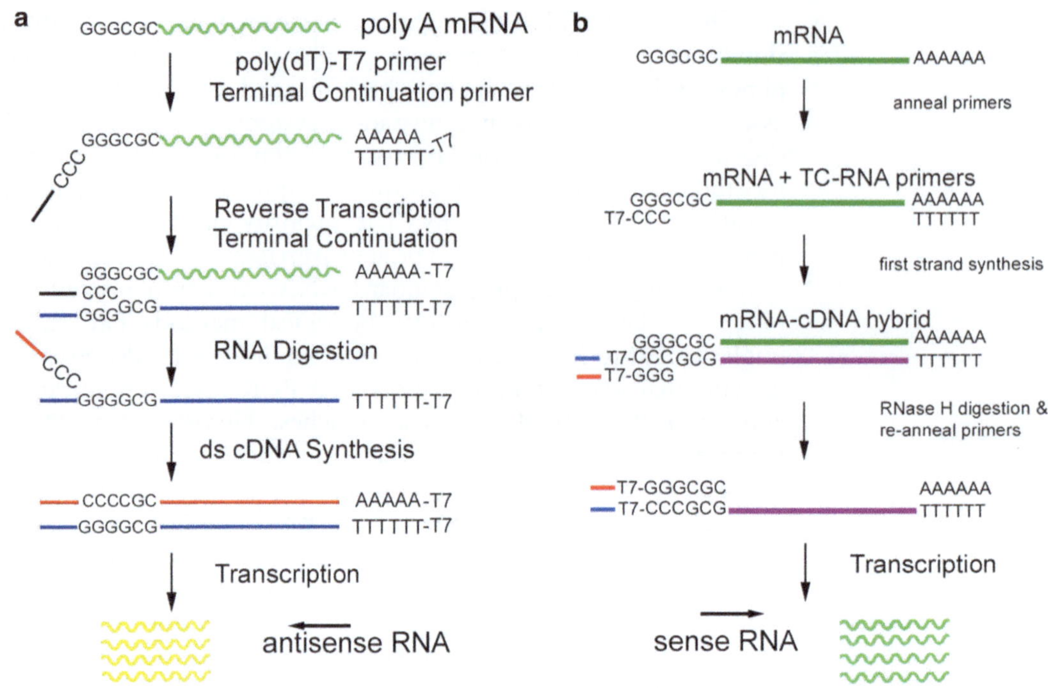

Fig. 1. Overview and analysis of the TC RNA amplification method. (**a**) Schematic representation of TC RNA amplification, antisense orientation. A poly d(T) primer (containing a bacteriophage promoter sequence for antisense orientation) and TC primer are added to the mRNA population to be amplified. First-strand synthesis that occurs as an mRNA–cDNA hybrid is formed after reverse transcription and terminal continuation of the oligonucleotide primers. Following RNase H digestion to remove the original mRNA template strand, second-strand synthesis is performed using Taq polymerase. The resultant double-stranded product is utilized as template for in vitro transcription, yielding high fidelity, linear RNA amplification of sense orientation (*rippled lines*). (**b**) Schematic representation of TC RNA amplification, sense orientation. mRNA to be amplified (*green line*) and the TC primer serve as templates for the first-strand synthesis, with poly d(T) acting as a primer. First-strand cDNA consists of three portions: the 5′ end comprised of the poly d(T), the mRNA complementary portion in the middle (*purple line*), and the 3′ end comprised of the TC primer complementary to the cDNA. The TC primer complementary sequence portion hybridizes with the TC primer present in the reaction and forms a double-stranded region without the need for further second-strand synthesis. Since the TC primer contains the T7 bacteriophage transcription promoter sequence, double-stranded TC primer regions provide a functional RNA synthesis promoter for IVT and subsequent robust RNA amplification. Adapted from (46, 48).

primers that exploit the poly(A) sequence present on the majority of mRNAs. The TC primer contains a span of three cytidine triphosphates (CTPs) or guanosine triphosphates (GTPs) at the 3′ terminus (31). Adenosine triphosphates (ATPs) or thymidine triphosphates (TTPs) do not perform well as constituents of the TC primer (5). As stated above, second-strand cDNA synthesis can be initiated for antisense RNA amplification (e.g., the T7 promoter sequence is on the poly d(T) primer) by annealing a second oligonucleotide primer complementary to the attached oligonucleotide (31), and is performed with robust DNA polymerases, such as Taq polymerase, or is avoided altogether when performing sense RNA amplification (e.g., when the T7 promoter sequence is on the TC primer) (47, 48). One round of amplification is sufficient for downstream genetic analyses (1, 31, 32, 46). The TC RNA amplification method is also being adjusted to enable amplification of noncoding RNAs including microRNAs (miRNAs) (50), which represents a new avenue for expression profiling studies that is just recently being recognized as a research endeavor. Below, this protocol describes the use of the TC RNA amplification procedure in detail for amplification of transcripts derived from small amounts of tissue, including neurons and other cellular populations microaspirated from tissues via laser capture microdissection (LCM) or related extraction procedures.

2. Materials

The methodology described herein is a step-by-step protocol for TC RNA amplification (47, 48). In terms of experimental orientation, this protocol begins at the point where cells are captured via LCM or a microaspiration strategy and follows through IVT using biotinylated, fluorescent, or radioactive methodologies to label the TC RNA amplified products. see Notes 1 and 2.

2.1. Isolation of RNA

1. Trizol reagent (Invitrogen, Carlsbad, CA) stored at 4°C.
2. Chloroform (Sigma, St. Louis, MO) stored at 22°C.
3. Isopropanol (Sigma) stored at 22°C.
4. 80% Ethanol (EtOH) stored at –20°C.
5. Linear acrylamide (5 mg/mL) (Applied Biosystems, Foster City, CA) stored at –20°C.

2.2. First-Strand Synthesis

1. First-strand synthesis primer {poly d(T); 100 ng/μl (IDT, Coralville, IA)} (Table 1).
2. Reverse transcription (RT) master mix: first-strand buffer (5×) (Invitrogen), dNTPs (10 mM) (Invitrogen), dithiothreitol (DTT) (0.1 M) (Sigma), and RNase inhibitor (Superase-In; Applied Biosystems; 20 U/μL) stored at –20°C.
3. Superscript III (Invitrogen, 200 U/μL) stored at –20°C.

2.3. Second-Strand Synthesis (For Antisense RNA Amplification Only)

1. 10× PCR buffer (Applied Biosystems; buffer includes 15 mM $MgCl_2$) stored at –20°C.
2. RNase H (Invitrogen, 10 U/μL) stored at –20°C.
3. Taq polymerase (Applied Biosystems; 5 U/μL) stored at –20°C.

2.4. Double-Stranded cDNA and cDNA/Primer Purification

1. 10,000 Molecular weight cut-off (MWCO) columns (Vivaspin 500; Sartorius Stedim, Goettingen, Germany).
2. 18.2 MΩ RNase-free water. see Note 1.

Table 1
Representative oligonucleotide sequences utilized for the poly d(T) and TC primers for the TC RNA amplification method

Antisense RNA orientation
First-strand synthesis primer (66 bp): 3′- AAA CGA CGG CCA GTG AAT TGT AAT ACG ACT CAC TAT AGG CGC TTT TTT TTT TTT TTT TTT TTT TTT -5′ *TC primer* (17 bp): 5′- TAT CAA CGC AGA GTC CC -3′
Sense RNA orientation
First-strand synthesis primer (18 bp): 3′- TTT TTT TTT TTT TTT TTT -5′ *TC-T7 primer* (51 bp): 5′- AAA CGA CGG CCA GTG AAT TGT AAT ACG ACT CAC TAT AGG CGC GAG AGG CCC -3′

2.5. IVT for TC RNA Amplification: Biotinylated/Fluorescent Probe Labeling

1. 10× Hybridization reaction buffer (Applied Biosystems) stored at –20°C.
2. 10× Biotin-labeled ribonucleotides (Enzo Life Sciences, Farmingdale, NY) stored at –20°C. see Note 3.
3. 10× RNase inhibitor mix (Enzo) stored at –20°C.
4. 10× DTT (Invitrogen) stored at –20°C.
5. T7 RNA polymerase (1,000 U/μL, Epicentre, Madison, WI) stored at –80°C.

2.6. IVT for TC RNA Amplification: Radioactive Probe Labeling

1. 5× Transcription reaction buffer (Epicentre) stored at –20°C.
2. 0.1 M DTT (Invitrogen) stored at –20°C.
3. 3NTPs (ATP, CTP, and GTP; 2.5 mM each) (Invitrogen) stored at –20°C.
4. UTP (100 μM) (Invitrogen) stored at –20°C.
5. RNase inhibitor (Superase-In) stored at –20°C.
6. ^{33}P-UTP (Perkin-Elmer, Boston, MA, 10 mCi/mL) stored at –80°C.
7. T7 RNA polymerase (1,000 U/μL, Epicentre).

3. Methods

3.1. Isolation of RNA

1. Add 500 μL of Trizol reagent to 0.7-mL thin-walled PCR tubes that will receive the microdissected regions/cells and/or profiles acquired via LCM, and keep on wet ice.
2. Invert the tubes so that Trizol reagent bathes microdissected material, and keep on wet ice. The samples can also be stored at –80°C at this juncture for future use.

3. Remove LCM cap and add 100 μL of chloroform to each sample. Vortex vigorously for 15 s and centrifuge the samples at 12,000 × *g*-force for 5 min at 4°C.
4. The mixture separates into an upper aqueous phase (clear) and a lower organic phase (red). Collect the aqueous phase containing RNA by aspirating with a pipette.
5. Add 250 μL of 100% isopropyl alcohol and 2 μL of linear acrylamide to precipitate the RNA from the aqueous phase. The samples can be stored at −80°C to precipitate RNAs if desired.
6. Mix vigorously and centrifuge the samples at 12,000 × *g*-force for 20 min at 4°C.
7. Decant the supernatant, by inverting the tube, being careful not to dislodge the pellet.
8. Add 500 μL of 80% EtOH to each sample to wash RNA pellet. Vortex vigorously for 15 s and centrifuge the samples at 12,000 × *g*-force for 5 min at 4°C.
9. Decant the supernatant, by inverting the tube, being careful not to dislodge the pellet.
10. Air dry the sample by inverting the tube for 5 min in the hood. see Note 4.
11. Resuspend the pellet in 6 μL of 18.2 MΩ RNase-free water.

3.2. First-Strand Synthesis

1. To each RNA sample, add 1 μL of first-strand synthesis primer (Table 1). Centrifuge for 10 s.
2. Heat the mixture for 2 min at 65°C, then for 1 min at 45°C, and place on ice. Total volume is 7 μL.
3. Prepare reverse transcription (RT) master mix (on wet ice): see Note 2.

5× first-strand buffer	4 μL
dNTPs (10 mM)	1 μL
0.1 M DTT	1 μL
RNase inhibitor	1 μL
18.2 MΩ RNase-free water	4 μL

4. Take an aliquot of TC primer (100 ng/μL) (Table 1), heat-denature for 2 min at 70°C, place on ice for several minutes, and then add to the RT master mix. see Note 5.
5. Aliquot Superscript III to the RT master mix. see Note 6. The RT master mix should comprise 13 μL per reaction.
6. Add the RT master mix to the 7 μL of sample, pipette vigorously, and centrifuge briefly. Incubate the mixture for 60 min at 50°C.

7. Inactivate the RT reaction by heating the sample for an additional 15 min at 65°C.
8. Centrifuge the reaction mixture briefly and cool immediately on wet ice. The samples can be stored at –20°C for short term or at –80°C for longer term storage.

3.3. RNA Removal and Second-Strand Synthesis

1. Prepare second-strand master mix (on wet ice):

10× PCR buffer	10 μL
RNase H	0.5 μL
18.2 MΩ RNase-free water	69 μL

2. Mix well with pipette tip and centrifuge briefly.
3. Distribute 79.5 μL of second-strand master mix to the appropriate number of 0.5 mL capacity thin-walled PCR tubes.
4. Add 20 μL of the sample to the second-strand master mix and mix thoroughly with a pipette tip.
5. Place the samples in a thermal cycler. see Note 7. Degrade the RNA for 30 min at 37°C. As soon as the block temperature reaches 95°C, pause the reaction (this will be the hot-start component). (Heat-denature RNase H for 3 min at 95°C and proceed to Sect. 3.4 if performing sense TC RNA amplification). Using a dedicated PCR pipette, add into each sample the following:

Taq polymerase	0.5 μL

6. Mix thoroughly with a pipette tip. see Note 8.
7. Continue the second-strand synthesis program by pushing the Continue function.
8. The second-strand synthesis program consists of the following:

Hot start denaturation	95°C	for 3 min.
Annealing	60°C	for 3 min.
Elongation	75°C	for 30 min.

9. Once the program is complete, remove the samples, centrifuge briefly, and store at –20°C for short term or at –80°C for longer term storage.

3.4. Double-Stranded cDNA or cDNA/Primer Purification

1. To purify double-stranded cDNA, use Vivaspin 500 columns (10,000 MWCO)

cDNA sample from Sect. 3.3	100 μL
18.2 MΩ RNase-free water	300 μL

2. Mix by pipetting vigorously and add to top of columns. Spin columns in their collection tubes in a microfuge at 10,000 × *g*-force for 7.5 min at 22°C and measure the resultant product in the column. The volume should be 13 μL. see Note 9.
3. Remove the purified cDNA from the column to a fresh 1.7-mL microfuge tube and store at −20°C for short term or at −80°C for longer term storage.

3.5. IVT for TC RNA Amplification: Biotinylated/ Fluorescent Probe Labeling

1. Prepare IVT master mix (on wet ice):

10× Hybridization reaction buffer	4 μL
10× Biotin-labeled ribonucleotides	4 μL
10× DTT	4 μL
10× RNase inhibitor mix	4 μL
18.2 MΩ RNase-free water	9 μL
T7 RNA polymerase	2 μL

2. The IVT master mix should comprise 27 μL per reaction.
3. Add 13 μL of TC sample to the IVT master mix (final volume of the reaction is 40 μL).
4. Mix thoroughly (very important!!). Centrifuge briefly and incubate for 5 h at 37°C.
5. TC RNA amplified products are now ready for purification, fragmentation, and application to cDNA or oligonucleotide arrays. see Note 10.

3.6. IVT for TC RNA Amplification: Radioactive Probe Labeling

1. Prepare IVT master mix (on wet ice):

5× RNA amplification buffer	8 μL
0.1 M DTT	1 μL
3NTPs (ATP, CTP, GTP; 2.5 mM each)	2 μL
UTP (100 μM)	1 μL
RNase inhibitor (20 U)	1 μL

2. The IVT master mix should comprise 13 μL per reaction.
3. Centrifuge briefly and add 13 μL of double-stranded TC sample to the IVT master mix (on wet ice).
4. To each sample, add ^{33}P-UTP (10 mCi/mL) 12 μL.
5. Add T7 RNA polymerase 2 μL.
6. Mix thoroughly (very important!!). Centrifuge briefly and incubate for 4 h at 37°C.
7. TC RNA amplified products are now ready to be hybridized to membrane-based array platforms. see Note 11.

4. Notes

1. All solutions throughout the entire protocol should be made with 18.2 MΩ RNase-free water (e.g., Nanopure Diamond, Barnstead, Dubuque, IA), and is referred to as 18.2 MΩ RNase-free water in the text.
2. Prepare enough master mix for each reaction step using this basic formula: tabulate the total number of samples plus two extra (i.e., calculate volumes of the mix for x # of samples + control + one to control for any volume loss and/or experimenter error).
3. This protocol is based on using the BioArray RNA transcript labeling kit (Enzo), although any biotinylated and/or fluorescent labeling protocols are suitable (minor modifications may apply).
4. Do not air dry RNA pellets longer than 5 min, as the pellet becomes difficult to resuspend.
5. Each sample in the RT master mix will get 1 μl of TC RNA primer; thus if you have eight samples, the RT master mix will be equivalent to ten samples and you will add 10 μL of the TC primer to the RT master mix prior to adding to the sample.
6. When taking aliquots of Superscript III from the tube, be extremely careful to not contaminate the vial of enzyme with primers or RNA samples. We recommend that laboratory investigators performing TC RNA amplification have their own aliquot of Superscript III to avoid cross-contamination of the enzyme stock.
7. For PCR cycling of the second-strand synthesis, it is a good idea to create a specific stepwise protocol to be programmed into the thermal cycler using final volume of 100 μL.
8. A crucial mistake that is commonly made is failure to mix the hot-start second-strand synthesis reaction when the Taq polymerase is added. Thorough mixing with a pipette tip will ensure a suitable reaction environment.
9. If more than 13 μL remains in the column after the first centrifugation spin, perform a repeat spin at 10,000 × *g*-force for 1 min until a volume of approximately 13 μL is recovered. If the volume in the column is below 13 μL, remove the product left in the top of the column to a fresh 1.7-mL microfuge tube and bring the volume to 13 μL with 18.2 MΩ RNase-free water.
10. There are numerous Internet sites and published protocols for hybridization of labeled probes to microarray platforms, subsequent washing, and imaging protocols (e.g., www.affymetrix.com and www.enzo.com (46, 49, 51)).

11. Procedures for membrane microarray synthesis, pre-hybridization, hybridization, washing conditions, and imaging are described in detail in several published papers emanating from the Ginsberg laboratory (5, 6, 31, 32, 35, 47, 48, 52).

Acknowledgments

The authors thank Irina Elarova, Shaona Fang, and Arthur Saltzman for their expert technical assistance. Support for this project comes from the NINDS (NS43939, NS48447), NIA (AG10668, AG14449, AG17617, AG09466), NICHD (HD057564), and the Alzheimer's Association.

References

1. Ginsberg, S. D., and Mirnics, K. (2006) Functional genomic methodologies, Prog Brain Res 158, 15–40.
2. Brown, P. O., and Botstein, D. (1999) Exploring the new world of the genome with DNA microarrays, Nat Genet 21, 33–37.
3. Counts, S. E., He, B., Che, S., Ikonomovic, M. D., Dekosky, S. T., Ginsberg, S. D., and Mufson, E. J. (2007) {alpha}7 Nicotinic receptor up-regulation in cholinergic basal forebrain neurons in Alzheimer disease, Arch Neurol 64, 1771–1776.
4. Counts, S. E., He, B., Che, S., Ginsberg, S. D., and Mufson, E. J. (2009) Galanin fiber hyper-innervation preserves neuroprotective gene expression in cholinergic Basal forebrain neurons in Alzheimer's disease, J Alzheimers Dis 18, 885–896.
5. Ginsberg, S. D., and Che, S. (2004) Combined histochemical staining, RNA amplification, regional, and single cell analysis within the hippocampus, Lab Invest 84, 952–962.
6. Ginsberg, S. D., and Che, S. (2005) Expression profile analysis within the human hippocampus: Comparison of CA1 and CA3 pyramidal neurons, J Comp Neurol 487, 107–118.
7. Ginsberg, S. D., Che, S., Counts, S. E., and Mufson, E. J. (2006) Shift in the ratio of three-repeat tau and four-repeat tau mRNAs in individual cholinergic basal forebrain neurons in mild cognitive impairment and Alzheimer's disease, J Neurochem 96, 1401–1408.
8. Ginsberg, S. D., Che, S., Wuu, J., Counts, S. E., and Mufson, E. J. (2006) Down regulation of trk but not p75 gene expression in single cholinergic basal forebrain neurons mark the progression of Alzheimer's disease, J Neurochem 97, 475–487.
9. Ginsberg, S. D., Alldred, M. J., Counts, S. E., Cataldo, A. M., Neve, R. L., Jiang, Y., Wuu, J., Chao, M. V., Mufson, E. J., Nixon, R. A., and Che, S. (2010) Microarray analysis of hippocampal CA1 neurons implicates early endosomal dysfunction during Alzheimer's disease progression, Biol Psychiatry, 68, 885–893.
10. Mufson, E. J., Counts, S. E., and Ginsberg, S. D. (2002) Single cell gene expression profiles of nucleus basalis cholinergic neurons in Alzheimer's disease, Neurochem Res 27, 1035–1048.
11. Sambrook, J., and Russell, D. W. (2001) Molecular cloning: a laboratory manual. Third edition, Cold Spring Harbor Laboratory Press, Cold Spring Harbor.
12. Madabusi, L. V., Latham, G. J., and Andruss, B. F. (2006) RNA extraction for arrays, Methods Enzymol 411, 1–14.
13. Van Deerlin, V. M., Gill, L. H., and Nelson, P. T. (2002) Optimizing gene expression analysis in archival brain tissue, Neurochem Res 27, 993–1003.
14. Kabbarah, O., Pinto, K., Mutch, D. G., and Goodfellow, P. J. (2003) Expression profiling of mouse endometrial cancers microdissected from ethanol-fixed, paraffin-embedded tissues, Am J Pathol 162, 755–762.
15. Lehmann, U., Bock, O., Glockner, S., and Kreipe, H. (2000) Quantitative molecular analysis of laser-microdissected paraffin-embedded human tissues, Pathobiology 68, 202–208.
16. Lewis, F., Maughan, N. J., Smith, V., Hillan, K., and Quirke, P. (2001) Unlocking the archive–gene expression in paraffin-embedded tissue, J Pathol 195, 66–71.
17. Penland, S. K., Keku, T. O., Torrice, C., He, X., Krishnamurthy, J., Hoadley, K. A., Woosley,

J. T., Thomas, N. E., Perou, C. M., Sandler, R. S., and Sharpless, N. E. (2007) RNA expression analysis of formalin-fixed paraffin-embedded tumors, Lab Invest 87, 383–391.

18. Rupp, G. M., and Locker, J. (1988) Purification and analysis of RNA from paraffin-embedded tissues, Biotechniques 6, 56–60.
19. Farrell Jr, R. E. (1998) RNA Methodologies, Second Edition, Academic Press, San Diego.
20. Blumberg, D. D. (1987) Creating a ribonuclease-free environment, Methods Enzymol 152, 20–24.
21. Stan, A. D., Ghose, S., Gao, X. M., Roberts, R. C., Lewis-Amezcua, K., Hatanpaa, K. J., and Tamminga, C. A. (2006) Human postmortem tissue: what quality markers matter?, Brain Res 1123, 1–11.
22. Weis, S., Llenos, I. C., Dulay, J. R., Elashoff, M., Martinez-Murillo, F., and Miller, C. L. (2007) Quality control for microarray analysis of human brain samples: The impact of postmortem factors, RNA characteristics, and histopathology, J Neurosci Methods 165, 198–209.
23. Fend, F., Emmert-Buck, M. R., Chuaqui, R., Cole, K., Lee, J., Liotta, L. A., and Raffeld, M. (1999) Immuno-LCM: laser capture microdissection of immunostained frozen sections for mRNA analysis, Am J Pathol 154, 61–66.
24. Goldsworthy, S. M., Stockton, P. S., Trempus, C. S., Foley, J. F., and Maronpot, R. R. (1999) Effects of fixation on RNA extraction and amplification from laser capture microdissected tissue, Mol Carcinog 25, 86–91.
25. Klimecki, W. T., Futscher, B. W., and Dalton, W. S. (1994) Effects of ethanol and paraformaldehyde on RNA yield and quality, BioTechniques 16, 1021–1023.
26. Su, J. M., Perlaky, L., Li, X. N., Leung, H. C., Antalffy, B., Armstrong, D., and Lau, C. C. (2004) Comparison of ethanol versus formalin fixation on preservation of histology and RNA in laser capture microdissected brain tissues, Brain Pathol 14, 175–182.
27. Qin, Y., Heine, V. M., Karst, H., Lucassen, P. J., and Joels, M. (2003) Gene expression patterns in rat dentate granule cells: comparison between fresh and fixed tissue, J Neurosci Methods 131, 205–211.
28. Mai, J. K., Schmidt-Kastner, R., and Tefett, H.-B. (1984) Use of acridine orange for histologic analysis of the central nervous system, J Histochem Cytochem 32, 97–104.
29. Vincent, V. A., DeVoss, J. J., Ryan, H. S., and Murphy, G. M., Jr. (2002) Analysis of neuronal gene expression with laser capture microdissection, J Neurosci Res 69, 578–586.
30. Ginsberg, S. D., Crino, P. B., Lee, V. M.-Y., Eberwine, J. H., and Trojanowski, J. Q. (1997) Sequestration of RNA in Alzheimer's disease neurofibrillary tangles and senile plaques, Ann Neurol 41, 200–209.
31. Che, S., and Ginsberg, S. D. (2004) Amplification of transcripts using terminal continuation, Lab Invest 84, 131–137.
32. Che, S., and Ginsberg, S. D. (2006) RNA amplification methodologies, in Trends in RNA Research (McNamara, P. A., Ed.), pp 277–301, Nova Science Publishing, Hauppauge.
33. Kacharmina, J. E., Crino, P. B., and Eberwine, J. (1999) Preparation of cDNA from single cells and subcellular regions, Methods Enzymol 303, 3–18.
34. Eberwine, J., Kacharmina, J. E., Andrews, C., Miyashiro, K., McIntosh, T., Becker, K., Barrett, T., Hinkle, D., Dent, G., and Marciano, P. (2001) mRNA expression analysis of tissue sections and single cells, J Neurosci 21, 8310–8314.
35. Ginsberg, S. D. (2008) Transcriptional profiling of small samples in the central nervous system, Methods Mol Biol 439, 147–158.
36. VanGelder, R., von Zastrow, M., Yool, A., Dement, W., Barchas, J., and Eberwine, J. (1990) Amplified RNA (aRNA) synthesized from limited quantities of heterogeneous cDNA, Proc Natl Acad Sci U S A 87, 1663–1667.
37. Tecott, L. H., Barchas, J. D., and Eberwine, J. H. (1988) In situ transcription: specific synthesis of complementary DNA in fixed tissue sections, Science 240, 1661–1664.
38. Wang, E., Miller, L. D., Ohnmacht, G. A., Liu, E. T., and Marincola, F. M. (2000) High-fidelity mRNA amplification for gene profiling, Nat Biotechnol 18, 457–459.
39. Zhumabayeva, B., Diatchenko, L., Chenchik, A., and Siebert, P. D. (2001) Use of SMART-generated cDNA for gene expression studies in multiple human tumors, BioTechniques 30, 158–163.
40. Dafforn, A., Chen, P., Deng, G., Herrler, M., Iglehart, D., Koritala, S., Lato, S., Pillarisetty, S., Purohit, R., Wang, M., Wang, S., and Kurn, N. (2004) Linear mRNA amplification from as little as 5 ng total RNA for global gene expression analysis, BioTechniques 37, 854–857.
41. Kurn, N., Chen, P., Heath, J. D., Kopf-Sill, A., Stephens, K. M., and Wang, S. (2005) Novel isothermal, linear nucleic acid amplification systems for highly multiplexed applications, Clin Chem 51, 1973–1981.
42. Matz, M., Shagin, D., Bogdanova, E., Britanova, O., Lukyanov, S., Diatchenko, L., and Chenchik, A. (1999) Amplification of cDNA ends based on template-switching effect and step-out PCR, Nucleic Acids Res 27, 1558–1560.

43. Iscove, N. N., Barbara, M., Gu, M., Gibson, M., Modi, C., and Winegarden, N. (2002) Representation is faithfully preserved in global cDNA amplified exponentially from sub-picogram quantities of mRNA, Nat Biotechnol 20, 940–943.
44. Xiang, C. C., Chen, M., Ma, L., Phan, Q. N., Inman, J. M., Kozhich, O. A., and Brownstein, M. J. (2003) A new strategy to amplify degraded RNA from small tissue samples for microarray studies, Nucleic Acids Res 31, E53.
45. Goff, L. A., Bowers, J., Schwalm, J., Howerton, K., Getts, R. C., and Hart, R. P. (2004) Evaluation of sense-strand mRNA amplification by comparative quantitative PCR, BMC Genomics 5, 76.
46. Ginsberg, S. D. (2005) RNA amplification strategies for small sample populations, Methods 37, 229–237.
47. Alldred, M. J., Che, S., and Ginsberg, S. D. (2008) Terminal continuation (TC) RNA amplification enables expression profiling using minute RNA input obtained from mouse brain, Int J Mol Sci 9, 2091–2104.
48. Alldred, M. J., Che, S., and Ginsberg, S. D. (2009) Terminal continuation (TC) RNA amplification without second strand synthesis, J Neurosci Methods 177, 381–385.
49. Ginsberg, S. D., Hemby, S. E., Mufson, E. J., and Martin, L. J. (2006) Cell and tissue microdissection in combination with genomic and proteomic applications, in Neuroanatomical Tract Tracing 3: Molecules, Neurons, and Systems (Zaborszky, L., Wouterlood, F. G., and Lanciego, J. L., Eds.), pp 109–141, Springer, New York.
50. Che, S., and Ginsberg, S. D. (2008) MicroRNA (miRNA) expression profiling using the miRNA signature sequence amplification (SSAM) technology in human postmortem brain tissues and in animal models of neurodegeneration, Proc. Soc. Neurosci. 34, 46.44.
51. Cheung, V. G., Morley, M., Aguilar, F., Massimi, A., Kucherlapati, R., and Childs, G. (1999) Making and reading microarrays, Nat Genet 21, 15–19.
52. Ginsberg, S. D., and Che, S. (2002) RNA amplification in brain tissues, Neurochem Res 27, 981–992.

Chapter 3

Expression Profiling in Brain Disorders

Peter J. Gebicke-Haerter

Abstract

The development of a variety of high-throughput (HT) technologies in molecular biology has markedly extended its analytical power and has given rise to an exponential jump of accumulation of data. The technologies can be roughly sorted into the ones used to analyze the transcriptome and the proteome. High-throughput DNA sequencing methods are only touched upon here in relation to transcription (epigenetics). Moreover, only a selection of new technologies can be covered within the scope of the article. It may give rise to more information upon new developments in this field to the reader. Evidently, the technologies can be used for many research applications. In general, they pave the way for a deeper understanding of biological systems. However, this fascinating potential requires substantial efforts in embedding the data in the biological context, presumably by aid of mathematical tools. Therefore, apart from describing the technologies, this pivotal aspect has been elaborated on in more detail here, in particular with respect to the complexity of the brain.

Key words: Brain disorders, Transcriptome, DNA microarrays, Epigenetics

1. Introduction

Research in biological systems has been hampered by the lack of technologies that allow obtaining an understanding of complex mechanisms of action. This is particularly true for cellular and molecular systems and even more so when it comes to understanding the functioning of the brain. Established technologies, therefore, had to rely on investigations on specific types of neural cells or on a single target molecule. This has resulted in a wealth of valuable data but, due to the lack of context, has not substantially broadened our understanding of interactions between cellular and molecular networks. The situation has changed in the last 10–15 years with the advent of a number of promising high-throughput

Yannis Karamanos (ed.), *Expression Profiling in Neuroscience*, Neuromethods, vol. 64,
DOI 10.1007/978-1-61779-448-3_3,

(HT) technologies in molecular biology. It has become possible to establish profiles of whole transcriptome in any given tissue or even in single cells within 24 h. This has led to a radically changed conception from the former reductionistic view of single components to a systems view of multiple, interacting components in closely interwoven networks of molecules. This view is able to explain seamlessly the mechanisms of action of single target approaches. It may not only elucidate that the impact on a target molecule in a certain metabolic pathway affects the function of that pathway or even has more far-reaching consequences on other pathways, but it may also easily reveal that quite some more molecules in those pathways would tentatively cause similar effects. Moreover, the view may tell us about tentative sources of side effects upon drug therapies. Furthermore, the broader view of understanding molecular systems can eventually guide us to more suitable targets for drug interventions. Finally, it may unveil early molecular changes that mark the beginning of a disease. In this way, molecular systems biology may be used as a powerful tool enabling very early diagnosis and subsequent very specific drug therapy.

The systems view has reminded us not only of the well-known fact that there is no privileged level in biological organisms, but also of the incomprehensible complexity on any given level including the molecular, and the brain appears to be the utmost challenge of all. While in peripheral organs like the skin or the liver the number of cell types is low and their functions typically serve a very restricted and specified task, cell type specializations in the brain are far more diverse, although there is a rough classification into neurons and glial cells. In general, it has to be kept in mind that within populations of only one cell type, metabolic activities even in neighboring single cells may be subtly different. This holds true especially for cells of the nervous system where neurons producing different neurotransmitters often are directly connected and, above all, fulfill different tasks depending on their location in distinct brain regions. Viewed from the molecular point of view, it is not unlikely that gene expression in a dopaminergic neuron in the substantia nigra is different from gene expression in a dopaminergic neuron in the ventral tegmental area. Additionally, there is the general problem of a sustained development of an organism from birth to death—its dynamics. This means that any sample taken from any brain region or from a single neural cell displays an expression pattern distinct from that of another sample taken from the same brain region or from just the neighboring neuron at any other time. This inherent dynamics of the whole system does not only hold for the normal development and aging of an organism, but also for the onset and progression of a disease. In that sense, each disease is unique, although major features can be identified. Therefore, there is hope that progression of a disease carries on in a rather stereotypic way which would be amenable by time-course studies. Facing this complex situation, it is obvious that the brain is extremely hard to analyze—and due to its limited accessibility, the human brain, in particular.

This gives rise to the question of how useful animal models are that show features of some human brain disorders. If the above hypothesis holds true, that disease development and progression are more or less comparable from one individual to the other—hence is disease specific—one could believe that certain basic features characterizing the trajectory toward disease can also be identified in those animals. Evidently, the three major advantages of investigations on animal models are (a) access to brain tissue, (b) genetic manipulations (e.g., knock out of gene functions), and (c) the opportunity to carry out time-course studies. Studies on human, postmortem brains have the advantage to permit a glance at the genuine human disorder, and even time-course studies appear to be feasible given the fact that the patients deceased at different ages. Obvious drawbacks include influences of medication, state of agony, or postmortem interval, which all can substantially impact on gene expression profiles. Therefore, it is probably the best to pursue both lines of research and try to find commonalities within the pool of data.

This leads us to some key issues pivotal for the retrieval of reliable data. First of all, and most importantly, a clearly circumscribed biological hypothesis has to be set up. Mere data collection does not lead anywhere. Then decisions have to be made as to what brain region or what specified cell type is to be investigated and which temporal frame is to be considered. Moreover, if a specific disease is to be studied, it has to be clarified beforehand what diagnostic markers are to be used and what exclusion criteria are to be observed. Then it has to be decided at which level the study is to be performed, which directly leads to the technologies available. This article does not claim to be comprehensive. It is meant to present some major techniques to analyze the transcriptome and the proteome, because they constitute the major levels of bottom-up approaches in systems biology, and to outline the contextual, biological framework that these techniques could make valuable contributions to.

2. Analyzing the Transcriptome (Fig. 1)

2.1. Preparation of Total RNA

Starting material for transcriptome analysis is messenger RNA. Because it is an extremely labile molecular species, extensive care has to be taken to protect its integrity from the tissue it is extracted from until the purified sample. Already in the tissue, its integrity is threatened as soon as cell damage occurs and RNAses are released. These enzymes can be found everywhere, are very robust, hence

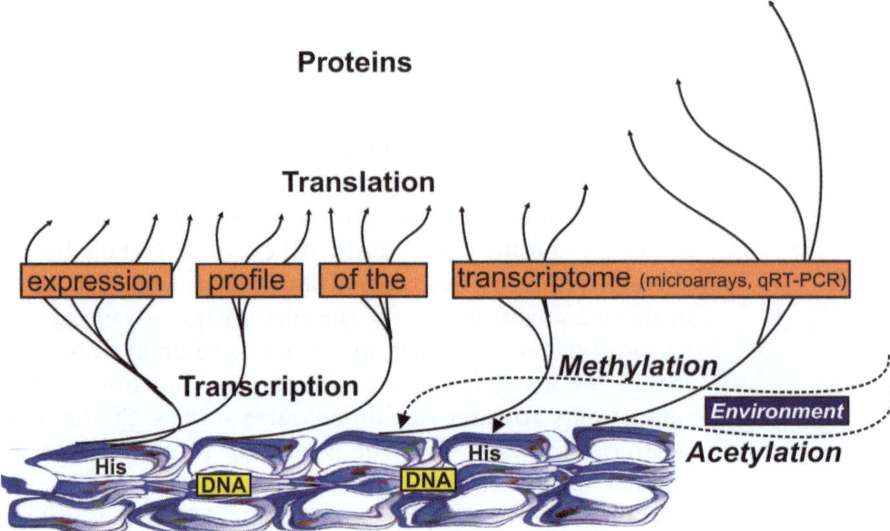

Fig. 1. *Analyzing the transcriptome.* Quantitative determinations of single transcripts or of the whole transcriptome reflect expression levels at the height of the *horizontal bar*. It is the sum of all the influences exerted by (endogenous) cooperative actions of transcription factors, by (exogenous) impact of differential DNA modifications (methylations, acetylations, and others), and by actions occurring at the level of RNA processing (alternative splicing, editing, etc.).

hard to inactivate, and work even at low temperatures. Therefore, removal of brain tissue has to be carried out fast if excised directly from brain at ambient temperature. Already at this point, the issue of reproducibility has to be taken into consideration. This means that each time interval—decapitation, brain dislocation, each cut until excision of the region of interest, and its immersion into a freezing liquid or fixative—has to be the same for each preparation. Typically, a well-experienced researcher or technician is required. The same is true for each subsequent step. Very often, specified brain regions are "punched" out from brain slices prepared in the cryostat. Different punching tools may be used in different laboratories. Moreover, various thicknesses of slices may be chosen—50, 100, or 150 μm. We are routinely using 120 μm. Also, it is highly dependent on the experimenter whether the excised area is exactly the same from brain to brain. Here, larger variations may be obtained when two or more people work on the same batch of brains. Variations may even be encountered with the same technician depending on the time he or she has been working on sectioning during the day (declining attention). With decreasing sizes of punched brain areas (e.g., hippocampus vs. amygdala) and animal species (rat vs. mouse), this work becomes increasingly more challenging and, hence, more apt to technical bias.

There is also a change in temperature from storage of brains (−80°C) to slicing (−15°C), which normally is given credit to by equilibration of the brains in the cryostat for approximately 1 h. Sometimes, after slicing, the remainder of the brain is frozen down again for additional applications. Although brain tissue is maintained frozen at all times, nobody has systematically analyzed tentative influences of higher or lower freezing temperatures.

The frozen tissues are then lysed on ice using Trizol. Most of the subsequent steps appear to be straightforward for a trained molecular biologist, but any slight changes, e.g., trituration 20× or 40× instead of 30×, or using a needle with another diameter, etc., may introduce technical bias. So, strict adherence to one protocol is essential for the quality of data. For a detailed protocol of RNA preparation as used in our laboratory, see **Protocol 1**.

Before reverse transcription of RNA into cDNA, RNA quality has to be evaluated. The NanoDrop UV–Vis spectrophotometer is very useful for a quick determination of total amounts of RNA, and estimations of DNA and protein contaminations (260/280 nm ratios: between 1.8 and 2.0 and 260/230 nm ratios, indicating the presence/absence of other organic compounds such as alcohol and phenol). It is no indication, however, of RNA integrity. This has to be assessed using the Bioanalyzer (Agilent), which works with capillary polyacrylamide electrophoresis performed in a chip that can analyze up to 12 samples per run. The printout displays a number of features, the most important one being the RNA Integrity Number (RIN), which is calculated from the 28S/18S ratio of ribosomal RNAs and RNA degradation products. It should be above 8.0, although sometimes with some tissues that are hard to prepare because of high RNAse content, compromises have to be made allowing also RIN values below 8.0.

RNA can be used for chip hybridizations or for quantitative, real-time PCR (qRT-PCR). In first expression profiling screens, genome-wide chips, like the SENTRIX bead chips from Illumina, are used, whereas custom-made chips can be used when profiles from genome-wide chips have provided sufficient information to restrict further searches to lower numbers of hypothesis-guided genes. If the number of these genes falls in the range of some dozens or even a few hundreds, qRT-PCR is the method of choice, because it is more precise and more sensitive than microarray analyses.

2.2. DNA Microarrays

For genome-wide expression profiling, RNA is reverse transcribed into cDNA, followed by an amplification/labeling step (in vitro transcription) to synthesize biotin-labeled cRNA according to the instructions of the MessageAmp II aRNA Amplification kit (Ambion, Inc., Austin, TX). Before hybridization, cRNA is purified and quality controlled using the RNA Nano Chip Assay on the Agilent 2100 Bioanalyzer. After hybridization, array signals are developed by 10-min incubation in 2 mL of 1 μg/mL Cy3–streptavidin, and scanning is done using a Beadstation array scanner. For a full protocol, see **Protocol 2**.

Often lists of up- and downregulated genes contain several hundred genes, which brings up considerations as to make hypothesis-driven chips. In this case, oligonucleotides (50–70mers) can be purchased and obtained already spotted and lyophilized into 96- or 384-well plates. To enable covalent binding to, for example, carboxy-modified chips, they should be elongated with a C6-spacer and a terminal NH group. For transfer of oligonucleotides to the chips, a variety of arrayers can be used. Typically, the quantities of provided oligonucleotides are sufficient for spotting 1,000 chips or more, depending on the oligonucleotide concentrations and on the number of array replicates on a chip. We routinely spot three replicates, and additionally add oligonucleotides for 20 "housekeeping" genes (1). RNA is reverse transcribed into cDNA using the Agilent Low Input Quick Amp Labeling kit that supports labeling of the cDNA with cyanine 3 or cyanine 5 (Cy3, Cy5) through in vitro transcription (http://www.chem.agilent.com/en-US/Search/Library/_layouts/Agilent/PublicationSummary.aspx?whid=55389&liid=3855).

Typically, dual labeling is done with one label attached to cDNA made from a universal RNA purchased from Stratagene (2). Hybridization is carried out using a Slide Booster SB401/800 device (Beckman Coulter, Advalytix, Munich) that ensures homogeneous, contact-free mixing of the hybridization solution. Washing of chips with increasing stringencies is extremely important to remove contaminating compounds like SSC or sodium dodecyl sulfate (SDS), and to reduce background. Specific fluorescence, however, can also be diminished. Because of photosensitivity, all washings have to be carried out in the dark. After the last washing, the chip is immediately dried by an oblique stream of air to minimize liquid drying on the chip. For the readout, a variety of scanners are available (Axon).

2.3. Quantitative (Real-Time) PCR

An option to hypothesis-driven oligonucleotide chips is qRT-PCR. RNA is reverse transcribed into cDNA using the High-Capacity cDNA Reverse Transcription kit (Applied Biosystems) and then diluted 1:10 or 1:100. QRT-PCR is performed in 96- or 384-well plates that are usually pre-spotted with specific primer pairs or TaqMan probes. If this job needs to be done in the own laboratory, it is highly recommended to have the pipetting done by a pipetting robot (Tecan, HP), especially in 384-well plates. We have routinely analyzed in one sample 180 distinct genes in duplicate plus 20 "housekeeping" genes (1) plus some background controls in one plate. There are companies offering pre-spotted, custom-made TaqMan plates (e.g., ABI) where only the reaction mix and the sample have to be added. In this case, it turned out that the precision between duplicates was so good that a difference of more than 0.5 Ct between duplicates was extremely rare. Moreover, next-generation multiwell plates (1,536 wells) combined with qRT-PCR are available, which would extend the number of genes to another fourfold and will make low-density, home-made chips completely dispensable.

The LightCycler® 1536 Instrument (Roche Applied Science) is one of the well-established instruments that supports mono- and dual-color applications for the detection of green intercalating dyes as well as hydrolysis probes. It works in the volume range of 0.5–2.0 μL/well. Other LightCyclers® work in the range of 5–20 μL (384-well format) and 10–100 μL (96-well format). Finally, it has to be mentioned that qRT-PCR is the standard method to confirm microarray results. Formerly, only few genes randomly picked or specifically selected were analyzed by qRT-PCR. The availability of the high-throughput formats depicted here permits to analyze all the differentially expressed genes revealed by microarrays. In this way, microarrays and qRT-PCR are strong, complementing partners to corroborate biological data.

2.4. DNA Modifications

Of course, expression profiles are the "final output" of a number of preceding events involved in the control and regulation of gene transcription and mRNA processing. After "zooming in" into tentatively important molecular networks, it may become interesting to learn more about mechanisms interfering with transcription of those genes. I only want to dwell here a little on DNA and histone modifications (epigenetics), although transcription factor analyses are equally important. DNA modifications entail methylations, acetylations, sumoylations, ADP-ribosylations, and a few more. Often, these modifications reduce gene transcription. Enzymes adding and maintaining (e.g., methyl transferases) or removing (e.g. demethylases) these residues are very sensitive to environmental influences. Apparently, there is a delicate balance between the activities of these two classes of enzymes. Too much methylation, acetylation, etc., may result in too much inactivation of a cell. In a few examples, methylation of genes, like methylation of the tumor suppressor gene p53, may result in an enhanced risk to develop cancer. Conversely, too much demethylation, and so on, may overactivate cells and may result in uncontrolled cell growth, which may also increase the risk to develop cancer. Similar functions are observed with chromatin (histone) modifications. Histone deacetylases (HDACs) mediate chromatin condensation and subsequent gene silencing, and histone acetyl transferases' (HAT) activities can lead to chromatin unfolding and increased gene transcription. Whether or not the increased or repressed transcriptional rates of genes surfacing in microarray investigations are influenced by these kinds of modifications can be closely analyzed by, for example, methylation-specific PCR (MSP).

2.4.1. Bisulfite Treatment of DNA

MSP involves the chemical conversion of all unmethylated cytosines to uracil. To this end, DNA is purified from brain tissue using the QIAamp DNA Mini kit. The DNA is treated with sodium bisulfite and fragmented. Methylated cytosines remain unaltered in the process. In subsequent PCRs, unmethylated cytosines/uracils are replaced by thymine. Therefore, the sequence of the DNA after bisulfite treatment will be different depending on the original methylation state of the DNA. The base exchange is revealed by the use of methylation-specific oligonucleotide primers for the genes of interest. Primers to the unmethylated and methylated sequences will be designed in such a way that mismatches are created depending on which sequence is present to prevent mispriming between the primer sets and the undesired target DNA. A typical experiment will involve performing two PCRs using the same bisulfite-treated template DNA. One reaction uses primers designed to anneal to the sequence present if the DNA is unmethylated. The other reaction will include primers designed to anneal to the sequence if the DNA is methylated. The PCR products can either be resolved in agarose gels or analyzed directly in quantitative real-time PCR (qPCR) setups. In this way, it is possible to establish a methylation pattern for each gene of interest in control and experimental samples and relate that to the expression pattern observed on the chip (**Protocol 3**). An optional, increasingly popular method to identify methylation patterns is sequencing of bisulfite-treated and of untreated DNA. If methylation patterns are to be investigated on a large or genome-wide scale, microarrays can be used for hybridization of bisulfite-treated, fragmented, and fluorescently labeled DNA. These arrays are composed of pairs of oligonucleotides of 17–25 bases synthesized on-chip. They represent sequences found in CpG islands and reflect either the unmethylated or the methylated versions (3). They may contain several hundred thousands of probes that cover about 25,000 gene promoters.

2.4.2. Chromatin Immunoprecipitation

Chromatin immunoprecipitation (ChIP) experiments can be used as a powerful tool to complement RNA transcription studies because they enable researchers to study the DNA–protein interactions that regulate gene expression. ChIP is a powerful tool for identifying proteins, including histone proteins and non-histone proteins, associated with specific regions of the genome by using specific antibodies that recognize a specific protein or a specific modification of a protein. This is also a drawback, because ChIP requires a priori knowledge of the existence of modifications (methylations, acetylations), because antibodies against the modifications have to be generated. Here, we present a simplified native chromatin immunoprecipitation (NChIP) protocol for frozen (never-fixed) human brain specimens. Starting with micrococcal nuclease digestion of brain homogenates, NChIP followed by qPCR can be completed within 3 days. The methodology presented here should be useful to elucidate epigenetic mechanisms of gene expression in normal and diseased human brain (**Protocol 4**). Often, the initial step of ChIP is the cross-linking of protein–protein and protein–DNA with formaldehyde, although this step can also be omitted. Then the tissue is dounced or sonicated to shear the DNA. Proteins associated with DNA are subsequently immunoprecipitated.

Protein–DNA cross-links in the immunoprecipitated material are then reversed and the DNA fragments are purified and PCR amplified.

Subsequently, targets are fragmented and labeled to hybridize onto GeneChip® tiling arrays. By comparing the hybridization signals generated by an immunoprecipitated sample versus an antibody-negative or nonspecific antibody control, the regions of chromatin protein interaction can be identified.

2.4.3. Histone Modifications HAT/HDAC Assay

Similar studies can be performed with modified or unmodified histones associated with the genes of interest. The nucleosomal organization of genomic DNA including DNA:core histone binding appears to be largely preserved in representative samples provided by various brain banks. Therefore, it is possible to study the methylation pattern and other covalent modifications of the core histones at defined genomic loci in postmortem brain (4).

Nuclear lysates are prepared from hippocampal tissue using the ProteoExtract® subcellular proteome extraction kit (Calbiochem 539790-1). For HAT assays, the protein of interest is immunoprecipitated and washed sequentially in lysis buffer followed by HAT assay buffer (50 mM Tris, pH 8.0; 10% glycerol; 0.1 mM EDTA; and 1 mM dithiothreitol). Immunoprecipitates are incubated in HAT assay buffer supplemented with acetyl-coenzyme A (100 μM), and either biotinylated histone H4 peptide or biotinylated peptide of protein of interest (0.5 μg) for 30 min at 30°C. An aliquot of the reaction mix is immobilized on a streptavidin plate, and acetylation is detected using a HAT enzyme-linked immunosorbent assay according to the manufacturer's instructions (Upstate Biotechnology, NY, or BioVision, K332-100) (5). HDAC assay is performed using the Fluorometric HDAC assay kit (BioVision, k330-100) according to the manufacturer's protocol. Briefly, 120 μg of nuclear extract is incubated for 4 h with the assay mix in Greiner 96 U-bottom transparent polystyrol plates. Analysis is performed using a TECAN Infinite 200 ELISA reader. This leads us directly to the analysis of the proteome.

3. Analyzing the Proteome (Fig. 2)

3.1. Western Blot–Immuno-PCR

Western Blotting is the standard procedure to investigate the occurrence and differential expression of specific proteins. All proteins of a given sample are separated by one- or two-dimensional polyacrylamide gel electrophoresis (PAGE) and electrophoretically transferred on nitrocellulose or other (e.g., PVDF or nylon) membranes. The protein of interest is detected either by a color reaction, e.g., through horseradish peroxidase-coupled secondary antibodies, or through fluorescence reactions (ECL) by exposure of the membrane to a photosensitive film.

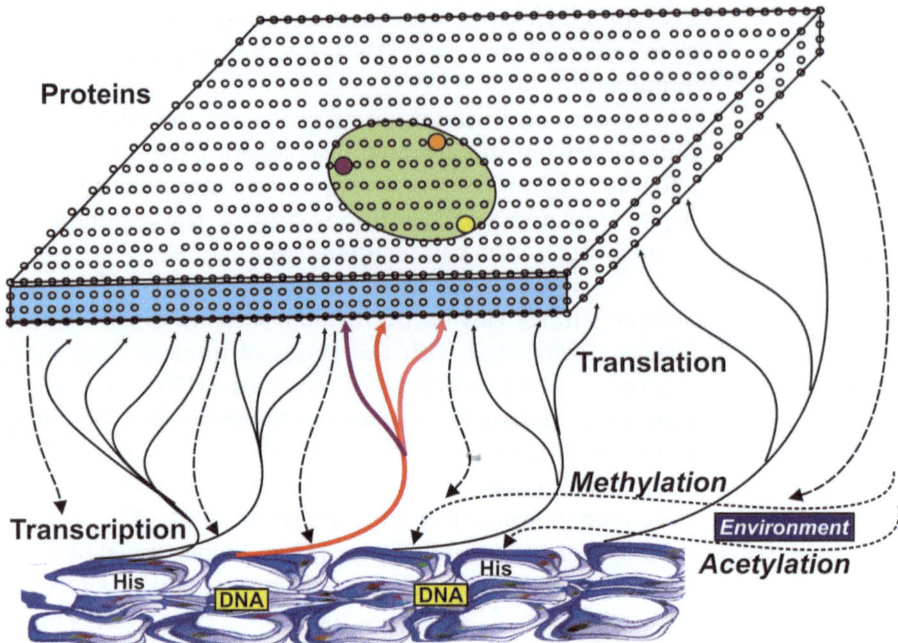

Fig. 2. *Protein expression profile upon transcription of one gene.* Analysis of the proteome is more complex than analysis of the transcriptome. The figure highlights the expression of only one gene. Several protein variants can emerge that may be expressed not only in different organs or cell types, but also simultaneously in one cell. These variants may have different tasks within metabolic (protein) networks, i.e., these may interact with distinct nearest neighbor partner proteins. Depending on the quantitative alterations of protein concentration due to changes on the transcriptional level, the whole protein network may be disturbed in its normal oscillatory behavior. Evidently, this is even more serious when expressions of several proteins are disturbed.

In keeping with epigenetic profiling as described above, the *detection of histone acetylation* by Western blotting is briefly summarized here. Brain tissue is homogenized in lysis buffer (50 mM Tris–HCl, 150 mM NaCl, 2 mM EDTA, 1% Triton-X 100, 1% NP-40, and 0.1% SDS) and subjected to Bioruptor (Diagenode) sonication for 15 min (High, 30 s ON, 30 s OFF) prior to centrifugation at 12,000 rpm for 10 min. The supernatant is used for immunoblotting. Protein concentrations are determined using Bradford protein assay reagent (Bio-Rad), and 10 μg from each sample is run on a 4–20% gradient SDS–polyacrylamide gel. Proteins are transferred to a polyvinylidene difluoride (PVDF) membrane by liquid (Tankblot from Biometra) or semidry blotting (e.g., the iBlot® Gel Transfer Device from Invitrogen). There are many advantages of semidry blotting, especially the very short transfer times (approximately 20 min), but the longer transfer times in tank blots may be more suitable when temperature-sensitive or native proteins are investigated. The longer-term and more gentle transfer may also be preferred for large proteins and for proteins that are generally hard to blot. After blotting, the membranes are probed with anti-H3Ac antibody (Upstate; 1:3,000 dilution), and either anti-pan-histone (Chemicon; 1:2,000 dilution) or anti-H4 antibody (Upstate; 1:2,000 dilution). After several washing steps, the histones are revealed with horseradish peroxidase-conjugated anti-rabbit antibody (Sigma; 1:5,000 dilution) and ECL^{+} chemiluminescent detection reagents (Amersham Biosciences). The signal is scanned and quantified on a Storm860 imaging system (Amersham Biosciences) or using an Odyssey Imager (Licor). It has to be made sure that signal intensities are within the linear range.

Although these blots are used for quantification of bands of interest, as described here, they are more suitable for qualitative, rather than quantitative determinations (6). Nevertheless, quantitations using Western blots have become very popular with the availability of numerous phosphorylation-specific antibodies where the changes are striking. But often differential changes are less pronounced, albeit still important. Quantitation becomes more difficult when proteins of lower abundance and showing more subtle quantitative changes, like receptors or enzymes, are to be studied. Because these changes not only depend on gene expression (changes on the transcript level), but are also subject to protein turnover and protein stability, more precise quantitative evaluations of the abundance of proteins have attained increasing interest. In these cases, the sensitivity of Western blots is not sufficient. Moreover, there is increasing demand of analyzing many samples within short periods of time. This would require running numerous gels and blots, which have to be compared with each other. For these reasons, Western blotting can easily become very laborious.

Optionally, for high-throughput purpose, immuno-PCR can be used. It is orders of magnitude more sensitive than Western blots and can be run in multiwell plates. Our protocol (**Protocol 5**) (7) closely adheres to a protocol published previously (8). The critical step is the biotinylation of the detection antibody. Azide and serum albumin often found in commercially available antisera interfere with biotinylation. Therefore, it is recommended to purify IgG from those antisera by protein-G agarose.

3.2. 2D-PAGE DIGE

The use of two-dimensional gel electrophoresis for differential analysis in proteomics was revolutionized by the introduction of 2D fluorescence difference gel electrophoresis (2D DIGE). This fluorescence-based technique allows for the use of multiplexed samples and an internal standard that virtually eliminates gel-to-gel variability, resulting in increased confidence that differences found between samples are due to real induced changes, rather than inherent biological variation or experimental variability. 2D DIGE has quickly become the "gold standard" for gel-based proteomics for studying tissues, as well as cell culture and bodily fluids such as serum or plasma.

The workflow for 2D DIGE consists of the following steps:

1. Labeling (e.g., Ettan™ DIGE technology from GE Healthcare)

   2D DIGE technology is based on the specific properties of spectrally resolvable dyes, the CyDye™ DIGE fluors. Two sets of dyes are available—Cy™2, Cy3, and Cy5 minimal dyes, and Cy3 and Cy5 saturation dyes—that have been designed to be both mass and charge matched. Therefore, identical proteins labeled with each of the CyDye DIGE fluors will migrate to the same position on a 2D gel.

2. Electrophoresis

   Protein separation in 2D DIGE is performed similarly to conventional 2D electrophoresis

3. Image acquisition

   Images are acquired on a multi-wavelength scanner capable of resolving the three CyDyes.

4. Image analysis

   Image analysis consists of the following processes:

   - Spot detection
   - Background subtraction
   - In-gel normalization
   - Gel artifact removal
   - Gel-to-gel matching
   - Statistical analysis

The drawbacks of 2D PAGE DIGE are that it is time consuming and not sufficiently reliable so that sometimes the results are difficult to reproduce. Low-molecular weight or relatively insoluble proteins are poorly separated. Moreover, spots often contain several proteins. Their advantage is that in one analysis, more than 1,000 proteins can be visualized and their further analysis can easily be interfaced to mass spectrometry.

3.3. Mass Spectrometric Methods to Analyze Protein Mixtures

Compared to 2D PAGE or 2D DIGE, mass spectrometry is a highly sensitive technology with recently increased sequencing speed. Similarly to PAGE, MS also separates proteins according to their mass-to-charge ratio. Because basically the proteins are prepared for MS to carry the same charge, they are separated by their molecular masses. There are two major methods to ionize proteins: electrospray ionization (ESI) and matrix-assisted laser desorption/ionization (MALDI). To achieve sufficient signals of all proteins, the samples should be composed of more or less equal quantities of constituents. Unfortunately, very often this does not hold for biological samples. Typically, certain protein species are highly abundant, like albumin in blood, and others may be very low. Therefore, ionization of these kinds of mixtures by ESI or MALDI results in strong signals preferentially arising from highly abundant components, concealing the signals from less abundant proteins. For this reason, methods for prefractionation, depletion, or enrichment of certain protein pools are required, which then results in a "subproteome." Typically, membrane proteins are underrepresented in total protein mixtures. Here, another problem arises in terms of what type of membrane is in the focus of interest (plasma membrane, nuclear, Golgi, ER, etc.) and subsequently the way to extract those proteins into an adequate buffer (9). Later on, it may happen that some of the buffer components required for solubilization interfere with MS technologies.

Often highly complex protein mixtures or proteins eluted from 2D PAGE are first digested by a protease, and peptides are separated by reverse-phase high performance liquid chromatography (HPLC). Then the peptides are transferred by ESI into a mass spectrometer (HPLC/MS = "shotgun proteomics"), where their masses and intensities are determined and selected peptides are isolated, fragmented again, and masses of the fragments are measured (MS/MS). Peptides can also be determined by MALDI—Time of Flight Mass (TOF) spectrometry. It is the most widely used method for peptide mass analysis. Several peptides per second can be processed in this way, and more than 1,000 proteins can be identified in such a single LC-MS/MS run. MALDI-TOF MS can resolve very small mass differences (approximately 0.1% of the total mass), which means that, for example, acetylated forms of proteins can be distinguished from nonacetylated ones. Hence, the technology can not only systematically map posttranslational modifications (PTMs) in a site-specific manner, and histone methylations or acetylations, as mentioned above, but also phosphorylations or ubiquitinations can be identified. However, determinations of amino acid sequences of the peptides do not yield absolutely safe results, because some amino acids have identical masses. Moreover, especially in mammalian organisms, peptide sequences can be shared by several proteins that are either isoforms or share extensive homologies. Therefore, the search of databases for deposited sequence information, which is required in any case, is not an easy task, although computer programs are available (e.g., MASCOT). Often, considering hundreds or thousands of low-scoring peptide matches, a cut-off score has to be set where it is believed that below that value there is no longer sufficient evidence for the occurrence of this protein in the sample. Usually, this cut-off score is set to 50. But this point is still being debated.

3.4. Antibody Microarrays

These arrays are becoming more and more popular the more antibodies are available or the less expensive it is to have them made. One reason for this popularity may be their capacity for simultaneous assessment of a very large variety of potential biomarkers in a high number of samples. The obvious drawback is their limitation to already known proteins and the availability of antibodies displaying high specificity. Therefore, those arrays are only as good as the antibodies they contain. Because of cross-reactivities that cannot be entirely excluded, only the highest quality antibodies should be used. There are presently more than 180 companies producing antibodies for research or the clinical market. The estimated number of 350,000 antibodies appears to be huge, but it is low considering the vast variety of protein variants including secondary or tertiary modifications. For antibodies, it has to be demonstrated that they are specific, selective, and reproducible. Often, the description of the antibody by the manufacturer is unsatisfactory. It may mention the starting dilutions and recommended applications, but may not show any typical data, like a Western blot or an immunocytochemical staining. Antibodies raised against a synthetic peptide may not work well when a protein is in its native conformation with intact 3D structure. So, they may fail in immunoprecipitations or immunohistochemistry. Also peptide competition experiments should be shown when a peptide has been used for immunization. Unfortunately, these data are also not fully convincing. It does not prove selectivity, because off-target binding of the antibody will also be inhibited by preadsorption with the blocking peptide. In case of phosphate-specific antibodies, results before and after phosphatase treatment should be provided. A negative control showing a blank when primary antibody is omitted is informative, but not sufficient. There is the additional issue of monoclonal vs. polyclonal antibodies. Polyclonal antibodies typically show a higher probability to detect the antigen in a range of different conditions (native vs. denatured). Monoclonal antibodies are more pure in that they are directed at only one epitope. However, the clones of fused cells may have been grown in host animals where the ascites fluid containing the secreted antibody is collected, which may result in contaminations of other antibodies. One option to solve this problem is a purification by use of the Nab Protein-G Spin kit (Thermo Scientific, Rockford, USA). Because of all these uncertainties, an initiative has recently been founded to set up rules for standards of quality of antibodies (10). A protocol of custom-made antibody arrays is provided below (**Protocol 6**).

Recently, antibody arrays have become commercially available. For instance, protneteomix (http://www.protneteomicx.com) has specialized in these kinds of arrays. However, arrays can also be home-made using rather inexpensive microarrayers and bulk packages of slides. TeleChem-ArrayIt (http://www.arrayit.com) is one of the companies established for a long time in this market. Although it was believed that protein arrays are too vulnerable to be stored for long term on glass slides, it has turned out that, for example, super-aldehyde-treated slides spotted with antibodies have a shelf life of at least 1 year—which compares well to DNA microarrays. For signal detection, CCD-based detectors are used in combination with filter sets to read fluorophores. Optionally,

ArrayIt recently offers a combination of CCD with complementary metal oxide semiconductor (CMOS) detection. Like with DNA microarrays, there is also the issue of one- or dual-color microchips. Apparently, dual-color measurements out-perform single-color approaches in terms of reproducibility and discriminative power, although it does not reduce the number of chips, because it is advisable to do dye swaps, too. It has been reported, too, that depletion of high-abundance proteins did not improve technical assay quality. Quite on the contrary, depletion of those proteins introduced a strong bias in protein representation.

4. Conclusions

Figure 3 is a simple display of tentative consequences following differential expression of one gene in a given brain region or cell type. It clearly shows that the situation becomes extremely compounding when more than one gene is affected by a disease. Moreover, the feedforward and feedback mechanisms encompass inherently

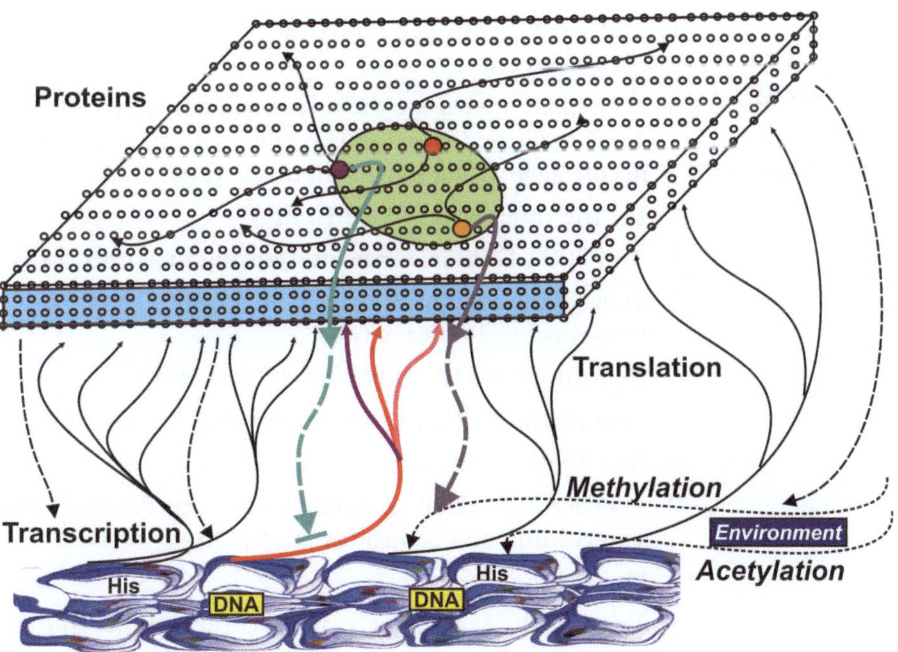

Fig. 3. *Tentative consequences of protein expression of one gene.* Spreading changes on the protein level (oscillatory cross-talk) may not only elicit crucial responses in remote regions of the networks, but also result in dampening, compensatory reactions. Furthermore, the changes on the protein level may feed back to the transcriptional level by either inhibiting (*green arrow*) (or enhancing—feedforward) transcription of its own gene, or by increasing (*magenta arrow*) (feedforward) or inhibiting transcription of another (other) gene(s). Here also, it has to be taken into consideration that typically more than one gene is differentially expressed in disease states of the brain. The figure also illustrates the considerable dynamics of the system.

dynamic processes that are spreading with time over large distances and not only affect more than one protein/metabolic pathway, but also circulate between the protein (cytoplasmic) and the gene (nuclear) compartments. Therefore, methods analyzing only events occurring on one level are insufficient to obtain deeper insights into these interactions. A better understanding is only achieved upon combining data collected from each level. Taken together, making predictions about the overall outcome of differentially expressed genes presently appears to be an insurmountable task. But a review of this short summary of new high-throughput technologies on both the transcript and the protein level may give rise to some hope that all data combined from all levels may yield a more comprehensive picture of specific molecular events going awry in disease. It is also feasible that we can obtain a better picture of the consequences of differential gene/protein expression by applying mathematical algorithms to the wealth of data that are available now and even more so along with more data in the future (11).

5. Detailed Protocols

5.1. Protocol 1: Isolation of Total RNA from Cryosections of Rat Brain and Quality Control

1. Add 1 mL Invitrogen's TRIzol® reagent to the tube containing the frozen tissue.
2. Immediately triturate the suspended tissue 30× through a 22G needle attached to a 1-mL syringe with the shaped side of the needle against the inner wall of the tube.
3. Let the tube sit at RT for 10 min.
4. Add 200 μL chloroform and thoroughly vortex for 30 s.
5. Transfer the content to a pre-spun (12,000–16,000× *g* for 2 min) PhaseLockGel™ heavy tube.
6. Centrifuge at >12,000× *g* for 15 min at 4°C.
7. Transfer the aqueous phase to a new RNAse-free tube (≈500 μL).
8. Add 1 vol of 70% ethanol and vortex.
9. Transfer 700 μL of content to a Qiagen RNeasy® MinElute™ spin column and centrifuge at >8,000× *g* for 15 s.
10. Discard the filtrate.
11. Transfer the remainder sample to the column and centrifuge again at >8,000× *g*.
12. Add 700 μL of Qiagen's buffer RW1 to the column and centrifuge at >8,000× *g* for 15 s.
13. Transfer the column to a new collection tube.
14. Add 500 μL of Qiagen's buffer RPE to the column and centrifuge at >8,000× *g* for 15 s.

15. Discard the filtrate.
16. Add 500 μL of 80% ethanol and centrifuge at >8,000× *g* for 2 min.
17. Transfer the column to a new collection tube and centrifuge at full speed for 5 min.
18. Place the column in a new 1.5-mL RNase-free tube.
19. Add 14 μL of H_2O directly on the membrane and incubate for 5 min.
20. Centrifuge at full speed for 1 min.
21. Take the filtrate, apply directly on the membrane, and incubate for 5 min.
22. Centrifuge at full speed for 1 min.
23. Take 1 μL and dilute with 9 μL of H_2O for OD-measurements and quality control.
24. Store the remainder at <–70°C until further use.

5.2. Protocol 2: Sample Preparation for SENTRIX Illumina Chips. Probe Labeling and Illumina Sentrix BeadChip Array Hybridization

Biotin-labeled cRNA samples for hybridization on Illumina Human/Mouse/Rat Sentrix-6/8/12 BeadChip arrays (Illumina, Inc.) are prepared according to Illumina's recommended sample labeling procedure based on the modified Eberwine protocol (12).

In brief, 250 ng total RNA is used for complementary DNA (cDNA) synthesis, followed by an amplification/labeling step (in vitro transcription) to synthesize biotin-labeled cRNA according to the MessageAmp II aRNA Amplification kit (Ambion, Inc., Austin, TX). Biotin-16-UTP is purchased from Roche Applied Science, Penzberg, Germany.

The cRNA is column purified using the TotalPrep RNA Amplification kit, and eluted in 60 μL of water. Quality of cRNA is evaluated using the RNA Nano Chip Assay on an Agilent 2100 Bioanalyzer and spectrophotometrically quantified (NanoDrop).

Hybridization is performed at 58°C, in GEX-HCB buffer (Illumina Inc.) at a concentration of 100 ng cRNA/μL, unsealed in a wet chamber for 20 h. Spike-in controls for low, medium, and highly abundant RNAs are added, as well as mismatch control and biotinylation control oligonucleotides.

Microarrays are then washed twice in E1BC buffer (Illumina Inc.) at room temperature for 5 min. After blocking for 5 min in 4 mL of 1% (wt/vol) Blocker Casein in phosphate-buffered saline Hammarsten grade (Pierce Biotechnology, Inc., Rockford, IL), array signals are developed by a 10-min incubation in 2 mL of 1 μg/mL Cy3–streptavidin (Amersham Biosciences, Buckinghamshire, UK) solution and 1% blocking solution. After a final wash in E1BC, the arrays are dried and scanned.

5.2.1. Scanning and Data Analysis

Microarray scanning is done in a Beadstation array scanner, with the setting adjusted to a scaling factor of 1 and PMT settings at 430. Data extraction is done for all beads individually, and outliers are removed when >2.5 median absolute deviation (MAD). All remaining data points are used for the calculation of the mean average signal for a given probe, and standard deviation for each probe is calculated.

Data analysis is done by normalization of the signals using the quantile normalization algorithm without background subtraction, and differentially regulated genes are defined by calculating the standard deviation differences of a given probe in a one-by-one comparisons of samples or groups.

5.3. Protocol 3: Bisulfite Treatment of DNA. Preparation of Nonmethylated Standard DNA

A 500–1,000-fold amplification of chromosomal DNA can be achieved by using phi29 DNA polymerase contained in the GenomiPhi DNA amplification kit (GE Healthcare /Amersham Biosciences). The amplified DNA is free of methylated cytosines and, hence, can be used as a standard for unmethylated DNA.

Typically, 1–10 ng DNA is dissolved in 10 μL sample buffer, denatured at 95°C for 3 min, and put back on ice.

1 μL phi29 DNA polymerase is diluted in 9 μL enzyme buffer and mixed in the denatured DNA sample. Polymerization is allowed to proceed for 18 h at 37°C and stopped at 65°C for 10 min.

5.3.1. Bisulfite Treatment of DNA

The methyl residue in the C5 position of 5-methyl cytosine protects the cytosine from deamination. Bisulfite-mediated conversion of unmethylated cytosine into uracil only occurs at single-stranded DNA. Therefore, double-stranded DNA has to be completely denatured before bisulfite treatment. After bisulfite treatment, two noncomplementary strands have been generated (bisulfite-1 and bisulfite-2). Bisulfite treatment results in cytosine-depleted DNA strands and, consequently, PCR-amplified bisulfite-treated DNA is guanine depleted.

1. Fragment the DNA with an appropriate restriction enzyme (e.g., EcoR1) and purify.
2. Up to 2 μg DNA is suspended in 100 μL H_2O.
3. 354 μL ($NaHSO_3$) bisulfite solution (5.89 mol/L) and 146 μL dioxane (98.6 mg/2.5 mL) are added.
4. After thorough mixing, the sample is denatured (strand separation) at 99°C for 3 min.
5. Further denaturations are carried out after 30, 90, and 180 min at 99°C for 3 min each.
6. After 5 h total incubation, 200 μL of H_2O is added and the sample is purified in two steps through a Millipore membrane filter (Microcon YM30).
7. To desulfonate the DNA, 100 μL of 0.2 M NaOH is added on the membrane and incubated for 1 min at room temperature.
8. Then the preparation is washed twice with 400 μL H_2O.
9. Finally, DNA is recovered in 50 μL H_2O at 50°C.

The DNA can be used for PCR, real-time PCR, MSP-PCR, combined bisulfite restriction analysis (COBRA), microarrays, pyrosequencing, and bisulfite sequencing.

For sequencing, the DNA is cut by a nonmethylation-specific restriction enzyme, like MspI.

The C′CGG ends are paired by GGCC with poly dT-tails, respectively poly dA-tails in the pairing strand. These DNA fragments are then subjected to high-throughput sequencing.

Tentative artifacts may occur through high salt concentrations in the bisulfite reaction, which favor reannealing of DNA strands. Moreover, some methylated cytosines may be converted to thymine. Additionally, DNA degradation may become significant through bisulfite treatment, especially at higher temperatures. In this case, the fragments may be too small to be analyzed. Facing low amounts of starting material in brain research or after microdissection, in particular, this loss of DNA may be an important issue. More DNA may be lost in subsequent desalting steps. Finally, the DNA has to be completely desulfonated. Otherwise, residual bisulphite inhibits the complete conversion of intermediate uracil sulfonates. These minor amounts of sulfonates impede DNA polymerases to replicate the templates.

DNA quantification in qRT-PCR is carried out using a concentration curve made from an *external standard DNA*. This DNA is made from bisulfite-treated, universally methylated DNA. DNA concentration is determined at 260 nm. A standard dilution series is done in the presence of poly-A (5 ng/μL, Roche Diagnostics) to minimize losses through adhesion of DNA to tube surfaces. Undiluted DNA is stored at –20°C. Repeated freezing and thawing have to be avoided.

5.3.2. Universally Methylated Standard DNA

CpGs in DNA can be completely methylated by *Sss*I DNA methyltransferase (New England Biolabs). 5 μg phi29-amplified, bisulfite-treated, or chromosomal DNA is incubated at 37°C in 20 μL reaction volumes containing 5 U *Sss*I and 160 mM *S*-adenosylmethionine (SAM). After 3 h of incubation, another 5 U *Sss*I enzyme and 160 mM SAM are added and incubation is continued at 37°C for 14 h. Then the enzyme activity is stopped at 65°C for 10 min.

5.4. Protocol 4: Chromatin Immunoprecipitation

- Dounce 50–500 mg of frozen postmortem gray matter tissue in 5× brain volume of douncing buffer (10 mM Tris–HCl, pH 7.5; 4 mM $MgCl_2$; and 1 mM $CaCl_2$) and place in 2.0-mL tube.
- Add 5 U/mL of micrococcal nuclease (Sigma-Aldrich), mix quickly by pipetting, and put on ice.
- Incubate the samples for 7 min at 37°C.
- Place the samples on ice and keep on ice during the next steps.
- Stop micrococcal nuclease by adding 0.5 M EDTA to a concentration of 10 mM.

- Place the samples into 15-mL tube and add 10× the sample vol of 0.2 mM EDTA, pH 8.0, 1/2,000 sample vol of 0.2 M benzamidine (0.1 mM), 1/1,000 sample vol of 0.1 M phenylmethanesulfonyl fluoride (PMSF) (0.1 mM).
- Incubate on ice for 1 h, and vortex every 10 min.
- Add 1/2,000 sample vol of 3 M DTT (1.5 mM).
- Vortex and centrifuge at 3,175 relative centrifugal force (RCF) for 10 min at 4°C.
- Take 500 μL of supernatant for Input Control and split the remainder into two tubes for ChIP. The Input Control is stored at −80°C until further use.
- Add 1:10 vol (160 μL) of 10× FSB (50 mM EDTA/200 mM Tris–HCl [pH 7.5]/500 mM NaCl) and 4 μg of antibody [-general H3 (1:20,000); -H3K4me1 (1:11,000; ab8895; Abcam, Cambridge, MA); -H3K4me2 (1:50,000; ab7766; Abcam); -H3K4me3 (1:50,000; ab8580; Abcam, marker for active transcription); -H3K27me1 (1:20,000;07-448; UBI/Millipore, Billerica, MA); -H3K27me3 (marker for silence loci) from Upstate; -H3k9 (Millipore); -H3K9/14Ac (1:3,500); -H4K12 (antibody from Abcam)] to each of the 2 ChIP samples and vortex.
- Rotate at 4°C overnight.
- Prepare protein G-agarose as follows:
- Add 1.6 mL 1× FSB to 245 μL protein G-agarose in a 2-mL tube.
- Split the solution into two 2-mL tube and fill up to 1.6 mL each with 1× FSB.
- Rotate at RT for 30 s and centrifuge at 0.1 RCF for 30 s.
- Remove supernatant.
- Add 1.6 mL 1× FSB, rotate for 30 s, and centrifuge at 0.1 RCF for 30 s.
- Remove supernatant and combine both tubes with 1.5 mL 1× FSB.
- Add 15 μL sonicated Salmon sperm DNA (10 mg/mL).
- Rotate at RT for 30 min and centrifuge at 0.1 RCF for 30 s.
- Remove supernatant.
- Add 200 μL 1× FSB.
- Now add two 90 μL aliquots of the slurry to the samples.
- Add 1 mL of 1× FSB to remaining agarose beads (negative control) and rotate at 4°C for 1 h.
- Centrifuge at 0.1 RCF for 30 s and discard the supernatant.
- Add 1 mL of ice-cold washing solution (2 mM EDTA/20 mM Tris–HCl) [pH 8.0]/500 mM.

- NaCl/0.1% SDS/1% Triton X-100 to the beads and rotate for 3 min at room temperature.
- Centrifuge at 0.1 RCF for 30 s.
- Discard supernatant.
- Repeat with 1 mL high-salt immune complex washing solution (2 mM EDTA/20 mM Tris–HCl [pH 8.0]/500 mM NaCl/1% SDS/1% Triton X-100).
- Repeat with LiCl buffer (250 mM LiCl/1 mM EDTA/10 mM Tris–HCl, pH 8.0/1% IGEPAL-CA360/1% deoxycholic acid).
- Wash twice in 1 mL 1× TE buffer (10 mM Tris–HCl, pH 8.0/1 mM EDTA).
- Add 250 μL freshly prepared elution buffer (100 mM $NaCO_3$/1% SDS).
- Rotate 15 min at room temperature, and centrifuge at 0.4 RCF for 1 min.
- Transfer supernatant in 2.0-mL tube.
- Add 250 μL elution buffer to each sample again and vortex manually for a few seconds.
- Vortex for 15 min on a vortexer.
- Centrifuge at 16 RCF for 4 min.
- Combine supernatant with aliquot in 2.0-mL tube (above).
- Add 10 μL 0.5 M EDTA, 25 μL 0.8 M Tris–HCl, pH 6.5, and 10 mg/mL proteinase K (1/200 sample) to each sample.
- Add lysis buffer for proteinase K digestion (1/10 of sample buffer) and 10 mg/mL proteinase K (1/200 of sample buffer) to each Input Control.
- Incubate Input and ChIP samples at 52°C for at least 3 h.
- Add 500 μL phenol–chloroform.
- Vortex several seconds and centrifuge at 13 RCF for 5 min.
- Transfer upper phase in 2.0-mL tube.
- Add a mixture of 2 μL glycogen, 50 μL 3 M sodium acetate, and 1.375 mL 100% ethanol.
- Vortex vigorously and precipitate at –80°C overnight.
- Place the samples on ice.
- Centrifuge at 15 RCF at 4°C for 10 min.
- Remove supernatant carefully (mind the pellet).
- Add 1 mL cold 75% ethanol and invert the tube 4–6×.
- Centrifuge at 18 RCF at 4°C for 5 min.
- Remove supernatant and allow pellets to dry.
- Dissolve pellets in 50 μL 4 mM Tris–HCl, pH 8.0, and store at –80°C until further use.

Optionally, the DNA-shearing kit and One-Day ChIP kit protocol from Diagenode (Diagenode, Belgium) have been used according to the manufacturer's instructions. The following modifications have been made to optimize the procedure for hippocampal tissue.

- Tissue is homogenized with twice the amount of the different buffers indicated.
- DNA shearing is performed by the Bioruptor (Diagenode) with the following settings (25 min, High, 30 s ON, 30 s OFF).
- Sheared chromatin samples are incubated for 60 min with 4 μL of antibody.
- Subsequently, the antibody–antigen complex is further incubated with pre-blocked beads and additional 500 μL of ChIP buffer for 60 min.
- Afterward, the beads are washed twice.
- Precipitated DNA is analyzed in a Bioanalyzer (Agilent).

5.4.1. Quantitative Real-Time PCR

For the ChIP experiments, qPCR for specific genomic DNA is performed using SYBR Green I Master kit (Roche). Primers for promoter regions are selected to be specific for the first 300 bp upstream of the TSS. Primers for coding regions are selected on the basis of ChIP-*Seq* results.

5.5. Protocol 5: Immuno-PCR

5.5.1. Coating of Microplates with Capture Antibody

1. Dilute the capture antibody to a concentration of 2 μg/mL in RB buffer. Titration range 20–0.2 μg/mL.
2. Add 30 μL per well of diluted capture antibody to TopYield modules (8-well strips) fixed to a frame.
3. Seal modules and incubate for 48 h (12–48) h at 4°C.
4. Wash three times for 1 min with 240 μL per well Tris-buffered saline (TBS) at room temperature with orbital shaking (500–600 rpm). Use a multichannel pipette for washing.
5. Add 240 μL per well of the blocking buffer BTBS (BSA*/Tris-buffered saline).
6. Seal the modules and incubate for at least 12 h at 4°C.

   The antibody-coated and blocked modules are stable and can be stored in blocking buffer for about 7 days at 4°C.

5.5.2. Antigen Immobilization

7. Wash antibody-coated and blocked modules using a multichannel pipette. Wash twice for 30 s and twice for 4 min with 240 μL per well Tween/EDTA/Tris-buffered saline buffer (TETBS) at room temperature with orbital shaking (500–600 rpm).

5.5.3. Standard Washing Step

8. For each assay, prepare a set of calibration standards to allow quantitation of antigen in "unknown" samples from serial dilutions of the antigen in matrix. Prepare antigen 10 μg/mL in buffer and dilute 1 μL in 99 μL Matrix to obtain 100 ng/mL

concentration "A." Add 10 μL "A" to 90 μL Matrix (concentration "B," 10 ng/mL). Continue with tenfold dilutions to prepare the concentrations "C" (1 ng/mL), "D" (100 pg/mL), "E" (10 pg/mL), "F" (1 pg/mL), and "G" (100 fg/mL). Also include a negative control (NC) of pure Matrix in each row of samples.

9. Add 30 μL of each sample (calibration curve, standards, controls, and unknowns) to each well of the plate.
10. Incubate for 30 min at room temperature with orbital shaking.
11. Perform a standard washing step.

5.5.4. Assembly of the Signal-Generating Immunocomplex

12. Dilute the ***biotinylated detection antibody*** to a concentration of 200 ng/mL in reagent dilution buffer (RDB). Titration range of 20 ng/mL–2 μg/mL; the antibody working dilution should be prepared immediately before use.
13. Add 30 μL of the antibody dilution to each well of the plate.
14. Incubate for 45 min at room temperature with orbital shaking.
15. Perform a standard washing step.
16. Dilute a stock solution of 100 mM recombinant streptavidin STV (stored at 4°C) 1:1,000 to a working concentration of 100 nM in TETBS. Titration range of 10 nM–1 mM; the working dilution should be prepared immediately before use.
17. Add 30 μL of the STV dilution to each well.
18. Incubate for 45 min at room temperature with orbital shaking.
19. Perform a standard washing step.
20. Dilute a stock solution of 10 mM biotinylated DNA marker (stored at −20°C) to a working concentration of 3 pM in TETBS. Titration range 0.01–100 pM; DNA must not contain any remaining biotinylated primers from the PCR synthesis.
21. Add 30 μL of the DNA dilution to each well. Keep the remaining DNA working solution for the positive control of the real-time PCR step.
22. Incubate for 45 min at room temperature with orbital shaking.

5.5.5. Real-Time PCR Signal Detection

23. Wash plate seven times with TETBS (4× 30 s, 3× 4 min) and 2× 1 min with TBS with orbital shaking (500–600 rpm).
24. Prepare the PCR master mix according to the manufacturer's instructions using the described concentrations of primer-1 and -2.
25. Pipette 30 μL of the PCR master mix into each well.
26. Seal the modules with adhesive foil. Remove modules from the frame.

27. Transfer sealed modules into the real-time PCR cycler and add the optical compression pad.
28. Close lid and start PCR. Make sure that the heated lid is operating and insert the correct well volume in the PCR program.
29. Carry out PCR as follows:

5.5.6. Time Temperature Repeats

5 min 95°C	1 × 30 s 50°C
30 s 72°C	28 × 30 s 95°C

Add melting curve temperature profile at will.

5.5.7. Control ELISA

1. Perform assay as described before in steps 1–12.
2. Add 30 μL of the STV/alkaline phosphatase conjugate dilution to each well.
3. Incubate for 45 min at room temperature with orbital shaking.
4. Wash seven times with TETBS (4× 30 s, 3× 4 min) and 2× 1 min with TBS.
5. Add 240 μL AttoPhos substrate to each well.
6. Incubate for 20 min at RT with orbital shaking.
7. Measure the fluorescence at 440 nm excitation and 550 nm emission wavelength.

5.5.8. Suggestions

(a) Buy/use for biotinylation NaN3- and BSA-free detection antibodies.
(b) Optimally, you buy/use a lyophilized or purified detection antibody for biotinylation.
(c) Use a commercially available biotinylation kit if possible to biotinylate the detection antibody. Read kit's instructions regarding the use of NaN3- and BSA-containing antibody solutions.
(d) Avoid the use of secondary antibody.
(e) Use fluorescent internal oligonucleotide as detection label instead of SybrGreen.

5.5.9. Reagent Setup

Prepare biotinylated DNA marker as follows: (*prepare a mono-biotinylated DNA marker by PCR using the pUC19 plasmid as the template and primer-1 and primer-2*).

1. For DNA preparation, use a 5′-biotinylated primer-1 and an unmodified primer-2.
2. Carry out PCR. Perform 40 cycles of amplification.
3. Purify the PCR amplification product (PCR Purification kit, Qiagen).
4. Determine the concentration by UV–spectrophotometry.

5. Concentrate *if necessary* the purified biotinylated DNA marker by ultrafiltration using a Centricon 100 microconcentrator tube (Millipore).
6. Store aliquots at 10 mM concentration at –20°C.

Prepare assay buffers as follows: (*ultrapure water for preparation of all buffers. Allow washing buffers to equilibrate to room temperature before use*).

– 10× TBS: 200 mM Tris-Cl, 1,500 mM NaCl, pH 7.5 (adjust pH with NaOH).

Store at room temperature up to 6 months.

– TBS, ready to use: 100 mL 10× TBS and 900 mL H_2O.

Store at RT up to 6 months.

– TE (Tris-buffered saline/EDTA buffer): 10 mM Tris-Cl, 1 mM EDTA, pH 7.5.

Store at RT up to 6 months.

– TETBS: 1× TBS, 5 mM EDTA, and 0.05% (vol/vol) Tween 20.

Store at RT for long periods of time.

– BTBS: 1× TBS, 0.5% (wt/vol) BSA*, 0.2% (vol/vol) NaN3, and 5 mM EDTA.

Store at –20°C. This buffer can be repeatedly frozen and thawed. The buffer should be placed on ice during assay procedure.

– RB (reagent buffer): 50 mM boric acid, pH 9.5; adjust pH with NaOH.

Store at 4°C for up to 6 months. Can be stored frozen.

– RDB: 1 mL BTBS and 9 mL TETBS.

Prepare fresh. The buffer should be placed on ice during assay procedure.

Prepare a solution of 100 mM recombinant streptavidin (STV) in TBS.

Prepare a working dilution of the fluorescence substrate AttoPhos according to the manufacturer's instructions.

*Recommendation: use biotin-free BSA.

5.6. Protocol 6: Antibody Microarrays

Only affinity-purified (protein-G) antibodies are used. If they are not commercially available, IgGs have to be purified.

– Adjust concentrations to 2 mg/mL by filtration through Microcon 100 kDa (Millipore).

– Dilute in 2× spotting buffer (20 mM sodium borate, pH 9.0, 250 mM $MgCl_2$, 0.01% (w/v) sodium azide, 0.5% (w/v) dextran, and 0.001% (w/v) [octylphenoxy]polyethoxyethanol in H_2O). Spot 10 μL of the 1 mg/mL antibody solutions on epoxysilane slides (Schott Nexterion, Jena, Germany) using a Microgrid microarrayer (BioRobotics, Cambridge, UK) with SMP3B pins

(TeleChem-ArrayIt). More than 1,000 arrays can be spotted with 10 μg of antibody (approximately 100× less antibody than required for respective ELISAs).

- Additionally, distribute antibodies against a variety of "housekeeping" proteins (1), polyspecific antibodies against serum proteins, and spotting buffer as negative control randomly across the array.
- After printing, wash slides 10× in 0.05% (w/v) Tween-20 and 0.05% (w/v) Triton-X 100 in PBS.
- Block overnight in 4% (w/v) skim milk powder and 0.05% (w/v) Tween-20 in PBS.
- Wash the slides 4× in PBSTT, 2× in 0.1× PBS, and then dry in a stream of air. In this condition, they can be stored in a humidity chamber at 4°C for up to 1 year.
- To control for proper arraying results, wash representative slides in PBSTT and stain with Sypro Ruby staining solution for 1 h (Sigma-Aldrich).
- Wash 4× with 10% methanol and 7% acetic acid in H_2O.
- Rinse 2× in H_2O and dry in a stream of air.
- Scan with Axon 4000XL scanner at 450 nm excitation and 610 nm emission wavelengths.
- To test for unspecific staining of secondary antibodies, incubate blocked slides for 2 h with 5 nM secondary antibodies conjugated with Cy3 or Cy5.
- Adjust protein samples to 4 mg/mL and label for 1 h with shaking at 4°C with 0.4 mg/mL of the NHS-esters of the dyes Dy-549 or Dy-649 (Dyomics, Jena, Germany) dissolved in 100 mM sodium bicarbonate buffer, pH 9.0, 1% Triton-X 100.
- Stop reactions by adding hydroxylamine to 1 M.
- After 30 min, remove unreacted dye and change buffer to PBS using Zeba Desalt columns (Thermo Scientific).
- Add Complete Protease Inhibitor Cocktail tablets (Roche).
- Store the samples at –20°C until use.
- Before incubation with labeled samples, block microarrays for 3 h in a casein-based blocking solution (Candor Biosciences) on a SlideBooster device (Advalytix, Munich, Germany).
- Dilute the samples 1:20 (600 μL total volume) in blocking solution containing 1% (w/v) Tween-20 and Complete Protease Inhibitor Cocktail, and incubate on the chip for 15 h.
- Wash the samples in PBSTT, in 0.1× PBS, and in H_2O, and then dry in a stream of air.

Acknowledgments

The most valuable comments by Dr. J. Hoheisel and the helpful suggestions by E. Röbel and F. Leonardi-Essmann are greatly acknowledged. The author was supported by NGFN+ alcohol (GS 08152 TP9), DFG Grant Ge 486/15-1, and DAAD 415-alechile/po-D/08/11685.

References

1. de Jonge, H.J., Fehrmann, R.S., de Bont, E.S., Hofstra, R.M., Gerbens, F., Kamps, W.A., de Vries, E.G., van der Zee, A.G., te Meerman, G.J., ter Elst, A. (2007) Evidence based selection of housekeeping genes. *PLoS ONE,* **2**, 898.
2. Novoradovskaya N, Whitfield ML, Basehore LS, Novoradovsky A, Pesich R, Usary J, Karaca M, Wong WK, Aprelikova O, Fero M, Perou CM, Botstein D, Braman J. (2004) Universal Reference RNA as a standard for microarray experiments. *BMC Genomics.* **5**, 20.
3. Beier V, Mund C, Hoheisel JD. (2007) Monitoring methylation changes in cancer. *Adv Biochem Eng Biotechnol.* **104**, 1–11.
4. Martins-de-Souza D, Maccarrone G, Wobrock T, Zerr I, Gormanns P, Reckow S, Falkai P, Schmitt A, Turck CW. (2010) Proteome analysis of the thalamus and cerebrospinal fluid reveals glycolysis dysfunction and potential biomarkers candidates for schizophrenia. *J Psychiatr Res.* **44**, 1176–1189
5. Sun Y, Xu Y, Roy K, Price BD. (2007) DNA damage-induced acetylation of lysine 3016 of ATM activates ATM kinase activity. Mol Cell Biol. **27**, 8502–8509.
6. Gassmann M, Grenacher B, Rohde B, Vogel J (2009) Quantifying Western blots: Pitfalls of densitometry. *Electrophoresis.* **30**, 1845–1855.
7. Leonardi-Essmann F., Lisboa F., Chavez-Peperkamp N., Roebel E., Gebicke-Haerter PJ (2010) Immuno-PCR: an improved laboratory protocol. Central Institute of Mental Health, Psychopharmacology, Mannheim
8. Niemeyer CM, Adler M, Wacker R.(2007) Detecting antigens by quantitative immuno-PCR. *Nat Protoc.* **2**, 1918–30.
9. Cordwell SJ, Thingholm TE. (2010) Technologies for plasma membrane proteomics. *Proteomics.* **10**, 611–627.
10. Gloriam DE, Orchard S, Bertinetti D, Björling E, Bongcam-Rudloff E, Borrebaeck CA, Bourbeillon J, Bradbury AR, de Daruvar A, Dübel S, Frank R, Gibson TJ, Gold L, Haslam N, Herberg FW, Hiltke T, Hoheisel JD, Kerrien S, Koegl M, Konthur Z, Korn B, Landegren U, Montecchi Palazzi L, Palcy S, Rodriguez H, Schweinsberg S, Sievert V, Stoevesandt O, Taussig MJ, Ueffing M, Uhlén M, van der Maarel S, Wingren C, Woollard P, Sherman DJ, Hermjakob H. (2010) A community standard format for the representation of protein affinity reagents. *Mol Cell Proteomics.* **9**, 1–10.
11. Tretter F, Gebicke-Haerter PJ, Mendoza ER, Winterer G. (Eds.) (2010) Systems Biology in Psychiatric Research. From High-Throughput Data to Mathematical Modeling. Wiley-VCH, Weinheim, Germany. ISBN: 978-3-527-32503-0
12. Eberwine J, Yeh H, Miyashiro K, Cao Y, Nair S, Finnell R, Zettel M, Coleman P.(1992) Analysis of gene expression in single live neurons. Proc. Natl. Acad. Sci. U. S. A. 89:3010–3014.

Chapter 4

Endothelial Cell Heterogeneity of Blood–Brain Barrier Gene Expression: Analysis by LCM/qRT-PCR

Tyler Demarest, Nivetha Murugesan, Jennifer A. Macdonald, and Joel S. Pachter

Abstract

The brain vasculature, and more specifically the microvasculature, represents a center of interest for the investigation of the brain. Indeed, this tissue at the interface between the circulatory system and the central nervous system (CNS) actively participates in the brain function either for the brain homeostasis or for the entrance of compounds or cells in the brain. One particular feature of the brain microvasculature, the blood–brain barrier (BBB), is the site of several restrictions. However, although representing a large exchange surface between blood and CNS, microvessels are still difficult to study due to their size and their position inside the CNS. The technological evolutions such as laser capture microdissection (LCM) give the opportunity to reach CNS microvessels for gene expression profiling analysis. This chapter presents the expression profiling of captured microvessels studied with qRT-PCR technology.

Key words: Blood–brain barrier, Endothelial cells, Laser capture microdissection

1. Introduction

1.1. Heterogeneity of the Cerebromicrovascular Endothelium

Far from being an invariant, inert tissue layer, the endothelium is functionally heterogeneous along the three major segments of the microvascular tree: *arterioles*, *capillaries*, and *venules*. Although this property is well established in the peripheral circulatory system (1–6), it has thus far received only scant attention in the central nervous system (CNS) (7, 8). As the CNS microvasculature is the site of the blood–brain barrier (BBB) (8–10), it is possible that genes regulating BBB activities also vary in their expression depending on CNS microvessel segment. Indeed, such heterogeneity might support a "division of labor" along the CNS microvascular

Yannis Karamanos (ed.), *Expression Profiling in Neuroscience*, Neuromethods, vol. 64,
DOI 10.1007/978-1-61779-448-3_4, © Springer Science+Business Media, LLC 2012

endothelium, such that stringent restrictions on solute transport are widely upheld in the face of opportunistic examples of cellular transendothelial migration (8, 11, 12). That leukocytes can cross CNS microvessels at early stages of inflammation—*before* any compromise in BBB integrity is evident—could, in fact, suggest the existence of specialized microvascular "domains" that discriminate between solute and cellular transport. Clarifying this issue holds tremendous significance for being able to appropriately model the *diverse* repertoire of CNS microvascular functions (13). Until recently, however, a means to characterize the full extent of this diverse endothelial behavior has not been available—bulk preparation of cerebral "microvessels," containing a mosaic of different segments, being the typical source of CNS endothelial cells for analysis (14). Laser capture microdissection (LCM) has provided a unique solution to the problem, enabling, for the first time, endothelial cells to be "plucked" from their in situ positions along the microvascular tree of the CNS. When coupled to a downstream analytical profile like qRT-PCR, this combined technology (LCM/qRT-PCR) is a powerful tool for exploring the depth of endothelial heterogeneity among the different CNS microvascular segments (15).

1.2. LCM/qRT-PCR: Proof of Principle

Key to any LCM approach is identifying both the specific cells targeted for capture, and those cells necessary to avoid. Failure to effectively eliminate the latter will lead to contamination of the desired cell type and, thereby not allow a faithful interpretation of the desired cell's profile. While "generic" histological stains such as hematoxylin and eosin are helpful for gross identification of tissues, they do not offer the resolution necessary to allow single cell types of the neurovascular unit (NVU) to be retrieved with precision. In this regard, the NVU is a functional syncytium of CNS microvascular endothelial cells and perivascular cell types comprised of astrocytes, neurons, and pericytes (16, 17). Of these perivascular cells, astrocytes are the most intimately and vastly associated with the endothelium and, hence, provide the biggest obstacle for retrieving highly purified populations of CNS microvascular endothelial cells by LCM. Double-label immunostaining, as demonstrated by this laboratory, has afforded a means to both clearly identify endothelial cells for LCM and, at the same time, avoid significant astrocyte contamination (18, 19). Combined immunohistochemistry and immunofluorescence enables images of endothelial cells and astrocytes to be viewed simultaneously by bright-field and epifluorescence optics, thus enabling the two cell types to be separated effectively in real time. Specifically, immunohistochemistry with anti-CD31 antibody has been used to tag brain microvascular endothelial cells (BMEC), as CD31 (PECAM-1) is expressed by *all* endothelial cells (20, 21)—CNS or peripheral, macrovascular, or microvascular. Such universal staining of endothelial cells by anti-CD-31 strongly contrasts with that obtained by lectins (22, 23), endogenous alkaline phosphatase activity (8, 23), or other endothelial "markers" such as Factor VIII-related antigen and angiotensin-converting enzyme (15, 24) which, while highlighting certain vascular beds, is highly irregular. Regarding development of the immunohistochemical precipitate, horseradish peroxidase (HRP) or alkaline phosphatase (Alk-Phos), with Diaminobenzide (DAB) or Nitro-Blue Tetrazolium Chloride/5-Bromo-4-Chloro-3′-Indolyphosphate p-Toluidine salt (NBT/BCIP) as substrates, respectively, has been used (15, 18, 19, 25, 26). But while both enzyme substrate combinations equally support efficient RNA extraction, Alk-Phos and NBT/BCIP were noted to foster higher protein recovery (25, 26), and thus this latter approach is now generally used. With regard to astrocyte detection, this is achieved by immunofluorescence using anti-GFAP conjugated directly to a fluorophore. Smooth muscle cell detection (for identifying arterioles) is analogously carried out using a fluorescently conjugated antibody to alpha smooth muscle actin (anti-SMA-α). Importantly, the combined immunostaining protocol must be brief (typically ≤10 min) in order to avoid RNA degradation, with inclusion of RNase inhibitor to maximize RNA yield.

Once immunostaining is complete, the tissue should be immediately dehydrated through graded alcohols and xylenes before commencing LCM. Upon completion of LCM, captured samples should be processed for RNA isolation/analysis without delay. This laboratory has enjoyed success with solubilizing LCM samples in TRIzol® and isolating RNA by a modification of the manufacturer's instructions, and the reader is directed to references (15, 19) detailing this protocol. Although highly efficient at preserving RNA, the numerous steps required for TRIzol® -based isolation of RNA can be cumbersome—particularly when dealing with large numbers of samples. Thus, a recent adaptation in this laboratory is to solubilize the LCM material in CL® Cell Lysis Buffer and proceed directly to reverse transcription without the need of RNA purification. This method is far quicker than utilizing TRIzol®, while providing equal or better sensitivity in downstream qRT-PCR (see Fig. 1), and is described below.

2. Materials

2.1. Preparation of Brain Tissue for Immunostaining

1. Compressed CO_2 in glass cylinders is used for euthanasia, as this allows the influx of gas to the induction chamber in a controlled manner. Cages are placed in the chamber, the chamber lid closed, and 100% CO_2 introduced at a rate of 10–20% of the chamber volume per minute. This rate of CO_2 introduction minimizes animal distress. After the animals become unconscious, the flow rate is increased to minimize the time to death. This euthanasia procedure is in accordance with measures

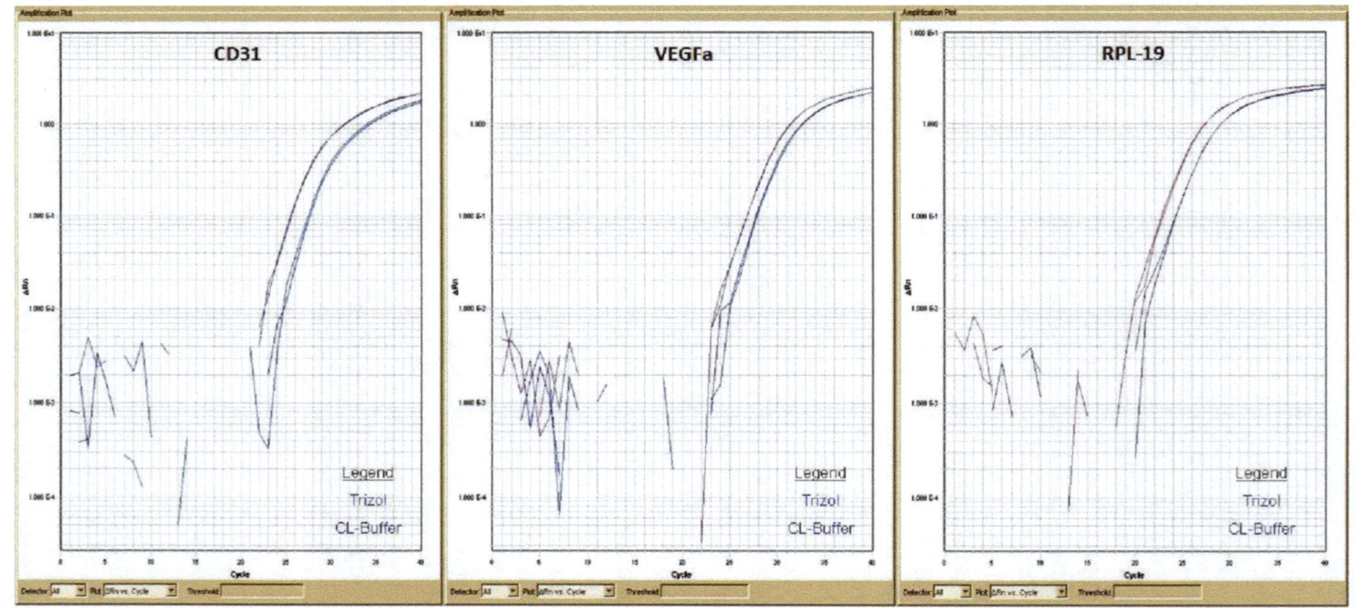

Fig. 1. LCM/qRT-PCR using CL® Lysis Buffer vs. TRIzol. Equal numbers of LCM "shots" (1,000) of cerebral capillary endothelium were obtained, and qRT-PCR then carried by one of two different methods. In one case, the tissue was solublized in CL® Lysis Buffer and qRT-PCR performed directly from this lysate—without need of RNA purification. In the other case, tissue was extracted with TRIzol, RNA isolated, and qRT-PCR performed. Equivalents of LCM material were processed in duplicate by both methods for the same three genes—**CD31**, **VEGF-A** and **RPL-19**—allowing for comparison in terms of sensitivity and reproducibility. Mean Ct values for each gene were consistently lower using CL® Lysis Buffer: Ct value for $\mathrm{CD31}_{\mathrm{CL®Lysis\ Buffer}}$ (26.37) *vs.* $\mathrm{CD31}_{\mathrm{TRIzol}}$ (28.39), a $\Delta CT = 2.02$; Ct value for $\mathrm{VEGFa}_{\mathrm{CL®Lysis\ Buffer}}$ (28.46) *vs.* $\mathrm{VEGFa}_{\mathrm{Trizol}}$ (29.50), $\Delta CT = 1.04$; $\mathrm{RPL\text{-}19}_{\mathrm{CL®Lysis\ Buffer}}$ (24.32) *vs.* $\mathrm{RPL\text{-}19}_{\mathrm{Trizol}}$ (25.80), $\Delta CT = 1.48$. The direct method of qRT-PCR, using CL® Lysis Buffer, was thus more sensitive than that using TRIzol-purified RNA, while both methods showed similarly high reproducibility between technical replicates.

stipulated by the Animal Care and Use Guidelines of the University of Connecticut Health Center (Animal Welfare Assurance # A3471-01).

2. Mouse brains.
3. Sterile surgical instruments: scalpel, forceps, and scissors.
4. Isopentane (2-methylbutane; Acros Organics, Morris Plains, NJ).
5. Shandon Cryomatrix-embedding medium (Thermo Fisher Scientific, Waltham, MA).
6. Microtome cryostat (Thermo Electron Corporation, San Jose, CA).
7. Shandon MX35 disposable microtome blades (Thermo Fisher Scientific, Waltham, MA).
8. RNase Away (Invitrogen, Carlsbad, CA).
9. Uncoated glass slides.

2.2. Immunostaining

1. Acetone (Fisher Scientific, Pittsburgh, PA).
2. Phosphate-buffered saline (PBS), pH 7.6.
3. Diethyl pyrocarbonate (DEPC; Sigma, St. Louis, MO).
4. Antibody diluting solution: All antibodies are diluted in 1× PBS containing 0.2% Tween-20
5. Pap-pen hydrophobic marker pen.
6. Rat antimouse CD31 antibody, 1:10 dilution (BD Biosciences, San Jose, CA).
7. Biotinylated rabbit antirat IgG antibody, 1:250 dilution (Vector Labs, Burlingame, CA).
8. Alexa 488-conjugated glial fibrillary acidic protein (GFAP) antibody, 1:5 dilution (Invitrogen, Carlsbad, CA), and fluorescein isothiocyanate-conjugated alpha smooth muscle actin (FITC-α-SMA) antibody, 1:5 dilution (Sigma, St. Louis, MO).
9. Alkaline phosphatase ABC kit (Vector Labs, Burlingame, CA).

   ABC Complex: Add one drop of reagent A to one drop of reagent B. Add 2.5 ml of 1× PBS. Mix well and incubate for 30 min at RT prior to use. Prewarm an aliquot to 37°C before use.
10. Alkaline phosphatase substrate (NBT/BCIP) kit IV, (Vector Labs, Burlingame, CA). NBT/BCIP substrate solution: Add one drop of reagent 1 (BCIP) to one drop of reagent 2 (NBT). Add one drop of reagent 3 ($MgCl_2$) and 1 ml of 0.1 M Tris-HCl, pH 9.5. Mix well.
11. RNasin Plus RNase inhibitor (Promega, Madison, WI): add to all antibody and substrate solutions to a final concentration of 0.4 U/μl (1:100 dilution of stock).

2.3. Dehydration of Tissue Sections Prior to Laser Capture Microdissection

1. Distilled water (DNase, RNase free; Molecular Devices, Sunnyvale, CA).
2. Ethanol (75%, 95%, and 100%; Fisher Scientific, Pittsburgh, PA).
3. Xylenes (Fisher Scientific, Pittsburgh, PA).

2.4. LCM

1. Model Pixcell IIe and XT LCM microscopes, each equipped with epifluorescent optics (Molecular Devices, MDS Analytical Technologies, Sunnyvale, CA).
2. Capsure HS caps (Molecular Devices, MDS Analytical Technologies, Sunnyvale, CA).
3. Sterile 0.5 ml microfuge tubes.

2.5. Tissue Solubilization

1. CL® Cell Lysis Buffer (Signosis, Inc., Sunnyvale, CA).

2.6. qRT-PCR

1. Superscript III Reverse Transcriptase kit (Invitrogen, Carlsbad, CA), including 0.1 M DTT, 5× First Strand Buffer and Superscript III RT.
2. Random Hexamers (Roche Diagnostics, Manheim, Germany).
3. dNTPs (Roche Diagnostics).
4. SYBR Green PCR Mix (ABI, Foster City, CA).
5. MicroAMP Fast Optical 96-well reaction plates (ABI).

3. Methods

3.1. Retrieval of Brain Tissue and Preparation of Tissue Sections

1. Kill the animal by euthanasia and spray the head with 75% ethanol.
2. Carefully remove the whole brain from the cranium using sterile surgical instruments.
3. Immediately immerse the brain tissue into isopentane freezing medium (precooled on dry ice) for 1–2 min.
4. Remove the frozen brain tissue from isopentane with a clean spatula.
5. Embed the frozen specimen in embedding medium, on dry ice.
6. Store the embedded tissue at –80°C until ready for sectioning.
7. Prepare 7 μm frozen brain sections of desired orientation (e.g., coronal, sagittal…etc).
8. Affix them onto uncoated clean glass slides, and keep on dry ice until cutting session is completed.
9. Place the tissue sections in a clean slide box precleaned with RNase Away.

10. Store at –80°C. Tissue sections are used for LCM within 1 week of sectioning in order to minimize degradation of RNA with increase in storage time. If it is desired to perform LCM at later times, it is best to leave brain tissue unsectioned at –80°C. Tissue left in this way has been analyzed by immuno-LCM coupled to qRT-PCR after being stored frozen for as long as 1 year.

3.2. Fixation of Frozen Tissue Sections and Quick Immunostaining for LCM

1. Remove tissue section from –80°C and quickly thaw to room temperature (see Note 1).
2. Fix the tissue section in 75% ethanol for 3 min.
3. Briefly air-dry the tissue and draw a water repellent circle around tissue section with Pap-pen hydrophobic marker pen.
4. Incubate with monoclonal rat antimouse CD31 antibody (1:10 dilution) for 3 min at room temperature (RT).

   All antibodies are diluted in 1× PBS + 0.2% Tween-20. Add RNasin Plus RNase inhibitor (1:100 dilution) to all staining reagents.
5. Wash briefly by dipping slide in 1× PBS for 5 s.
6. Incubate with biotinylated rabbit antirat antibody (1:250 dilution) for 3 min at RT.
7. Wash briefly by dipping slide in 1× PBS for 5 s.
8. Add prepared avidin-biotinylated enzyme complex (ABC), prewarmed to 37°C, to the tissue for 3 min.

   Preparation of ABC: Add one drop of reagent A + one drop of reagent B to 2.5 ml of 1× PBS. Mix well and incubate for 30 min at RT prior to use. Prewarm an aliquot to 37°C before using in step 8 (see Notes 2, 3).
9. Wash briefly by dipping slide in 1× PBS for 5 s.
10. Add prepared NBT/BCIP substrate solution and incubate until purple color develops (7 min).

    Preparation of NBT/BCIP solution: Add one drop of reagent 1 (BCIP) + one drop of reagent 2 (NBT) + one drop of reagent 3 ($MgCl_2$) to 1 ml of 0.1 M Tris-HCl, pH 9.5. Mix well.
11. Wash briefly by dipping slide in 1× PBS for 5 s.
12. Add a 1:5 dilution of Alexa 488-conjugated GFAP antibody or FITC-conjugated α-SMA antibody for 8 min.
13. Wash briefly by dipping slide in 1× PBS for 5 s.

3.3. Dehydration of Tissue Sections Prior to LCM

1. Dip the immunostained slide in 75% ethanol for 10 s (see Note 4).
2. Transfer to 95% ethanol for 30 s.
3. Immerse in first 100% ethanol for 60 s.
4. Immerse in second 100% ethanol for 90 s.
5. Transfer to first xylene wash for 2 min.

6. Transfer to second xylene wash for 3 min.
7. Air-dry the slide for 5 min after the final xylene wash.

3.4. LCM

1. The dehydrated double-immunostained slide is placed on the microscope stage.
2. Vacuum seal is used to keep the slide in place.
3. An HS Capsure LCM cap is placed in position, in the path of the laser beam, above the tissue section.
4. Both bright field and epifluorescence optics is used to simultaneously visualize the CD31+ microvessels (purple) and fluorescent (green) GFAP+ astrocyte processes or α-SMA+ smooth muscle cells (see Figs. 2 and 3) Different microvessel segments are identified as follows: capillaries, <10 μm in diameter; venules, >10 μm in diameter and α-SMA–; and arterioles, >10 μm in diameter and α-SMA+.
5. A 7.5-μm laser spot size is used at a power range of 65–80 mW and pulse duration of 550–800 ms. *This combination of parameters allows for efficient retrieval of microvessel sample and limits contamination from perivascular cells* (see Note 5).
6. Specific microvessel segments are targeted for LCM. The cap containing LCM-captured material is fit onto a clean 0.5 ml microfuge tube and stored at –80°C until further analysis (see Note 6).

3.5. Tissue Solubilization

1. Heat 22 μl of CL® Cell Lysis Buffer to 75°C for 5 min, and place directly on cap containing LCM-captured tissue. Scrape buffer along the cap for 2–3 min, then remove entire liquid contents with dislodged cap polymer gel and transfer to a fresh 0.5 ml sterile microfuge tube.
2. Incubate solubilized tissue/polymer gel at 75°C for an additional 15 min. Sample can be stored frozen at –80°C until cDNA synthesis.

3.6. qRT-PCR

1. Perform reverse transcription and first strand cDNA synthesis using SuperScript III Reverse Transcriptase and manufacture's protocol, with a modification in the extension temperature to 42°C for 60 min. cDNA may be stored at –20°C until used for qRT-PCR.
2. Measurement of cDNA levels by qRT-PCR is carried out using an ABI PRISM 7900 Sequence Detection System Version 2.3 and SYBR green fluorescence to quantify relative amplicon amounts. A separate cocktail of specific primer pair (final concentration of 250 nM), SYBR Green Mix (2× stock) and water is made for each gene to be analyzed, and 23 μl of this cocktail dispensed into respective wells of a 96 well Microamp qRT-PCR reaction plate.

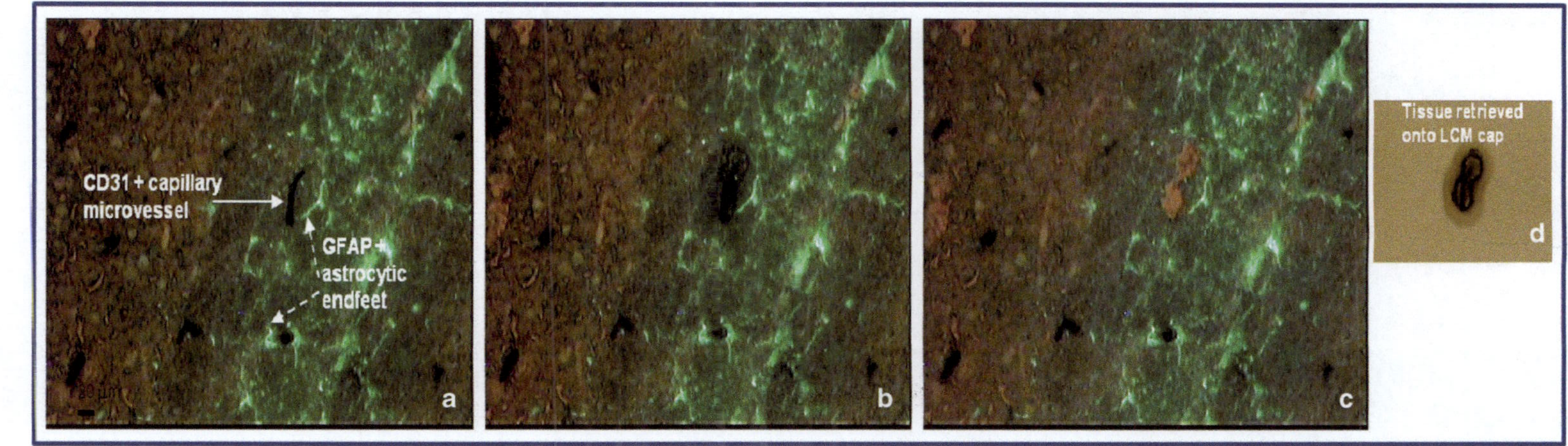

Fig. 2. Retrieval of brain capillary endothelial cells by immuno-LCM. A 7-μm thick frozen coronal section of mouse brain double-immunostained with antibodies against CD31 and GFAP is shown. LCM is performed using the PixCell IIe. (**a**) Tissue stained with anti-CD31 (*dark purple*) and anti-GFAP (*green*), viewed under bright-field and epifluorescence optics simultaneously, prior to LCM. *Arrows* indicate the CD31+ stained capillary endothelial cells (*dark purple*) and the perivascular distribution of astrocytic endfeet (*green*). (**b**) Tissue section immediately after the laser pulse, with the LCM cap placed over the CD31+ capillary microvessel. (**c**) Tissue after LCM, showing the entire endothelial layer was removed and the fluorescent distribution of astrocytic end-feet was not disturbed. (**d**) Tissue transferred to cap after LCM, showing intact endothelium, was retrieved, with no detection of astrocytic endfeet. A similar process is used to retrieve brain venular endothelial cells.

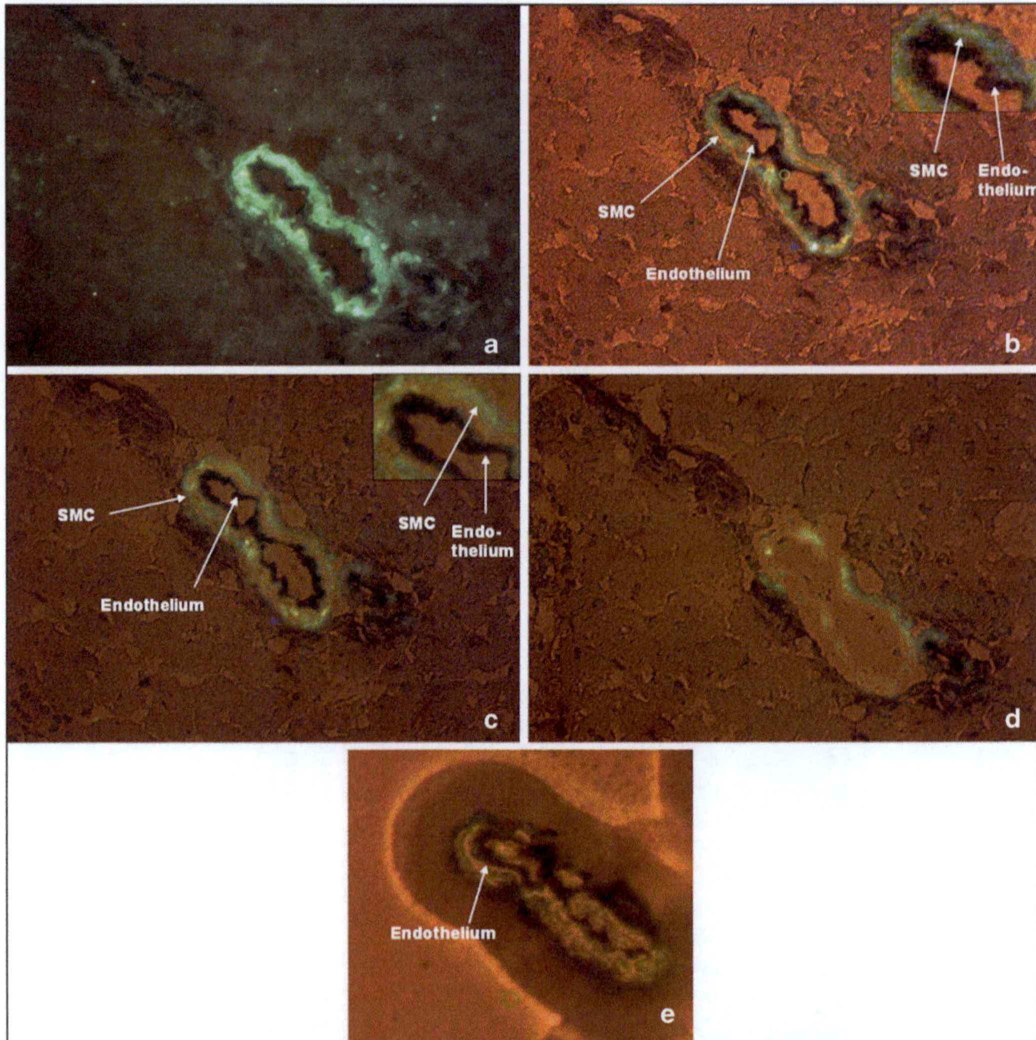

Fig. 3. Retrieval of brain arteriolar endothelial cells by immuno-LCM. Because of the close apposition of a dense muscle layer of α-SMA+ cells near the arteriolar endothelium, an alternative LCM procedure is employed to retrieve endothelial cells from brain arterioles. For this purpose, an Arcturus Model XT LCM is used. (**a**) Immunofluorescent detection of thick *tunica media* of α-SMA+ smooth muscle cells (*green*). (**b**) Same section after double-labeling, showing CD31 expression (*dark brown*) by the endothelium, and closely associated layer of α-SMA+ cells; the close proximity of these cells is highlighted in the *box* in the *upper right*. (**c**) The same tissue section after "burning" away part of the *tunica media* with an ultra-violet (UV) laser; *box* at *top right* now shows "gap" between endothelial cells and α-SMA+ smooth muscle cells. (**d**) The section following "lifting" of the endothelium using an infra-red (IR) laser; remnants of the burned *tunica media* are seen left behind. (**e**) Only the endothelium is deposited on the cap (for qRT-PCR analysis). Adjustments in the parameters of UV and IR lasers (e.g. spot size, power and duration) are dictated by specific tissue conditions and the degree of cell enrichment desired.

3. cDNA is diluted based on both the number of LCM "shots" performed and total genes to be analyzed (see Note 7). A volume of 3.5 μl of diluted cDNA sample (depending on initial RNA sample volume) is added to each well, resulting in a total volume of 26.5 μl/well.

4. The following qRT-PCR controls are also used: no template control, and no reverse transcriptase control. Each primer pair's efficiency (E) is calculated by $E=(10^{(-1/\text{slope})}-1)$, where slope refers to the slope of the standard curve from the log serial fold dilutions of template cDNA against their Ct values (27). Relative quantification is calculated using the formula: $(1+E\text{target})^{\text{Ct(target)}}/(1+E\text{ref})^{\text{Ct(ref)}}$ and ref: RPL-19, target: gene of interest, Ct: threshold cycle.
5. The amplification protocol used is as follows:
   (a) 50°C for 2:00 min
   (b) 95°C for 10:00 min
   (c) 95°C for 0:15 min, 60°C for 1 min, 95°C for 0:15 min; repeat 40×
   (d) 95°C for 0:15 min, 60°C for 0:15 min, 95°C for 0:15 min.

4. Notes

1. Tissue sections should be used within 1 week of sectioning in order to minimize degradation of RNA. Unsectioned brain tissue (embedded in cryomatrix) has been kept frozen at −80°C for up to 1 year without significant loss of qRT-PCR signal.
2. Immunohistochemistry using alkaline phosphatase rather than horseradish peroxidase is preferred as the NBT/BCIP reaction product of the former enzyme is more soluble in a variety of solvents, allowing for greater RNA and protein recovery (15, 24, 26). Additionally, endogenous alkaline phosphatase activity in endothelial cells amplifies the immune signal.
3. Prewarm the ABC (avidin-biotinylated enzyme complex) reagent to 37°C prior to use. This will shorten the time for colorimetric development of the reaction product.
4. Extensive dehydration of tissue sections is critical for high efficiency capture during LCM. Alcohol solutions and xylenes should be changed frequently, and use of a room dehumidifier is highly recommended.
5. A single pulse of the laser is referred to as an LCM "shot." This terminology is used to define comparative amounts of LCM tissue being analyzed. As tissue sections only contain fractions of cellular material, it is inappropriate to refer to such LCM captures as "cells."
6. Loosely adhered tissue, nonspecifically stuck to the cap, can be removed using a clean Post-it™, prior to storage.

7. Based on the total number of genes analyzed, cDNA volume is diluted accordingly. For the examples used herein, 1,000 LCM shots were solubilized in 22 μl of CL® buffer and this was dispensed to analyze 3 genes in duplicate. However, the cDNA may be further diluted to analyze >3 genes. Using a similar LCM/qRT-PCR approach, this laboratory has described heterogeneity in expression of a catalog of 87 BBB-associated genes in brain capillary *vs.* venular endothelial cells (15).

Acknowledgments

This work was supported by grants RO-1-MH54718 and R21-NS057241 from the National Institutes of Health, grant PP-1215 from the National Multiple Sclerosis Society and grant 2011–0143 from the Connecticut Dept. of Public Health to J.S.P., and grant 0815733D from the American Heart Association to N.M.

References

1. Garlanda, C., and Dejana, E. (1997) Heterogeneity of endothelial cells. Specific markers. *Arterioscler. Thromb. Vasc. Biol.* 7, 1193–1202.
2. Boegehold, M.A. (1998) Heterogeneity of endothelial function within the circulation. *Curr. Opin. Nephrol. Hypertens.* 7, 71–78.
3. Thurston, G., Baluk, P., and MacDonald, D. (2000) Determinants of endothelial phenotype on venules. *Microcirculation* 7, 67–80.
4. Aird, W.C. (2007) Phenotypic heterogeneity of the endothelium: I. Structure, function and mechanisms. *Circ. Res.* **100**, 158–173.
5. Aird, W.C. (2007) Phenotypic heterogeneity of the endothelium: II. Representative vascular beds. *Circ. Res.* **100**, 174–190.
6. Tse, D., and Stan, R.V. (2010) Morphological heterogeneity of endothelium. *Semin. Thromb. Hemost.* **36**, 236–245.
7. Spatz, M., Kawai, N., Merkel, N., Bembry, J., and McCarron RM. (1997) Functional properties of cultured endothelial cells derived from large microvessels of human brain. *Am. J. Physiol.* **41**, C231–C239.
8. Ge, S., Song, L., and Pachter J.S. 2005. Where is the blood-brain barrier…really? *J. Neurosci. Res.* **79**, 421–427.
9. Abbott, N.J., Patabendige, A.A., Dolamn, D.E., Yusof, S.R., and Begley, D.J. (2010) Structure and function of the blood-brain barrier. *Neurobiol. Dis.* **37**, 13–25.
10. Cardoso, F.L., Brities, D., and Brito, M.A. (2010). Looking at the blood-brain barrier: molecular anatomy and possible investigation approaches. *Brain Res. Rev.* **64**, 328–363.
11. Bechmann. I., Galea, I., and Perry, V.H. (2007) What is the blood-brain barrier (not)? *Trends Immunol.* **28**, 5–11.
12. Engelhardt, B. (2010) The blood-central nervous system barriers actively control immune cell entry into the central nervous system. *Curr. Pharm. Des.* **14**, 1555–1565.
13. Cucullo, L., Marchi, N., Hossain, M., and Janigro, D. (2010) A dynamic in vitro model for the study of immune cell trafficking into the central nervous system. J. Cereb. Blood Flow Metab. doi:10.1038/jcbfm.2010.162.
14. Ribiero, M.M., Castanho, M.A., and Serrano, I. (2010) In vitro blood-brain barrier models -- latest advances and therapeutic applications in a chronological perspective. *Mini. Rev. Med. Chem.* **10**, 262–270.
15. Macdonald, J., Murugesan, N., and Pachter, J.S. (2009) Endothelial cell heterogeneity of blood-brain barrier gene expression along the cerebral microvasculature. *J. Neurosci. Res.* **88**, 1457–1474.
16. McCarty, J.H. (2009) Cell adhesion and signaling networks in brain microvascular units. *Curr. Opin. Hematol.* **16**, 209–214.

17. del Zoppo, G.J. (2010) The neurovascular unit in the setting of stroke. *J. Intern. Med.* **267**, 156–171.
18. Kinnecom, K., and Pachter, J.S. (2005) Selective capture of endothelial cells and perivascular cells from brain microvessels by laser capture microdissection. *Brain Res. Protoc.* **16**, 1–9.
19. Macdonald, J., Murugesan, N., and Pachter, J.S. (2008) Validation of immuno-laser capture microdissection coupled with quantitative RT-PCR to probe blood-brain barrier gene expression in situ. *J. Neurosci. Method.* **174**, 219–226.
20. Feng, D., Nagy, J.A., Pyne, K., Dvorak, H.F., and Dvorak, A.M. (2004) Ultrastructural localization of platelet endothelial adhesion molecule (PECAM-1, CD-31) in vascular endothelium, *J. Histochem. Cytochem.* **52**, 87– 101.
21. Ilan, N., and Madri, J. PECAM-1: old friend, new partners. (2003) *Curr. Opin. Cell Biol.* **15**, 515– 524.
22. Smolkova, O., Zavadka, A., Benkston, P., and Lutsyk, A. (2001) Cellular heterogeneity of rat vascular endothelium as detected by HPA and GS I lectin-gold probes. *Med. Sci. Monit.* 7, 659–668.
23. Vorbrodt, A.W. (1988) Ultrastructural cytochemistry of blood–brain barrier endothelia. *Prog. Histochem. Cytochem.* **18**, 1–99.
24. Jandeleit-Dahm, K., Jackson, B., Paxton, D., Perich, R., and Johnston, C.I. (1995) Characterization of angiotensin converting enzyme from different vascular beds. *Blood* **4**, 170–176.
25. Lu, C., Murugesan, N., Macdonald, J., Wu, S.-L., Pachter, J.S., and Hancock, W.S. (2008) Analysis of brain microvascular endothelium using immuno-laser capture microdissection coupled to a hybrid LTQ-FT-MS proteomics platform. *Electrophoresis* **29**, 2689–2695.
26. Murugesan, N., Wu, S.-L., Hancock, W.S., and Pachter, J.S. (2011) Analysis of mouse brain microvascular endothelium using laser capture microdissection coupled with proteomics. *Methods Mol. Biol.* **686**, 297–311.
27. Pfaffl, M.W. (2001). A new mathematical model for relative quantification in real-time RT-PCR. *Nuc. Acids Res.* **29**, e45.

Chapter 5

Gene Expression Profiling Using 3′ Tag Digital Approach

Yan W. Asmann, E. Aubrey Thompson, and Jean-Pierre A. Kocher

Abstract

Massive parallel sequencing will become the method of choice for transcriptome profiling. Two protocols have been developed to quantify level of expressions: full-length RNA sequencing (RNA-SEQ) and 3′ tag digital gene expression (DGE). We have studied the performance of 3′ tag DGE profiling and used this protocol to compare the expression profiles of brain RNA to universal human reference RNA. This comparison highlighted that DGE is highly quantitative with excellent correlation of differential expression with quantitative real-time PCR. Our analysis also showed that when compared to microarray, one lane of 3′ DGE sequencing had wider dynamic range for transcriptome profiling and was able to detect expressed genes that are below the detection threshold of microarray. We conclude that 3′ tag DGE profiling is highly sensitive and reproducible for transcriptome profiling. It outperforms microarray platforms in detecting lower abundant transcripts.

Key words: Parallel sequencing, Transcriptome profiling, Bioinformatics

1. Background and Introduction

High-throughput analysis of transcriptome profiles has become a standard tool in clinical research (1). The prevailing wisdom holds that one should be able to interpret patterns of gene expression in diseased individuals, with a view toward using these data to elucidate etiology (2), refine diagnosis (3), predict prognosis (4), and improve treatment (5) of diseases based upon unique associations between cohorts of genes that are over- or underrepresented in the transcriptome of pathological samples. The recent development of the next-generation sequencing (NGS) technologies is providing means to quantify genomics variations and gene expression at an unprecedented level of resolution. In the near future, NGS will enable the sequencing of Giga-bases of DNA or RNA sequences at

Yannis Karamanos (ed.), *Expression Profiling in Neuroscience*, Neuromethods, vol. 64,
DOI 10.1007/978-1-61779-448-3_5, © Springer Science+Business Media, LLC 2012

an affordable cost in reasonable time. Although currently still expensive, these novel approaches attract a high level of interest driven by the potential for discoveries and clinical applications. NGS technologies are also versatile, enabling the interrogation of various genomics features such as SNPs, mutations, CNV regions, fusion genes, and transcript splice variants. The application of NGS to the profiling of transcripts is particularly interesting since this technology presents significant advantages over traditional quantification methods including hybridization-based technologies such as microarray and beadarray (6, 7), sequencing-based technologies such as SAGE (8), and clone of cDNA libraries sequencing (9, 10). Microarrays have been the technology of choice for the quantification of gene expression while suffering from well-known limitations including insufficient sensitivity for quantifying lower abundant transcripts, narrow dynamic range, and nonspecific hybridizations. Microarrays are also limited to the measure of known/annotated transcripts. Sequencing-based methods such as SAGE rely upon cloning and sequencing cDNA fragments. The count of cDNA fragments associated with a transcript provides an estimate of the mRNA abundance, with the assumption that cDNA fragments sequenced contain sufficient information to identify a transcript.

The NGS technology (11–13) eliminates some of these barriers. First, it enables massive parallel sequencing, which simultaneously sequences tens to hundreds of millions of DNA fragments and generates giga-bases of sequence information from a single experiment. The DNA/RNA is randomly fragmented into segments of few hundred nucleotides in length. These are amplified by a process that retains spatial clustering of the PCR products, and sequenced by one of several technologies. Second, besides providing a more effective approach to SAGE and traditional cDNA sequencing (14, 15), NGS offers a number of technical advantages over hybridization-based microarray methods. The output from sequence-based protocols is digital, rather than analog, simplifying data normalization and summarization, and eliminates the need for background correction. The dynamic range is essentially infinite, if one accumulates enough sequence tags. Finally, no prior knowledge of the transcriptome is required, making NGS a useful technology for the discovery and annotation of novel transcripts as well as for analysis of poorly annotated genomes.

One application of NGS is the quantification of transcriptome by sequencing 20–21 base cDNA tags located in the 3′ region of genes. This approach, referenced as 3′ tag digital gene expression (DGE), sequences cDNA libraries enriched in the 3′ untranslated regions of polyadenylated mRNA. Although less informative than full-length RNA sequencing (RNA-Seq), DGE offers a cost-effective approach to the quantification of transcripts with poly-A tails.

In theory, one run of DGE using Illumina Genome Analyzer (GA) II can generate sequencing depth of coverage extensive enough to quantify single copy of gene expression. If one assumes that on average, about 350,000 transcripts are expressed per cell, and that the Illumina GA II produces sequence tags at least 10 million reads per run, about 30 tags per transcript expressed at one copy per cell could be detected. Based on this assumption, DGE can be used to interrogate low-abundance transcripts, which may comprise as much as half of the nonstructural RNA within a cell.

To validate the potential of DGE and assess the use of 3′ tag DGE as an alternative to microarrays or RNA-Seq, several experiments were designed. Two RNA samples from commercially available samples used by the MicroArray Quality Control (MAQC) project (16), Human Brain Reference RNA (HBRR), and Universal Human Reference RNA (UHRR) were used for this purpose. These samples have been well characterized by microarray (repeatedly assayed by 7 microarray platforms) and quantitative real-time PCR (qPCR, 3 quantitative PCR platforms). The sensitivity, dynamic range, reproducibility, and accuracy of 3′ tag DGE sequencing were assessed by comparing the gene expression profile obtained from 3′ tag DGE with the ones obtained from microarray and qPCR experiments.

2. Material and Methods

A detailed description of the methods that are summarized in the following brief section is provided in the published work by Asmann et al. (17). The UHRR and HBRR samples were, respectively, purchased from Stratagene (catalog no.740000) and Ambion/Applied Bioscience (catalog no. AM6051). Eight HBRR libraries (L1–L8) and 1 UHRR library (L9) were independently prepared. The libraries L3–L8 were prepared at Mayo Clinic, Rochester, and the libraries L1–L3 and L9 were prepared at Mayo Clinic Florida. The libraries were sequenced in 6 runs (6 flow cells) for a total of 38 flow cell lanes (35 lanes for HBRR and 3 lanes for UHRR) using Illumina Genome Analyzer (GA) I and II (Table 1a, b). One lane from each flow cell was used to run the bacteriophage ΦX174 DNA control sample.

3. Results

Each flow cell was processed independently. One lane of sequencing typically generated 3–5 million reads on the GA I and 6–8 million reads on the GA II. On average, 60% of the reads aligned to the

Table 1
Library and run summary. (a) Summary of 3′ tag digital gene expression libraries including RNA samples, locations of the laboratory that prepared the library, and the library ID. (b) Flow cell layout for individual sequencing runs. A total of 7 runs were performed on Genome Analyzer I (GA I) and GA II. There are 8 lanes on each flow cell named as Lanes 1–8, one of which was used in each run for a control sample PhiX, the bacteriophage DNA

(a) RNA	Lab location	Library ID
HBRR	Florida	L1
HBRR	Florida	L2
HBRR	Florida	L3
HBRR	Minnesota	L4
HBRR	Minnesota	L5
HBRR	Minnesota	L6
HBRR	Minnesota	L7
HBRR	Minnesota	L8
UHRR	Florida	L9

		Lane numbers							
(b) RNA #	Sequencer	1	2	3	4	5	6	7	8
1	GA I	L1	L1	L2	L2	PhiX	L3	L3	L3
2	GA I	L4	L5	L4	L5	PhiX	L4	L6	L4
3	GA II	L4	L5	L4	PhiX	L5	L4	L6	L4
4	GA II	L4	L4	L5	PhiX	L6	L7	L8	L5
5	GA II	L4	L7	L4	PhiX	L8	L6	L7	L8
6	GA II	L9	–	–	L9	PhiX	–	–	L9

canonical, mitochondrial, or rRNA tag tables. The remaining reads aligned to repeat (~7%), noncanonical (~3%), noncoding (~0.02%), and intergenic (~14%) tag tables. Please refer to the Stowers Institute's website for the definitions of the tag tables (http://research.stowers-institute.org/microarray/tag_tables/index.html). Approximately `15% of reads could not be aligned. Gene expression was normalized to tag *C*ounts *P*er *M*illion total *T*ags (CPMT) of the lane to enable comparison between sequence counts between lanes within or across flow cells.

3.1. Reproducibility of 3′ Tag DGE

The sequencing results obtained from the two RNA samples gave us the opportunity to assess both the technical and the biological reproducibility of the 3′ tag DGE technology.

3.1.1. Technical Replications

The pair-wise analyses of gene expression levels from strict technical replicates (same library, same run, and different lanes) showed high correlations ($r > 0.999$) within most of the libraries with the exception of L8 ($0.93 < r < 0.95$).

3.1.2. Biological Replications

This analysis was performed on the HBRR libraries (L1–L8, 35 lanes). We observed high correlations ($r > 0.93$) of expression between biological replicates (same sample, different libraries, and same or different runs). Although having different read counts, different libraries sequenced on GA I and GA II also showed good correlation ($r > 0.93$). The same observations were made between libraries prepared between two Mayo locations (Minnesota and Florida). We further evaluated the concordance/reproducibility of gene identification. A gene is defined as identified/expressed when associated with at least one tag, or 1 CPMT. A total of 18,000 genes with at least one tag were identified from the combined analysis of the reads included in the 35 lanes. Of these genes, 12,825 (71.25%) are repeatedly identified in all lanes from all libraries. The remaining 5,175 genes (28.75%) are identified in fewer numbers of lanes (Fig. 1a). Figure 1b, c illustrates, respectively, the correlations between gene expression levels and numbers of lanes in which the gene was identified (Fig. 1b), and the histogram of gene expressions (Fig. 1c). All genes not consistently identified in all 35 lanes had lower expression levels between 1 and 2 CPMT. The small peak in the histogram (Fig. 1a) around genes identified in 1–5 lanes is related to a group of extremely low-abundant transcripts whose expression levels are around the detection sensitivity threshold of DGE in the current experiment.

These results are, of course, depending upon the sequencing throughput (the number of tags sequenced per lane). Since the time when these libraries were sequenced, the throughput of one lane of sequencing on GA II has increased to ~30 million tags. We estimated that with this sequencing throughput, more than 90% of the 18,000 genes would be detected in all 35 lanes.

3.2. Accuracy of 3′ Tag DGE

3.2.1. Protocol-Dependent Bias and Gene Summarization

The enzymatic digestion protocol used in the 3′ tag DGE approach is designed to capture the digestion site most proximal to a polyadenylation site. If one assumes complete *DpnII* digestion during library preparation and assumes a single polyadenylylation site per gene, there should be theoretically only one type of tag per gene. However, according to the run summary data in Table 1, there were on average more than three unique tags (reads) per gene identified in the HBRR and UHRR samples. The analysis of three libraries (L1–L3) of HBRR indicates that the number of unique tags per gene ranges

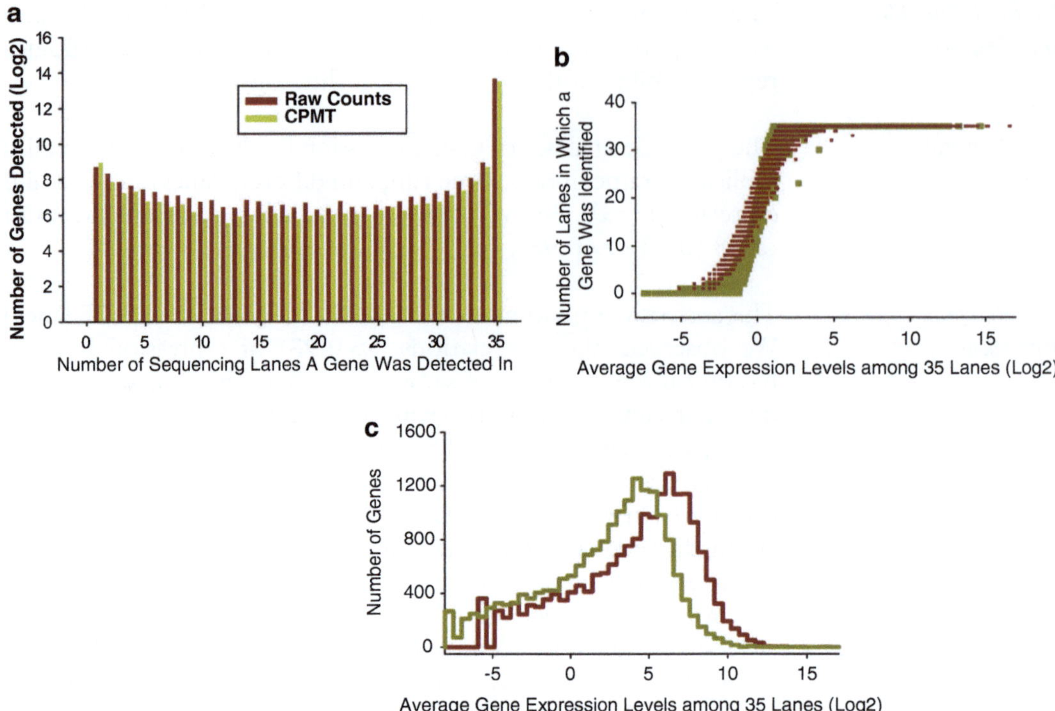

Fig. 1. Reproducibility of 3′ tag digital gene expression profiling. (**a**) Concordance of gene detection. Gene expression levels are represented by the raw number of reads (*dark red*) and number of reads per million (*yellow*). More than 70% of the genes were repeatedly detected in all 35 lanes. (**b**) The relationship between gene detection and expression levels. Genes that were detected in less than 35 lanes were lower expressed at levels of 1–2 CPMT, or 0.35–0.7 copies per cell. (**c**) The histogram of gene expression levels.

from 1 to 46 tags, with a mean of 4 and a median of 3 (data not shown). Figure 2 shows frequency distribution plots of tags per digestion site (referenced relative to the transcript 3′ end), as a function of the total number of reads analyzed. On average, in both HBRR and UHRR samples, the 3′ most *DpnII* digestion site (position 1) accounts for 70–80% of the total mapped tags, with an exponential decrease in the number of tags observed as the digestion site becomes closer to the 5′ end region. In individual genes, this exponential profile may not be observed. For instance, the two most abundant tags of the PGK1 gene (Fig. 3) were the 1st and 4th from the 3′ end. Analysis indicates that the different transcripts from the same gene with different poly-A sites resulted in different 3′ most *DpnII* digestion sites for each transcript, and therefore generated several unique tags per gene. In addition, sequence analysis of the genes with multiple types of tags showed that some of these tags originate from alternative *DpnII* enzymatic digestion sites within the gene without the presence of multiple poly-A sites, suggesting that *DpnII* enzymatic digestion was not always 100% efficient.

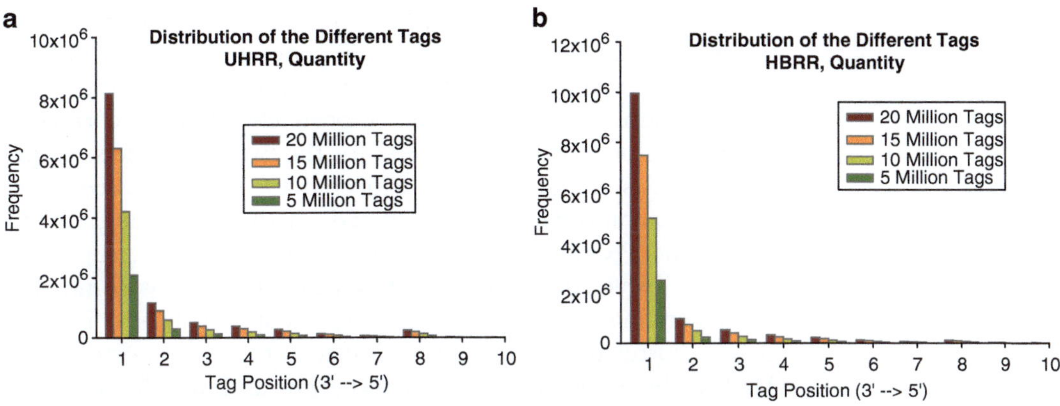

Fig. 2. Distribution of tag frequency per DpnII digestion site (ordered 3′ to 5′) for samples UHRR (**a**) and HBRR (**b**).

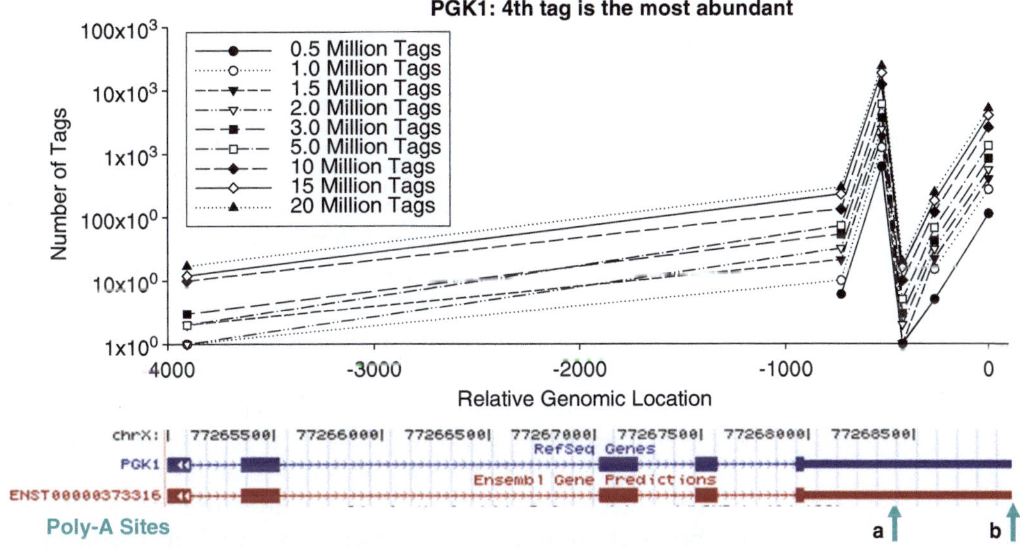

Fig. 3. Tag Count Distribution of PGK1. PGK1 has 11 known exons. Among randomly selected 0.5, 1.0, 1.5, 2.0, 3.0, 5.0, 10. 15, and 20 million tags from UHRR sample, we have consistently observed 5 tags from the 3′ most exon, and 1 tag from the 5th exon from the 3′ end. As shown in the figure above, the most abundant tag was always the 4th tag from the 3′ end which is directly upstream of the poly-A site a. The second most abundant tag was the 3′ most tag and is directly upstream of poly-A site b.

The gene expression levels in the current study are represented by the total tag counts per gene. When there were multiple types of tags aligned to different locations of the same gene, the gene expression levels are represented by the summation of all.

3.2.2. Comparison Between 3′ Tag DGE and Quantitative Real-Time PCR

To assess the accuracy of 3′ tag DGE, DGE gene expression levels were compared to TaqMan qPCR data produced by the MAQC project. The gene expression levels from the qPCR data are reported in number of cycles of amplification. Since large cycle value is representative of low level of expression, cycle numbers and expression levels are inversely correlated.

Data on both DGE and qPCR were available on 815 Ensembl genes in the UHRR samples and 800 genes in the HBRR samples. The direct comparison between the level of expressions and amplification cycles highlights, respectively, Pearson correlation coefficients of −0.289 and −0.302 for the two sample sets. Largest discrepancies are observed as expected for low-expressed genes. When the comparison is done based on the fold changes of the genes, the Pearson Coefficient increased to −0.902, suggesting that the lower correlations observed in the previous comparison are depending upon technology platform-specific biases. This fold change correlation coefficient is similar to the one reported by the MAQC between Affymetrix microarray and qPCR (r=~0.92) (16). The high correlation of differential expression measured by NGS and gold-standard qPCR validates and demonstrates the values in using NGS DGE for gene expression quantitation.

3.2.3. Comparison Between 3′ Tag DGE and Affymetrix U133 Plus 2.0 Microarray

Levels of expression from DGE were also compared to those obtained from the Affymetrix U133 Plus 2.0 microarrays. The number of ENSEMBL genes for which both data are available was, respectively, 17,303 and 17,187 for the UHRR and HBRR samples. About 63% of these genes were commonly shared, 33% were only observed by 3′ tag DGE without a minimum count threshold, and 4% were only observed by Affymetrix microarray. Smoothed histogram plots shown in Fig. 4 a–d demonstrate that most of the genes detected only with DGE are in very low abundance, while the genes identified by Affymetrix microarray but not by DGE are uniformly distributed across different levels of expression. If a detection count threshold of minimum 5 is applied, the total number of observed genes decreased by ~18% (Fig. 4e), but the observation that 3′ DGE detected more less abundance genes still holds.

The correlations between differentially expressed genes identified by DGE and Affymetrix microarray were also studied. The analysis was performed on 10,980 ENSEMBL genes in the UHRR sample, and 10,856 in the HBRR sample. There are 9,512 ENSEMBL genes shared by both UHRR and HBRR and were used to study differential expressions between these two sample sets. The Pearson correlation coefficient for the differential expression genes between UHRR and HBRR was 0.895, a correlation comparable to the one observed between DGE and qPCR. Discrepancies are emphasized on low-expressed genes as well, and the DGE expression levels spanning a broader range than the ones measured by Affymetrix microarray.

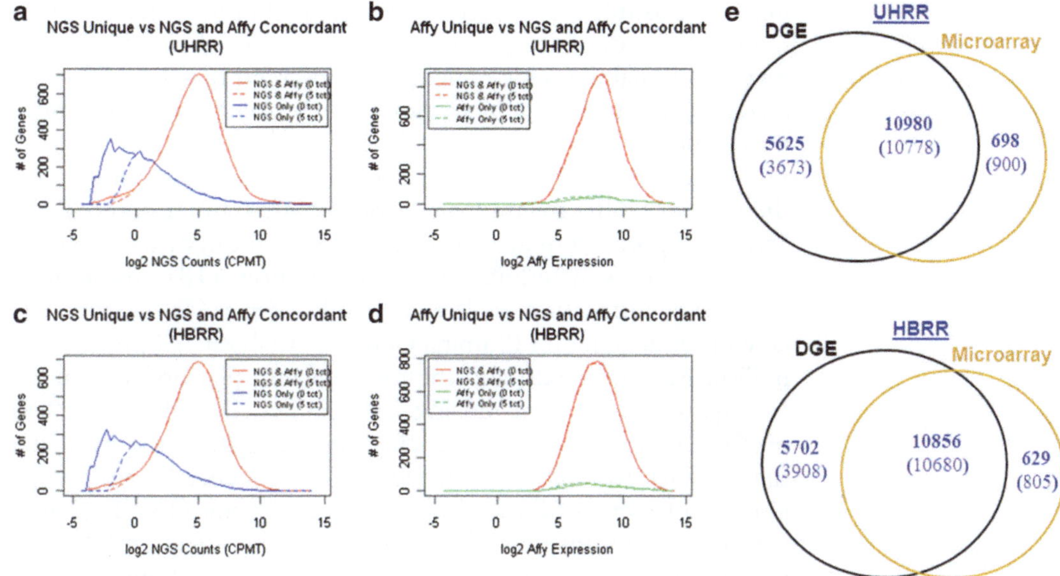

Fig. 4. ENSEMBL gene expression distribution separated by method of detection and sample analyzed: expression distribution in DGE counts for genes identified in (**a**) UHRR and (**c**) HBRR, by both DGE and Affymetrix microarray (*red line*), as well as by DGE alone (*blue line*); expression distribution in Affymetrix microarray expression values for genes identified in (**b**) UHRR and (**d**) HBRR, by both DGE and Affymetrix microarray (*red line*), as well as by Affymetrix alone (*green line*). The differential expression between UHRR and HBRR for genes identified by (**e**) both DGE and Affymetrix, as well as NGS alone, in NGS counts; both NGS and Affymetrix, as well as Affymetrix only, in Affymetrix microarray expression values. Comparison of the number of expressed genes detected by DGE and microarrays. Values for relaxed (at least one read) and stringent (at least five reads) DGE parameters are in bold or in brackets, respectively.

4. Discussion

Our Study highlights the high reproducibility and accuracy of 3′ tag DGE. The correlation coefficient between technical replicates was higher than 0.999 comparable to the coefficients reported by the MAQC project for microarrays (>0.99). The study of biological replicates also showed good correlations between experiments to the exclusion of library L8. This variance seems to be related to library construction instead of sequencing issues as evident by the high correlation coefficients computed for the other library that were higher than 0.99. The 3′ tag DGE is also very accurate. Fold changes in gene expressions between HBRR and UHRR that correlate well with qPCR data are comparable to the correlations observed between microarray and qPCR. Although comparable in accuracy, 3′ tag DGE is more sensitive than microarrays. 3′ tag DGE detected about 20% more transcripts in the HBRR and UHRR libraries. Most of these transcripts were expressed at levels below the detection threshold of microarrays, suggesting that 3′ tag DGE can be used to study low-expressed transcripts. Low-expressed transcripts could account for nearly half of all the functional transcripts in a cell, possibly playing critical roles in pathology and physiology. 3′ tag DGE could also be the methodology of choice to discover new functional genomic regions. We observed that 13–15% of the reads aligned to intergenic regions which may be related to novel transcripts.

Other sequencing approaches, providing more information than 3′ tag DGE, are available to profile the transcriptome. RNA-Seq is an approach of growing interest since it can be used to study RNA variations including expressed mutations (18), fusion genes (19), or quantification of alternative splice forms. At this time, one lane of sequencing on Illumina Genome Analyzer IIX produces 30 million pair-end reads, and the most current platform (Illumina's HiSeq) is capable of producing 250 Gb per run, which is at least fivefold increase in the sequencing throughput. It was estimated that at least 40 million reads need to be sequenced from a single library to achieve 90% coverage of the transcriptome (13). Therefore, with the fast advancement in sequencing technologies, mRNA-Seq will become a more and more attractive approach compared to 3′ tag DGE.

Acknowledgments

This work was supported by Donna Foundation for breast cancer research, NIH Grants AI 33144, AI 48793, AI 40065, and 1 UL1 RR024150-01 from the National Center for Research Resources (NCRR), a component of the National Institutes of Health, and the NIH Roadmap for Medical Research.

References

1. Trevino V, Falciani F, Barrera-Saldana HA: DNA microarrays: a powerful genomic tool for biomedical and clinical research. *Molecular medicine (Cambridge, Mass* 2007, **13**(9–10):527–541.

2. Mootha VK, Lindgren CM, Eriksson KF, Subramanian A, Sihag S, Lehar J, Puigserver P, Carlsson E, Ridderstrale M, Laurila E, Houstis N, Daly MJ, Patterson N, Mesirov JP, Golub TR, Tamayo P, Spiegelman B, Lander ES, Hirschhorn JN, Altshuler D, Groop LC: PGC-1alpha-responsive genes involved in oxidative phosphorylation are coordinately downregulated in human diabetes. *Nat Genet* 2003, **34**(3):267–273.

3. Golub TR, Slonim DK, Tamayo P, Huard C, Gaasenbeek M, Mesirov JP, Coller H, Loh ML, Downing JR, Caligiuri MA, Bloomfield CD, Lander ES: Molecular classification of cancer: class discovery and class prediction by gene expression monitoring. *Science* 1999, **286** (5439):531–537.

4. van 't Veer LJ, Dai H, van de Vijver MJ, He YD, Hart AA, Mao M, Peterse HL, van der Kooy K, Marton MJ, Witteveen AT, Schreiber GJ, Kerkhoven RM, Roberts C, Linsley PS, Bernards R, Friend SH: Gene expression profiling predicts clinical outcome of breast cancer. *Nature* 2002, **415**(6871):530–536.

5. Potti A, Dressman HK, Bild A, Riedel RF, Chan G, Sayer R, Cragun J, Cottrill H, Kelley MJ, Petersen R, Harpole D, Marks J, Berchuck A, Ginsburg GS, Febbo P, Lancaster J, Nevins JR: Genomic signatures to guide the use of chemotherapeutics. *Nat Med* 2006, **12**(11):1294–1300.

6. Schena M, Shalon D, Davis RW, Brown PO: Quantitative monitoring of gene expression patterns with a complementary DNA microarray. *Science* 1995, **270**(5235):467–470.

7. Lockhart DJ, Dong H, Byrne MC, Follettie MT, Gallo MV, Chee MS, Mittmann M, Wang C, Kobayashi M, Horton H, Brown EL: Expression monitoring by hybridization to high-density oligonucleotide arrays. *Nat Biotechnol* 1996, **14**(13):1675–1680.
8. Velculescu VE, Zhang L, Vogelstein B, Kinzler KW: Serial analysis of gene expression. *Science* 1995, **270**(5235):484–487.
9. Adams MD, Kerlavage AR, Fleischmann RD, Fuldner RA, Bult CJ, Lee NH, Kirkness EF, Weinstock KG, Gocayne JD, White O, Sutton G, Blake JA, Brandon RC, Chiu M, Clayton RA, Cline RT, Cotton MD, Hughes JE, Fine LD, Fitzgerald LM, FitzHugh WM, Fritchman JL, Geoghagen NSM, Glodek A, Gnehm CL, Hanna MC, Hedblom E, Hinkle Jr. PS, Kelley JM, Klimek KM, Kelley JC, Liu L, Marmaros SM, Merrick JM, Moreno-Palanques RF, McDonald LA, Nguyen DT, Pellegrino SM, Phillips CA, Ryder SE, Scott JL, Saudek DM, Shirley R, Small KV, Spriggs TA, Utterback TR, Weidman JF, Li Y, Barthlow R, Bednarik DP, Cao L, Cepeda MA, Coleman TA, Collins E, Dimke D, Feng P, Ferrie A, Fischer C, Hastings GA, He W, Hu J, Huddleston KA, Greene JM, Gruber J, Hudson P, Kim A, Kozak DL, Kunsch C, Ji H, Li H, Meissner PS, Olsen H, Raymond L, Wei Y, Wing J, Xu C, Yu G, Ruben SM, Dillon PJ, Fannon MR, Rosen CA, Haseltine WA, Fields C, M. FC, Venter JC: Initial assessment of human gene diversity and expression patterns based upon 83 million nucleotides of cDNA sequence. *Nature* 1995, **377**(6547 Suppl):3–174.
10. Boguski MS, Tolstoshev CM, Bassett DE, Jr.: Gene discovery in dbEST. *Science* 1994, **265**(5181):1993–1994.
11. Mardis ER: The impact of next-generation sequencing technology on genetics. *Trends Genet* 2008, **24**(3):133–141.
12. Mardis ER: Next-generation DNA sequencing methods. *Annual review of genomics and human genetics* 2008, **9**:387–402.
13. Wold B, Myers RM: Sequence census methods for functional genomics. *Nat Methods* 2008, **5**(1):19–21.
14. Cloonan N, Forrest AR, Kolle G, Gardiner BB, Faulkner GJ, Brown MK, Taylor DF, Steptoe AL, Wani S, Bethel G, Robertson AJ, Perkins AC, Bruce SJ, Lee CC, Ranade SS, Peckham HE, Manning JM, McKernan KJ, Grimmond SM: Stem cell transcriptome profiling via massive-scale mRNA sequencing. *Nat Methods* 2008, **5**(7):613–619.
15. Sultan M, Schulz MH, Richard H, Magen A, Klingenhoff A, Scherf M, Seifert M, Borodina T, Soldatov A, Parkhomchuk D, Schmidt D, O'Keeffe S, Haas S, Vingron M, Lehrach H, Yaspo ML: A global view of gene activity and alternative splicing by deep sequencing of the human transcriptome. *Science* 2008, **321**(5891):956–960.
16. Shi L, Reid LH, Jones WD, Shippy R, Warrington JA, Baker SC, Collins PJ, de Longueville F, Kawasaki ES, Lee KY, Luo Y, Sun YA, Willey JC, Setterquist RA, Fischer GM, Tong W, Dragan YP, Dix DJ, Frueh FW, Goodsaid FM, Herman D, Jensen RV, Johnson CD, Lobenhofer EK, Puri RK, Schrf U, Thierry-Mieg J, Wang C, Wilson M, Wolber PK, Zhang L, Amur S, Bao W, Barbacioru CC, Lucas AB, Bertholet V, Boysen C, Bromley B, Brown D, Brunner A, Canales R, Cao XM, Cebula TA, Chen JJ, Cheng J, Chu TM, Chudin E, Corson J, Corton JC, Croner LJ, Davies C, Davison TS, Delenstarr G, Deng X, Dorris D, Eklund AC, Fan XH, Fang H, Fulmer-Smentek S, Fuscoe JC, Gallagher K, Ge W, Guo L, Guo X, Hager J, Haje PK, Han J, Han T, Harbottle HC, Harris SC, Hatchwell E, Hauser CA, Hester S, Hong H, Hurban P, Jackson SA, Ji H, Knight CR, Kuo WP, LeClerc JE, Levy S, Li QZ, Liu C, Liu Y, Lombardi MJ, Ma Y, Magnuson SR, Maqsodi B, McDaniel T, Mei N, Myklebost O, Ning B, Novoradovskaya N, Orr MS, Osborn TW, Papallo A, Patterson TA, Perkins RG, Peters EH, Peterson R, Philips KL, Pine PS, Pusztai L, Qian F, Ren H, Rosen M, Rosenzweig BA, Samaha RR, Schena M, Schroth GP, Shchegrova S, Smith DD, Staedtler F, Su Z, Sun H, Szallasi Z, Tezak Z, Thierry-Mieg D, Thompson KL, Tikhonova I, Turpaz Y, Vallanat B, Van C, Walker SJ, Wang SJ, Wang Y, Wolfinger R, Wong A, Wu J, Xiao C, Xie Q, Xu J, Yang W, Zhong S, Zong Y, Slikker W, Jr.: The MicroArray Quality Control (MAQC) project shows inter- and intraplatform reproducibility of gene expression measurements. *Nat Biotechnol* 2006, **24**(9):1151–1161.
17. Asmann YW, Klee EW, Thompson EA, Perez EA, Middha S, Oberg AL, Therneau TM, Smith DI, Poland GA, Wieben ED, Kocher JP: 3' tag digital gene expression profiling of human brain and universal reference RNA using Illumina Genome Analyzer. *BMC genomics* 2009, **10**:531.
18. Wang Z, Gerstein M, Snyder M: RNA-Seq: a revolutionary tool for transcriptomics. *Nature reviews* 2009, **10**(1):57–63.
19. Maher CA, Kumar-Sinha C, Cao X, Kalyana-Sundaram S, Han B, Jing X, Sam L, Barrette T, Palanisamy N, Chinnaiyan AM: Transcriptome sequencing to detect gene fusions in cancer. *Nature* 2009, **458**(7234):97–101.

Chapter 6

Sharing Expression Profiling Data with Gemma

Anton Zoubarev and Paul Pavlidis

Abstract

We describe a method for analyzing and sharing gene expression data using a Web-based system, "Gemma." Gemma is designed to support meta-analysis in a collaborative framework. Gemma allows users to log on, upload their data, annotate and analyze it, and share the data with others. Users can then compare results across public data sets or other private data sets. We also briefly review alternative approaches for sharing and meta-analyzing gene expression data.

Key words: Microarray, Bioinformatics, Data mining, World Wide Web, Databases, Internet, Epilepsy

1. Introduction

Expression profiling experiments generate large quantities of data that are complex to manage, analyze, and interpret. As these data accumulate and increasingly enter public repositories, there is an increased interest in reusing the data (1). Data reuse can take the form of reanalysis or meta-analysis. An interesting case is the situation where one would like to enhance the interpretation of new expression profiling data by leveraging published data. For example, one might want to ask if there are relevant patterns of expression already identified that should be sought in the new data set. One could try to combine the new data set with similar published data sets to increase power, or take the results of the new data set and search existing data for anything that looks similar. All of these cases require a means of efficiently locating relevant published data, accessing it, and analyzing and querying it in various ways. Despite the exciting possibilities that would be opened up by rich data reuse resources, currently the main way such data reuse is accomplished is through "hand-crafted" reanalyses that require case-specific effort

Yannis Karamanos (ed.), *Expression Profiling in Neuroscience*, Neuromethods, vol. 64,
DOI 10.1007/978-1-61779-448-3_6, © Springer Science+Business Media, LLC 2012

and bioinformatics expertise (for further discussion and review see (2–5)). These types of analyses would be greatly facilitated by the availability of software tools specifically designed for the task.

While the main topic of this chapter is sharing expression data, sharing and reanalysis/meta-analysis are closely related. By definition, sharing is an act of giving access to the data for a second party's use, and the obvious use of the data is to analyze it. Data can be shared in two basic ways. First, the data could be made public. This is commonly a requirement of publication, and normally requires submission into a public repository such as the Gene Expression Omnibus (GEO) (6). Of more interest for our current purposes is the process of sharing data with one or more other specific individuals, without making the data public.

Currently, the main means of sharing expression data prior to publication is the use of ad hoc mechanisms such as e-mail, internal networks, and physical data transfer. Of course, there are more sophisticated mechanisms. Users of commercial or open-source expression database solutions usually have remote access to data, for example, the Web interfaces provided by the free and open-source system BASE (7). Tools like BASE are primarily designed to serve a small community of users, perhaps those on a single research team, and typically are focused on data archiving and single dataset analyses, not meta-analysis. Interestingly, BASE can be connected with the "Multi-Experiment Viewer" analysis tools (8), by installing the "MeV Launcher" (http://baseplugins.thep.lu.se/wiki/net.sf.basedb.mev). MeV is an extensive set of tools for analysis and visualization of gene expression profiles with a graphical user interface, and the combination with BASE is powerful. Because installing the BASE server requires systems administration expertise and access to adequate hardware resources, it is not a quick solution to short-term data sharing needs.

Another example of a powerful way to leverage existing data is to use packages from the Bioconductor (9) system, which is part of the R statistical computing environment (10). Data from GEO can be imported into R using GEOQuery (11), or from ArrayExpress (12), after which it can be analyzed using all the tools available in R/Bioconductor. Of particular relevance are Bioconductor meta-analysis packages such as RankProd (13) and MADAM (14) and GeneMeta (15). While powerful and very flexible, these tools require facility with command-line tools and statistical programming. While such expertise is increasingly available, there is room for user-friendly graphically-oriented tools.

In this chapter, we describe sharing and analysis of expression data using Web-based tools developed in our laboratory, "Gemma." Gemma is an open-source software and database system designed to support the reuse and meta-analysis of publicly available expression data, with a strong element of permitting introduction and sharing of novel data. The goal of Gemma is to make it easier for researchers to combine and compare their data to public data or to that of their collaborators. A key feature of Gemma is that it allows private storage of data and sharing of that data with other users selected by the data submitter. In this way, Gemma is a platform for collaboration as well as a meta-analysis tool.

Gemma currently has data from over 3,000 expression studies, mostly from mouse, human, and rat. Each data set in Gemma is analyzed for differential expression (changes in expression levels related to an experimental parameter of interest such as genotype) and coexpression (correlations of expression profiles of use for understanding regulatory networks or inferring functional relations). The results of these analyses are stored for later retrieval.

Gemma's software has features that are common among expression data resources such as the ArrayExpress Atlas (16) and other features that are apparently unique. What Gemma has the most in common with other systems is in regard to features available to anonymous nonregistered users. Such guest users can search, visualize, and download public data. Importantly, Gemma allows users to query across data sets, so that genes which are coexpressed or differentially expressed in multiple studies can be identified. The most unique features of Gemma have to do with its focus on neuroscience-related data and the ability of registered users to have their own "workspace" including their own data or simply customized "slices" of the public data. To our knowledge, no other Web exposed expression data resource allows users to jointly analyze their own private data with hundreds of publicly available data sets.

Gemma is under active development and new features are being released regularly. More complete instructions for how to use Gemma are available from our Web site. In the remainder of this chapter, we briefly describe how a user can upload their own data into Gemma, share it with other users, and explore the data in various ways.

2. Materials

1. Expression profiling data in tab-delimited text format. This format is fairly straightforward, with one row per probe, and each column containing the data for one sample. Values in the matrix are expression levels expressed as ratios as might be output in a two-channel experiment, or signal intensities as might be output by single channel studies. More details on the format expected are available at http://www.chibi.ubc.ca/Gemma/static/expressionExperiment/upload_help.html. The means of preparing this file will depend on how the data were preprocessed. Currently Gemma only supports a single matrix of data per uploaded experiment. That is, you cannot

upload "absent–present" calls in addition to your expression values. Thus missing values should be indicated with "NA" or blanks in the submitted file. Gemma is also intended to be used with experiments that have reasonably large numbers of samples, ideally 10 or more.

2. A modern Web browser (e.g., Mozilla Firefox 3+, Microsoft Internet Explorer 7+, Google Chrome) and a reasonably fast Internet connection. Because Gemma uses client-side methods for many features, we expect that very old computers may display Gemma pages sluggishly. Gemma is accessed via http://www.chibi.ubc.ca/Gemma.

3. Methods

In the next section, we walk through some of Gemma's functionality through a hypothetical usage scenario. In the scenario, we have two researchers: Alice, who wants to upload her private data and share it with her collaborator, Bob. Bob, in turn, needs this data to compare it with publicly available studies. To accomplish this, Alice uploads her data, annotates it and runs differential expression analysis. She later shares the data set with Bob who runs differential expression queries over a group of public and private data sets focusing on a few genes of interest.

The data for the example is based on a subset of the data described in (17). A meta-analysis involving this experiment was previously described in (18). In the study, rats were injected with kainate to induce seizures in an experimental model of epilepsy, and brain samples were assayed for gene expression as compared to controls. In our hypothetical example, Bob and Alice are interested in comparing gene expression changes in their study to published work and in looking at genes which were previously shown to be differentially expressed in this model.

3.1. Setting Up Data in Gemma

Submitting data to Gemma involves uploading, annotating, and analyzing it with tools provided by the system. These steps can be done by registered Gemma users. The registration process should be familiar to anybody who has registered for any Web site. The user name and password chosen at registration are used to identify the user in future sessions.

3.1.1. Uploading Data

Figure 1 shows part of the "Data Upload" page (http://www.chibi.ubc.ca/Gemma/expressionExperiment/upload.htm). Gemma accepts a tab-delimited file with probe identifiers, sample names, and gene expression values (see Sect. 2 for more details). This form asks, among other things, for a study name, a brief description of the

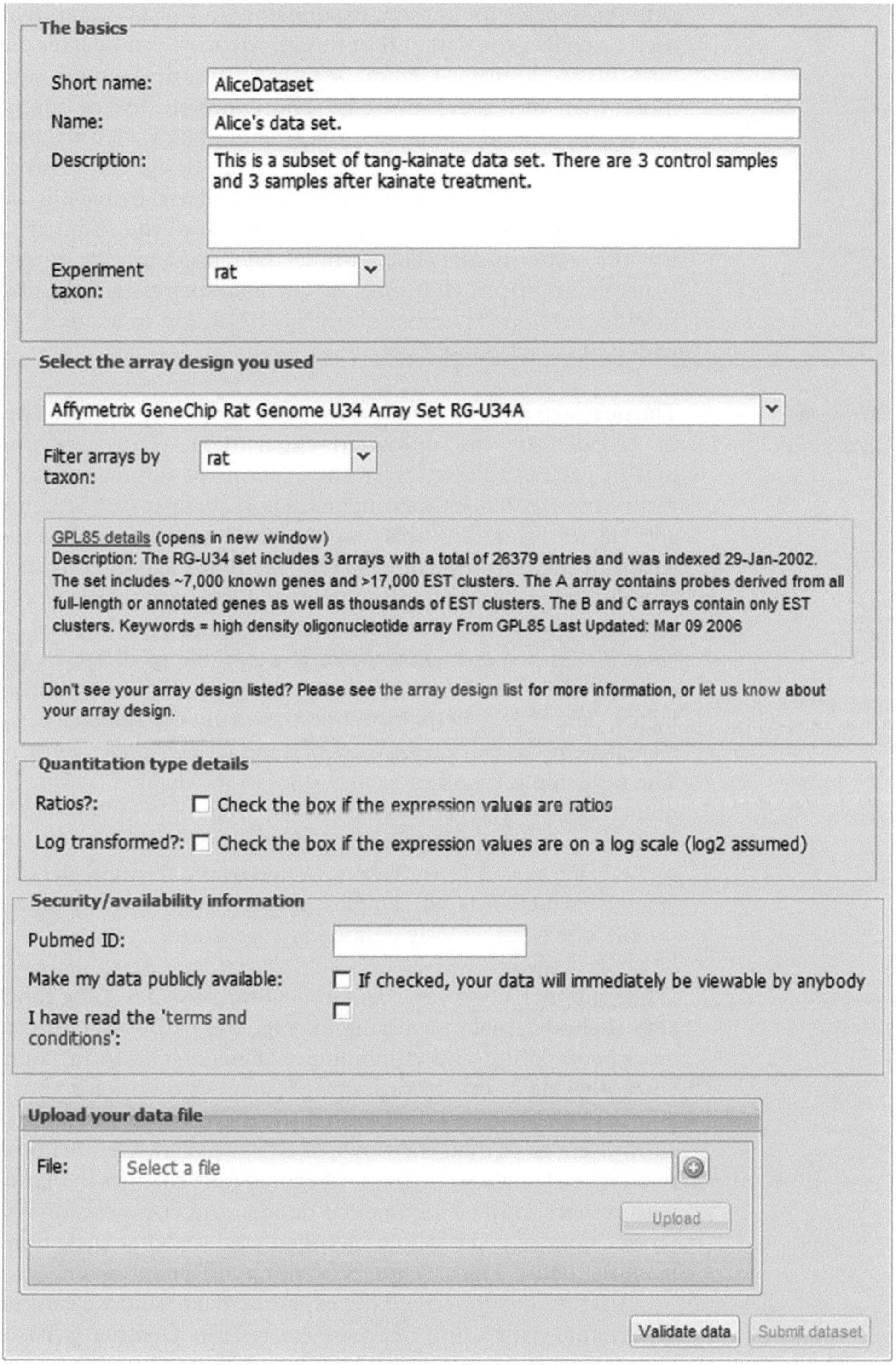

The basics

Short name: AliceDataset

Name: Alice's data set.

Description: This is a subset of tang-kainate data set. There are 3 control samples and 3 samples after kainate treatment.

Experiment taxon: rat

Select the array design you used

Affymetrix GeneChip Rat Genome U34 Array Set RG-U34A

Filter arrays by taxon: rat

GPL85 details (opens in new window)
Description: The RG-U34 set includes 3 arrays with a total of 26379 entries and was indexed 29-Jan-2002. The set includes ~7,000 known genes and >17,000 EST clusters. The A array contains probes derived from all full-length or annotated genes as well as thousands of EST clusters. The B and C arrays contain only EST clusters. Keywords = high density oligonucleotide array From GPL85 Last Updated: Mar 09 2006

Don't see your array design listed? Please see the array design list for more information, or let us know about your array design.

Quantitation type details

Ratios?: ☐ Check the box if the expression values are ratios

Log transformed?: ☐ Check the box if the expression values are on a log scale (log2 assumed)

Security/availability information

Pubmed ID:

Make my data publicly available: ☐ If checked, your data will immediately be viewable by anybody

I have read the 'terms and conditions': ☐

Upload your data file

File: Select a file

Upload

Validate data Submit dataset

Fig. 1. The "Data Upload" form is used to submit data into the Gemma database. The submitter enters the data set name and a brief description of the data set, as well as indicating the array design that was used. Clicking the "Validate data" button runs checks on your data to ensure it meets the format accepted by Gemma.

data set, species used in the experiment, and array design that was used to collect the data. All of this information can be later used to find this study using Gemma search. Currently, Gemma supports more than 300 array designs. The complete list of supported platforms can be found at http://chibi.ubc.ca/Gemma/arrays/showAllArrayDesigns.html. If your array design is missing you can contact the Gemma administrator and ask to have it added. Once the data are uploaded, you are taken to the summary page for the experiment. The sample data used in this tutorial is available at http://chibi.ubc.ca/Gemma/expressionExperiment/showExpressionExperiment.html?id=1944; log in as user "guest" with password "guest."

3.1.2. Defining the Experimental Layout

The next step in our example is to configure the experimental design, or layout, for the uploaded experiment. The experimental design page is accessed from the experiment summary page. This form is used to define experimental factors (in this case "Treatment") and "factor values" (in this case "control" and "kainate") and then to assign them to the samples. Our example data set consists of six samples: three of which are controls and three are tissues from mice treated with kainate. First, we need to define experimental factors (Fig. 2). An experimental factor is a variable of interest that was controlled or measured in the experiment. In our example, the factor describes whether or not the drug was given. We enter "Treatment" as our factor category and a descriptive factor name. The next step is to assign factor values to the defined factors. In our simple case, we need just two such values: "control" and "kainate." Now that these are defined, we can annotate our samples with experimental factors and factor values. We go to the "Sample details" tab (Fig. 3) which lists all samples in our data set and assign each sample one of previously defined factor values.

3.1.3. Running a Data Analysis

While simply having your data in Gemma provides some functionality such as visualization and sharing, the real power of Gemma lies in comparing and combining analysis results from your data with other data sets. In order to get analysis results for your data, you have to analyze it in Gemma (future versions of Gemma will allow users to upload their own analysis results). Gemma provides two types of analysis: differential expression (changes in expression with respect to the experimental design) and coexpression analysis (correlations of genes with each other, used to define gene networks from the data). Here we focus on differential expression.

After we have specified the experimental design we can run the differential expression analysis. Analysis in Gemma is based on linear models (e.g. *t*-tests, ANOVA/ANOCVA). Depending on the number of factors in the experimental design, Gemma picks the appropriate model. You can also manually choose factors you want to be included in the analysis. The analysis itself takes a few minutes

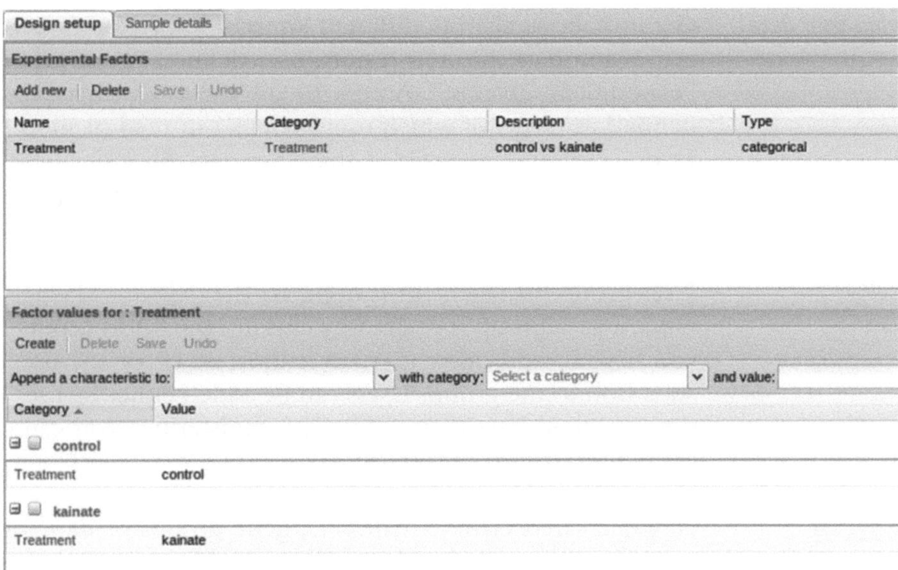

Fig. 2. The "Experimental Design" page is used to define the experimental design or layout, for your data. In the top half of the "Design setup" tab, experimental factors are defined. For example, clicking the "Add new" button creates a new factor. You then need to specify name, category, and description for it. Factor values are created in a similar way in the bottom half of the window.

Design setup | Sample details

Save | Undo | Bulk changes: select a factor | Select a factor value

BioMaterial	BioAssay	control vs kainate
Control1_AliceDataset	Control1	control
Control2_AliceDataset	Control2	control
Control3_AliceDataset	Control3	control
Kainate1_AliceDataset	Kainate1	kainate
Kainate2_AliceDataset	Kainate2	kainate
Kainate3_AliceDataset	Kainate3	kainate

kainate
control

Fig. 3. This is the "Sample details" tab of the "Experimental Design" form. This form is used to assign factor values to each sample, selected from the factors that you have defined using "Design setup" tab (Fig. 2).

to complete, after which the experiment details page is updated with a summary of the results.

The analysis results are stored in Gemma's database and are therefore available via differential expression queries, explained later. It is also possible to download these results in a tab-delimited format for further processing outside of Gemma. Analogous facilities are available for coexpression analysis.

At this point, Alice's data is annotated, analyzed, and ready to be used for meta-analysis and sharing.

3.2. Sharing Your Data

Gemma allows sharing different kinds of information such as gene expression data, analysis results, user-defined "gene groups," and "experiment groups." Access to all these Gemma resources is defined in terms of which user groups can read or modify them. In our example, in order to share her data Alice needs to create a user group containing Bob and her, and then give this group access to her data.

3.2.1. Creating a User Group

User group management is done via the "Group Manager" page. Once a user group is created by clicking the "Add group" button, you have to add other users to it for it to be of any use, so Alice adds Bob to the group. Users can be added either by user name or the e-mail address they used when registering. You can create as many user groups as you like.

3.2.2. Making Data Available to Your Group

Newly created Gemma data sets, gene groups or data set groups are accessible and editable only by you. However, it is possible to change these permissions and make them available either publicly or to specific groups of users. Using the data set details page, Alice can modify access rights for her data. Once she grants access to her study to the user group she previously created, other group members can use her data. Most likely you will want to make the data "read-only" by group members other than yourself, but you can allow other group members to edit your data if you like. Your group members will be able to do most anything you can do with data: view it, download it, add it to experiment groups—but not delete it, even if they have "write" access. Deletion of a data set from Gemma is reserved to the owner of the data set or administrators. In a similar fashion you can edit the access permissions of your gene groups.

3.2.3. Making Data Public

At some point, presumably after publication, you might want to make your data public. Making data public through Gemma might not satisfy the requirements of the stakeholders, but it does provide another access point and is easy to do. You simply check the "Public" box on the data set permissions form.

3.3. Exploring Data

The main way of data exploration in Gemma is through queries. To help users narrow down their search space, Gemma offers two convenient mechanisms: gene groups and data set groups. They provide a concise way to specify queries such as: "Find differentially expressed genes from my gene group by looking only at data sets in my data set group."

3.3.1. Gene Groups

Since data sets typically contain a large number of genes, researchers often concentrate on a set of genes they are interested in. Gemma allows users to define "gene groups" and use them later in different parts of Gemma interface. Gene groups offer a simple way to add a "preset" of gene information to Gemma. Suppose our researcher

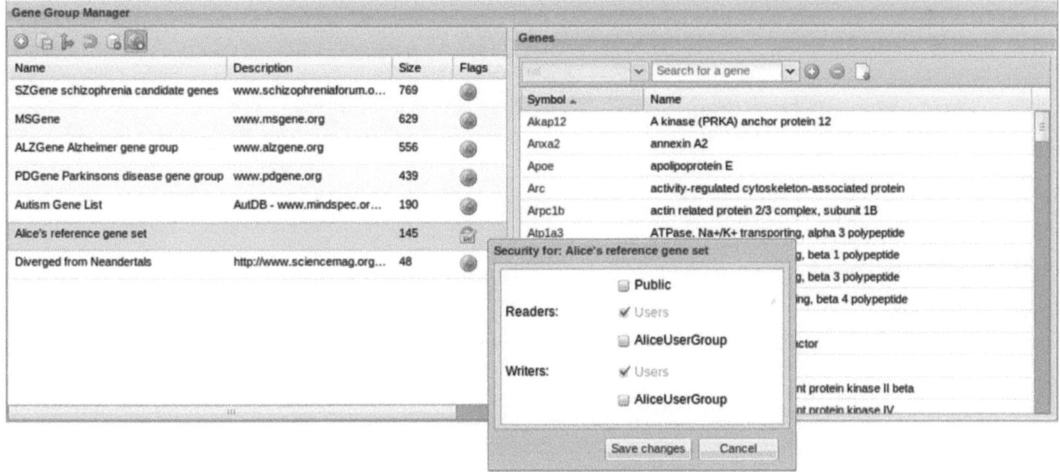

Fig. 4. The "Gene Group Manager" allows creation and management of groups of genes. You can create an empty group or create a copy of existing one and then modify it. Modification is done using "+" (add) and "–" (remove) buttons. It is also possible to add genes in bulk. Also shown here is the pop-up window that appears when the security icon is clicked (the "world" or "lock" icon shown in the *left panel*). This is how sharing of data is set.

wants to look at expression data for genes on a reference list for epilepsy. Figure 4 shows the Gemma page for managing gene groups. We give a name for our gene group and add the genes we are interested in. Also shown in Fig. 4 is an interface to change the security settings on a gene group. In this case, Alice is changing the gene group to be shared with her user group.

You can have as many gene groups as you like. At the time of this writing, gene groups are unadorned with any meta-data: for example, you cannot indicate which genes are upregulated or provide an associated *p*-value, but we expect to add this feature soon.

Alice can now provide her gene group to Gemma's visualization tool and look only at genes in her data set that she is interested in. Currently, Gemma supports heat maps and line plots as possible visualization choices.

3.3.2. Data Set Groups

Gemma also has a facility for creating data set groups, which is like the gene set concept applied to experiments. You might want to create an experiment group that includes your own data as well as data sets you determine are relevant to your analysis. Similar to gene groups, experiment groups show up as choices in relevant points of the interface.

3.3.3. Queries Over Multiple Data Sets

Having your data set analyzed in Gemma might be useful in itself, but things get more interesting when you compare your data to other results in the system.

We return to our hypothetical scenario. Bob now has access to a privately shared data set and is going to include it into his own data set group in order to run differential expression queries. Using the "Dataset Group Editor" (Fig. 5), Bob adds Alice's shared data

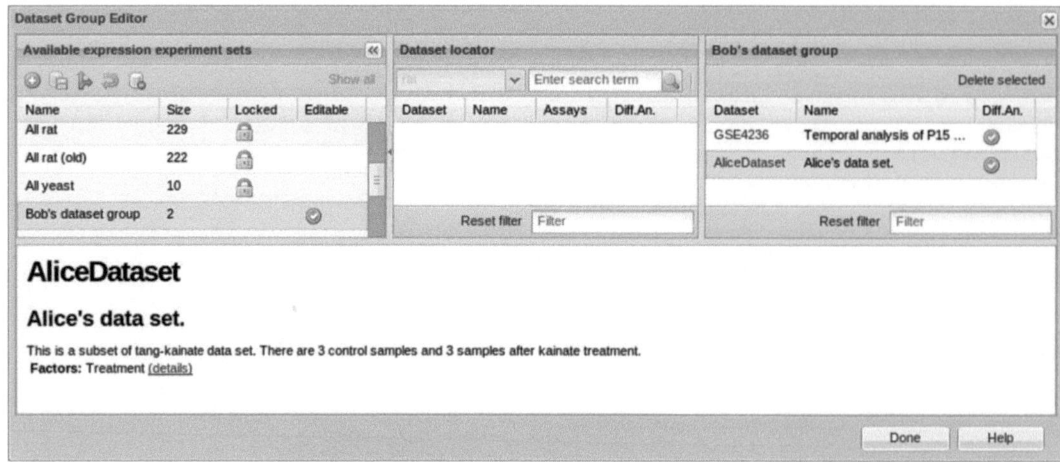

Fig. 5. This form is used to create and manage dataset groups. The *left panel* contains a list of the dataset groups to which you have access. When creating or editing a group, the middle panel is used to find datasets and add them to the selected group. The *right panel* shows datasets in currently selected group. The area at the bottom shows a brief description of the selected study.

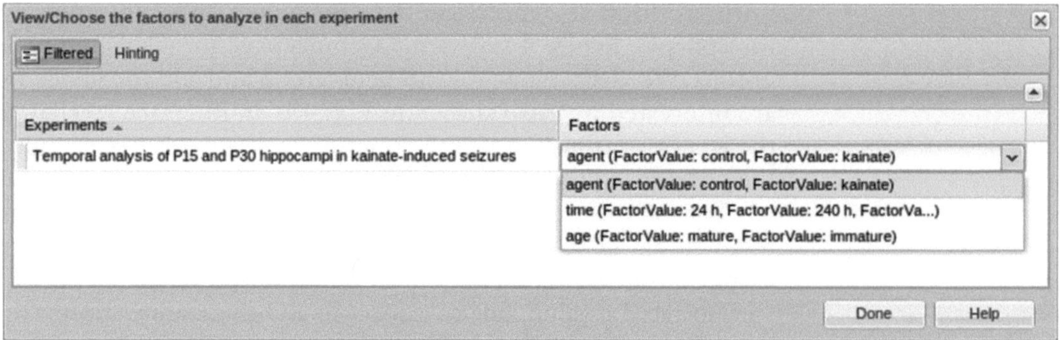

Fig. 6. While the example we used has only one factor, many data sets have multiple factors that were used in their differential expression analyses. When using differential expression query "factor chooser" form pops up where so you can specify which factors you are interested in for this particular search.

set to his data set group. He now has a mixture of private and public data in his group of data sets.

Bob is interested in differential expression of a few specific genes across two experiments. He uses the differential expression search page to construct his query and restricts search to his data set group. Since some data sets have more than one factor, Bob has to select the one he is interested for each study (Fig. 6). In this case, that means choosing the "Agent" factor and ignoring "Age" and "Time." The query results may help him better understand how genes of interest behave in those data sets. In our example, Bob finds that gene Hspb1 is differentially expressed in both data sets (Fig. 7).

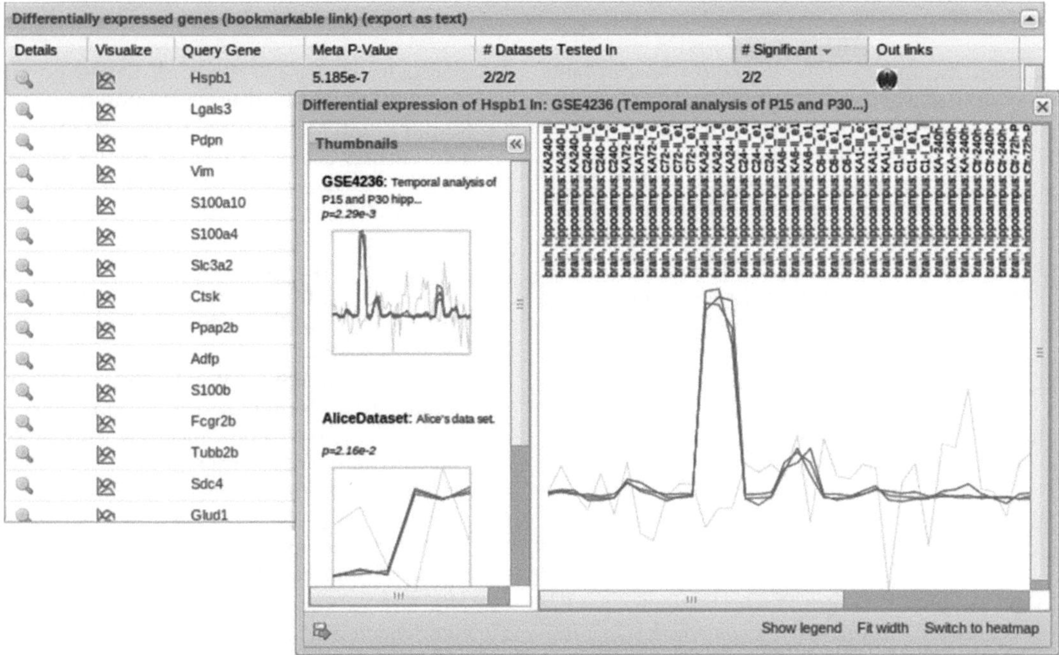

Fig. 7. The "Differential expression search results" form (window in the *background* of this *screenshot*) shows how many data sets contain each gene and how many have this gene differentially expressed, subject to the specified threshold. Clicking the "Visualize" icon brings up a window (shown here in the *foreground*) that shows a line plot (or heat map) of the expression levels for selected gene across samples of the study. *On the left side*, thumbnails of the line plots are displayed for each data set found in the search.

4. Conclusion

Gemma is a work in progress, and at this writing not every feature we would like is implemented. We hope that by introducing the capabilities of system we will encourage users to try it and provide feedback about future development. Questions, bug reports, and feature requests should be addressed to gemma@chibi.ubc.ca.

Acknowledgments

We thank the members of the Pavlidis lab for their contributions to Gemma. Development of Gemma is supported by NIH Brain Research Project grant GM076990. Hardware support is provided by a grant from the Canadian Foundation for Innovation. PP is a Michael Smith Foundation for Health Research Career Scholar and a CIHR Young Investigator award holder.

References

1. Wan, X., and Pavlidis, P. (2007) Sharing and reusing gene expression profiling data in neuroscience, *Neuroinformatics* 5, 161–175.
2. Fierro, A. C., Vandenbussche, F., Engelen, K., Van de Peer, Y., and Marchal, K. (2008) Meta Analysis of Gene Expression Data within and Across Species, *Curr Genomics* 9, 525–534.
3. Ramasamy, A., Mondry, A., Holmes, C. C., and Altman, D. G. (2008) Key issues in conducting a meta-analysis of gene expression microarray datasets, *PLoS Med* 5, e184.
4. Cahan, P., Rovegno, F., Mooney, D., Newman, J. C., St Laurent, G., 3 rd, and McCaffrey, T. A. (2007) Meta-analysis of microarray results: challenges, opportunities, and recommendations for standardization, *Gene* 401, 12–18.
5. Suarez-Farinas, M., and Magnasco, M. O. (2007) Comparing microarray studies, *Methods Mol Biol* 377, 139–152.
6. Barrett, T., Troup, D. B., Wilhite, S. E., Ledoux, P., Rudnev, D., Evangelista, C., Kim, I. F., Soboleva, A., Tomashevsky, M., and Edgar, R. (2007) NCBI GEO: mining tens of millions of expression profiles--database and tools update, *Nucleic acids research* 35, D760–765.
7. Vallon-Christersson, J., Nordborg, N., Svensson, M., and Hakkinen, J. (2009) BASE--2nd generation software for microarray data management and analysis, *BMC Bioinformatics* 10, 330.
8. Howe, E., Holton, K., Nair, S., Schlauch, D., Sinha, R., and Quackenbush, J. (2010) MeV: MultiExperiment Viewer, in *Biomedical Informatics for Cancer Research*, pp 267–277, Springer, New York.
9. Gentleman, R. C., Carey, V. J., Bates, D. M., Bolstad, B., Dettling, M., Dudoit, S., Ellis, B., Gautier, L., Ge, Y., Gentry, J., Hornik, K., Hothorn, T., Huber, W., Iacus, S., Irizarry, R., Leisch, F., Li, C., Maechler, M., Rossini, A. J., Sawitzki, G., Smith, C., Smyth, G., Tierney, L., Yang, J. Y., and Zhang, J. (2004) Bioconductor: open software development for computational biology and bioinformatics, *Genome Biol* 5, R80.
10. Team, R. D. C. (2010) R: A Language and Environment for Statistical Computing.
11. Sean, D., and Meltzer, P. S. (2007) GEOquery: a bridge between the Gene Expression Omnibus (GEO) and BioConductor, *Bioinformatics* 23, 1846–1847.
12. Kauffmann, A., Rayner, T. F., Parkinson, H., Kapushesky, M., Lukk, M., Brazma, A., and Huber, W. (2009) Importing ArrayExpress datasets into R/Bioconductor, *Bioinformatics* 25, 2092–2094.
13. Hong, F., Breitling, R., McEntee, C. W., Wittner, B. S., Nemhauser, J. L., and Chory, J. (2006) RankProd: a bioconductor package for detecting differentially expressed genes in meta-analysis, *Bioinformatics* 22, 2825–2827.
14. Kugler, K. G., Mueller, L. A., and Graber, A. MADAM - An open source meta-analysis toolbox for R and Bioconductor, *Source Code Biol Med* 5, 3.
15. Gentleman, R., Ruschhaupt, M., Huber, W., and Lusa, L. (2008) Meta-analysis for microarray experiments, Bioconductor.
16. Parkinson, H., Kapushesky, M., Kolesnikov, N., Rustici, G., Shojatalab, M., Abeygunawardena, N., Berube, H., Dylag, M., Emam, I., Farne, A., Holloway, E., Lukk, M., Malone, J., Mani, R., Pilicheva, E., Rayner, T. F., Rezwan, F., Sharma, A., Williams, E., Bradley, X. Z., Adamusiak, T., Brandizi, M., Burdett, T., Coulson, R., Krestyaninova, M., Kurnosov, P., Maguire, E., Neogi, S. G., Rocca-Serra, P., Sansone, S. A., Sklyar, N., Zhao, M., Sarkans, U., and Brazma, A. (2009) ArrayExpress update--from an archive of functional genomics experiments to the atlas of gene expression, *Nucleic acids research* 37, D868–872.
17. Tang, Y., Lu, A., Aronow, B. J., Wagner, K. R., and Sharp, F. R. (2002) Genomic responses of the brain to ischemic stroke, intracerebral haemorrhage, kainate seizures, hypoglycemia, and hypoxia, *Eur J Neurosci* 15, 1937–1952.
18. Rogic, S., and Pavlidis, P. (2009) Meta-analysis of kindling-induced gene expression changes in the rat hippocampus, *Front Neurosci* 3, 53.

Chapter 7

Two-Dimensional Protein Analysis of Neural Stem Cells

Martin H. Maurer

Abstract

The current protocol describes in the first part the isolation and culture of neural stem and progenitor cells. In the second part, the two-dimensional gel electrophoresis of neural stem and progenitor cells is explained, consisting of sample preparation, first-dimensional isoelectric focusing, second-dimensional acrylamide gel electrophoresis, protein staining, and image digitalization. The third part concentrates on the software analysis of two-dimensional gel images and the bioinformatical processing.

Key words: Neural stem cells, Neural progenitor cells, Two-dimensional gel electrophoresis, Proteomics

1. Introduction

1.1. Neural Stem Cells

Neural stem cells are stem cells isolated from the central nervous system (brain and spinal cord). They have been isolated from various regions of the adult mammalian brain (1–3). In the embryo, neurogenesis is common, whereas neurogenesis in the adult mammalian organism is found to a lesser extent. At least in two regions, spontaneous neurogenesis persists until late adulthood, i.e., the dentate gyrus of the hippocampus, and the subventricular zone. Neural stem cells can give rise to the cells of the neural and glial linage, which are neurons, astrocytes, and oligodendrocytes. The generation of other cell types from neural stem cells is still under investigation.

During recent years, neural stem cells have been discussed as potential new therapies for a number of neurodegenerative diseases, such as Alzheimer's Disease, Parkinson's Disease, Amyotrophic Lateral Sclerosis (ALS, Lou Gehrig's disease), spinal cord injury, or ischemic stroke (1). The loss of neurons in these diseases is thought to be replaced by novel cells which differentiate and integrate into existing biological networks, and replace the function of the diseased cells.

Yannis Karamanos (ed.), *Expression Profiling in Neuroscience*, Neuromethods, vol. 64,
DOI 10.1007/978-1-61779-448-3_7,

When isolated from the brain, neural stem cells can be propagated in culture for prolonged periods of time (several passages) without losing their ability to proliferate, to self-renew, and to differentiate. After harvesting from specific brain regions, neural stem cells can be grown in vitro as neurospheres, large networks of some hundreds, or even thousands of cells. In this article, the technique of the neurosphere culture is described (4). When neurospheres are passaged for longer time periods (about >10 passages), they seem to change their properties: sometimes, they become independent from growth factor stimulation, or they change their gene expression profile, and show altered proliferative kinetics. In consequence, only short-term neurosphere cultures (about <10 passages) should be used for gene and protein expression studies, unless one is interested in molecular mechanisms during cellular aging. On the other hand, stem cells have to be expanded in culture in order to exclude contamination by other types of neural cells and to yield sufficient amounts of DNA, RNA, or protein, which are necessary to specify gene and protein expression.

1.2. Large-Scale Protein Expression Analysis Using Two-Dimensional Gel Electrophoresis

Large-scale screening tools have become major technologies in the biomedical sciences. For the analysis of DNA or RNA expression, genomic and transcriptomic approaches based on nucleic acid microarrays have been developed, respectively. In analogy, the analysis of protein expression has been named proteomics (5, 6).

Proteomics can be used to (1) validate genomic sequences, (2) identify novel proteins, (3) characterize regulating (stimulating and inhibiting) proteins, (4) detect posttranslational modifications, (5) monitor expression patterns of large sets of proteins, (6) purify proteins in high-resolution, (7), detect immunogenic proteins, e.g., in vaccine studies, (8) analyze mechanisms of action of therapeutic agents and their toxicological relevance, and (9) identify novel drug targets.

The rationale of a proteomics experiment can be twofold: In a typical "profiling" approach, the major aim of the study is to identify ideally all proteins in a given cell, compartment, organ, or even the whole organism. On the other hand, the "functional" approach aims at identifying all proteins with differential expression comparing two (or more) states. This approach is used, for example, in differentiation, aging, or pharmaceutical studies.

Independently from the approach chosen, proteomics always rely on two technical principles, the separation of proteins, and their identification. Major technologies for the separation of proteins involve two-dimensional gel electrophoresis (2DGE) (7, 8), liquid chromatography (LC), capillary electrophoresis (CE), and others. For the identification of proteins, mass spectrometry methods have become state of the art.

In the current protocol, I will concentrate on the two-dimensional gel electrophoresis, which can be performed in the laboratory

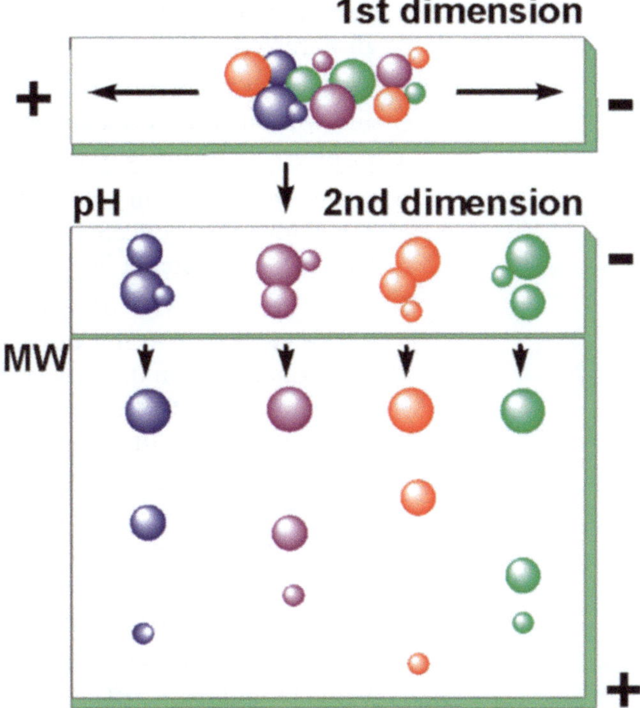

Fig. 1. Principle of the two-dimensional gel electrophoresis. In the first dimension, the protein mixture is applied to a gel containing an immobilized pH gradient for separation by isoelectric focusing (IEF). In the electric field, the proteins migrate to the point where their electric charge is zero. Then the gel strip of the first dimension is attached to the second dimensional polyacrylamide gel. This gel sieve separates the proteins according to their respective molecular weights in the second electric field at right angles to the first electric field.

without specific needs. I will not describe protein identification protocols by mass spectrometry, since nowadays it is easy to find a commercial provider for mass spectrometry services, e.g. at the URL http://www.ionsource.com/links/proteolinks.htm.

The typical 2DGE experiment involves the following steps (9), which are outlined in the present chapter (1) sample preparation, (2) first dimension: isoelectric focusing (IEF) (Fig. 1), (3) second dimension: polyacrylamide gel electrophoresis (PAGE) (Fig. 1), (4) in-gel staining of proteins, (5) digitizing of gel images, (6) image analysis, (7) protein identification, and (8) bioinformatics.

2. Materials

2.1. Neural Stem Cell Isolation and Culture

1. Laminar flow cell culture hood.
2. Glass- and plasticware: Sterilized 100 mL glass flasks, sterile 15 mL and 50 mL polypropylene tubes (Falcon), 1 mL plastic

Pasteur pipettes. Petri dishes. Scissors, scalpel, and 2–3 forceps, all sterilized.

3. Dulbecco's Phosphate-Buffered Saline (D-PBS), without Mg^{2+} and without Ca^{2+}, cell culture grade (Invitrogen).
4. D-PBS-Glc: Prepare 1 mL stock solutions of 450 mg/mL D-glucose (Sigma) in 1× DPBS (100×). Store at −20°C. Dilute 1:100 for a final concentration of 4.5 g/L D-glucose in D-PBS. Pass through a sterile filter.
5. PPD: Dissolve 370 μL papain [Sigma; final concentration 0.01% (w/v)], 100 mg dispase II (neutral protease) [Sigma; final concentration 0.1% (w/v)], 10 mg DNase I (Sigma); final concentration 0.01% (w/v) in 100 mL of Hank's Buffered Salt Solution (HBSS, Invitrogen). Then add 149 mg manganese sulfate ($MgSO_4$), pass through a sterile filter, and store 10 mL aliquots at −20°C.
6. L-Glut: Prepare a 200 mM stock solution of L-glutamine (Sigma). Store 1 mL aliquots at −20°C.
7. Pen/Strep: Prepare a 10,000 units/mL penicillin (Sigma) and 10,000 μg/mL streptomycin (Sigma) stock solution. Store 1 mL aliquots at −20°C.
8. Mix 100 mL of Dulbecco's Modified Eagle's Medium-Ham's F-12 medium (Invitrogen) with 1 mL of the L-Glut stock solution (final concentration 2 mM), and 1 mL of the Pen/Strep stock solution (final concentrations 100 units/mL penicillin, 100 μg/mL streptomycin).
9. Prepare 100 μL stock solutions for the following growth factors: 20 μg/mL bFGF in DPBS (final concentration 20 ng/mL); 2 mg/mL heparin (final concentration 2 μg/mL); and 20 μg/mL EGF in ddH_2O (final concentration 20 ng/mL). Store aliquots at −20°C. Use 100 μL of the stock solution per 100 mL final solution.
10. NB medium: To 100 mL of the Neurobasal media (Invitrogen) add 1 mL of the L-Glut stock solution (final concentration 2 mM), 1 mL of the Pen/Strep stock solution (final concentrations 100 units/mL penicillin, 100 μg/mL streptomycin), 2 mL of the 50× B27 supplement (Invitrogen), 100 μL of each of the growth factor stock solutions for bFGF, EGF, and heparin.

    Note: for mouse neurosphere cultures, add additionally 1 mL of a 300 mg/mL D-glucose stock solution (final concentration 6 g/L).

    After addition of the growth factors, use media within 2 weeks, and keep refrigerated at 4°C.
11. Table centrifuge for 50 mL tubes.
12. Accutase (Sigma) or AccuMax (Sigma).
13. DMSO (Sigma).

2.2. Two-Dimensional Gel Electrophoresis

All fine chemicals are from Sigma, but similar products also may be used.

2.2.1. Sample Preparation

1. Detergent lysis buffer containing 7 M urea, 2 M thiourea, 4% (w/v) CHAPS, 0.5% (v/v) Triton X-100, 0.5% (v/v) IPG buffer pH 3–10 (Amersham Biosciences, Uppsala, Sweden), 100 mM DTT and 1.5 mg/mL Complete protease inhibitor (Roche, Mannheim, Germany).
2. Tissue sample buffer consisting of 40 mM tris, 7 M urea, 4% 3-[(3-Cholamidopropyl)dimethylamino]-1-propanesulphonate (CHAPS), 10 mM 1,4-dithiothreitol, and 1 mM EDTA.
3. Glassware for tissue homogenization: Potter-Elvehjem homogenizer.
4. Table centrifuge, or alternatively, ultracentrifuge.

2.2.2. First Dimension: Isoelectric Focussing

1. Sample buffer: 7 M urea, 2 M thiourea, 4% (w/v) CHAPS, 0.5% (v/v) IPG buffer pH 3–10, trace of bromophenol blue.
2. Immobiline DryStrip pH 3–10 NL, 18 cm, Amersham Biosciences, Uppsala, Sweden. Other pH ranges of the gel strips may also be used, but then the IPG buffer in the sample buffer(s) must be adjusted, respectively.
3. IPGphor apparatus (Amersham Biosciences, Uppsala, Sweden) with strip holders.

2.2.3. Second Dimension: Polyacrylamide Gel Electrophoresis

1. DTT equilibration buffer: 1% (w/v) DTT in 50 mM Tris-HCl, pH 8.8, 6 M urea, 30% (v/v) glycerol, 2% (w/v) SDS, trace of bromophenol blue.
2. IAA equilibration buffer: 2.5% (w/v) iodoacetamide in 50 mM Tris-HCl, pH 8.8, 6 M urea, 30% (v/v) glycerol, 2% (w/v) SDS, trace of bromophenol blue.
3. Electrophoresis chamber with 20×20 cm^2 glass plates, e.g., PerfectBlue Vertical Double Gel System Twin L (Peqlab).
4. Electrophoresis power supply, e.g., peqPower E250 (Peqlab).
5. Monomer solution: 30% acrylamide, 0.8% *N,N'*-methylenebisacrylamide. The monomer powder is extremely hazardous. Therefore, I recommend purchasing the complete monomer solution in liquid form.
6. 4× Resolving gel buffer, 1 L: To 181.7 g Tris base (final concentration 1.5 M) add 750 mL double distilled H_2O, stir until complete solution. Adjust pH with HCl to 8.8. Fill with ddH_2O to 1 L. Store at 4°C.
7. 10% (w/v) Sodium dodecyl sulfate (SDS) solution. Dissolve 5 g SDS in double distilled H_2O, fill to 50 mL total volume. Store at room temperature.

8. 10% (w/v) Ammonium persulfate (APS) solution. Dissolve 0.1 g APS in 1 mL double distilled H_2O. This solution should be prepared fresh. Alternatively, you can prepare a larger volume, and store 1 mL aliquots at –20°C.
9. Gel casting solution: Assemble the gel casting apparatus. Make sure the gel chamber is tightly sealed. Calculate the volume necessary for the gel(s). The acrylamide percentage of the gel determines its resolution of the molecular weight of the proteins analyzed. Here, the casting of a 12.5% acrylamide gel is described, typically resolving proteins in the range of 14–100 kDa. For other protein ranges, see Table 1. For 100 mL of a 12.5% acrylamide gel, combine 41.7 mL of the monomer solution, 25 mL of the 4× resolving gel buffer, 1 mL 10% SDS, 500 μL 10% APS, 31.8 mL double distilled H_2O. Stir until a clear solution is reached. Then start the polymerization process by adding 33 μL TEMED solution. Cast the gel(s) immediately. Overlay the solution with water-saturated *n*-butanol. This will give the gel a flat surface and prevent oxidation. The polymerization process will take up to 1 h. Store the polymerized gel at 4°C.
10. SDS electrophoresis buffer consisting of 25 mM Tris-HCl, pH 8.3, 192 mM glycine, 0.1% SDS. Combine 30.3 g Tris-base,

Table 1
Acrylamide concentrations for the resolving gel (modified from (9))

Acrylamide percentage in resolving gel	Separation size range (kDa)
Single percentage gels	
5%	36–200
7.5%	24–200
10%	14–200
12.5%	14–100[a]
15%	14–60[a]
Gradient gels	
5–15%	14–200
5–20%	10–200
10–20%	10–150

[a]Larger proteins may fail to enter the gel

144.0 g glycine, and 10 g SDS in a 10 L jar. Fill with double distilled H_2O to 10 L. You will need large quantities of this buffer. Store at room temperature.

11. Agarose sealing solution: Add 0.5 g agarose, a trace of Bromophenol blue, and 100 mL of double distilled H_2O into a 500 mL Erlenmeyer flask. Swirl to disperse. Heat in a microwave oven until the agarose is completely dissolved, but do not allow the solution to boil over. Let the solution cool down until it is hand-warm. Use immediately to seal the gel strip. The solution gets a jelly-like consistence at room temperature. Reuse the solution by heating in the microwave oven. Store at room temperature.

2.2.4. Protein Visualization and Image Digitization

Silver Staining

1. Fixation: 40% (v/v) methanol, 10% (v/v) acetone.
2. Washing: 30% (v/v) abs. ethanol.
3. Sensitizing: 0.02% (w/v) $Na_2S_2O_3{\cdot}5H_2O$ (0.2 g/L).
4. Staining: 0.2% (w/v) $AgNO_3$ (2 g/L); 0.02% (v/v) HCOH 37% (200 μL).
5. Developer: 3% (w/v) Na_2CO_3 (30 g/L); 0.05% (v/v) HCOH 37% (500 μl/L); 0.0005% (w/v) $Na_2S_2O_3{\cdot}5H_2O$ (50 μL of a 10% solution).
6. Stop solution: 0.5% (w/v) glycine (5 g/L).

Destaining of Silver-Stained Gels

1. Farmer's reagent component (A): 30 mM $K_3[Fe(CN)_6]$, potassium hexacyanoferrate(III), MW = 329.25, 0.99 g/dL.
2. Farmer's reagent component (B): 100 mM $Na_2S_2O_3$, sodium thiosulfate, MW = 158.11, 1.58 g/dL.
3. Stop solution: 100 mM NH_4HCO_3.

"Blue Silver" Sensitive Coomassie Gel Stain

1. Fixation: 40% (v/v) methanol, 10% (v/v) acetone.
2. "Blue silver" staining stock/working solution: For 1 L of total volume, add 100 mL phosphoric acid (H_3PO_4) to 100 mL ddH_2O. Then add 100 g ammonium sulfate (NH_4SO_2). Dissolve under constant stirring. Add 1.2 g Coomassie Blue G-250. Add ddH_2O to 800 mL. Add 200 mL methanol (MeOH). Fill with double distilled H_2O to 1 L.

2.3. Software Analysis

1. Software: Phoretix 2D Expression, version PG200 (Nonlinear Dynamics, Newcastle-upon-Tyne, U.K.).
2. Desktop scanner, through-light.

3. Methods

3.1. Neural Stem Cell Isolation and Culture

3.1.1. Preparation and Primary Culture

1. Sacrifice 6 of about 6–8 week old male Wistar rats (body weight 180–200 g). Alternatively, you can use mice (body weight 20 g), but adjust glucose concentrations in the media.
2. Remove brains, collect brains in 30 mL ice-cold D-PBS + 4.5 g/L Glucose (D-PBS-Glc), wash once in same medium.
3. Dissect brain areas of spontaneous neurogenesis: Hippocampus, subventricular zone, and olfactory bulb, collect in a 50 mL tube in D-PBS-Glc.
4. Centrifuge for 5 min at 800× *g* in a table centrifuge at 4°C.
5. Remove supernatant with Pasteur pipette, mechanically dissect tissue into small pieces in Petri dish with scissors and scalpel, place back into tube and fill to 20 mL with D-PBS-Glc. Mix well by inversion.
6. Centrifuge for 5 min at 800× *g* in a table centrifuge at 4°C.
7. Remove supernatant and resuspend pellet in 10 mL PPD. Use plastic pipettes, single use, for triturating cells (no adhesion at wall, shear forces). Incubate about 30–40 min, RT. Mix well every 10 min with 10 mL pipette.
8. Centrifuge, 5 min, 800× *g*, 4°C. Resuspend pellet in 1 mL DMEM-Ham's F12 medium. Mix well, triturate with 1 mL blue tip, and then add to 10 mL with media. Repeat this step 2–3 times.
9. (optional:) Percoll separation (see Note 1).
10. Resuspend cells in 1 mL NB medium.
11. Plate cells in a primary culture: To 200 μL cell suspension, add 8 mL B27-Neurobasal (NB) per small cell culture bottle (=25 cm^2 growth area), working volume per bottle is 8 mL. Change media once per week, keep 2 mL conditioned medium, add 6 mL new medium.
12. Passage every 2–2½ weeks (see Sect. 3.1.2), after first passage, replate cells in 6-well plates, working volume per well is 4 mL. To change media in 6-well plates, replace 3 mL by new medium, keep 1 mL of conditioned medium.

3.1.2. Passaging

1. Collect all neurospheres from a 6-well plate in a 50 mL tube.
2. (optional:) wash all wells with about 1 mL D-PBS, add to tube.
3. Centrifuge 5 min, 1,000× *g*, at room temperature.
4. Remove supernatant.
5. Wash pellet once in about 15–20 mL D-PBS.
6. Centrifuge 5 min, 1,000× *g*, at room temperature.
7. Remove supernatant.

8. Make single cell suspension by resuspending the pellet in the appropriate enzyme solution (see Note 2). Incubate for the given time.
9. Count the cell number using the trypan blue stain (see Note 3) and replate to 150,000–200,000 cells per well.

3.1.3. Cryoconservation

1. Passage cells (see **Protocol 3.1.2).**
2. Wash pellet in about 5–10 mL medium to remove enzyme (e.g. Accutase), 5 min.
3. Centrifuge at 1,000× *g*, remove and discard supernatant.
4. Resuspend pellet in 1 mL fresh medium, determine cell number.
5. Dilute cell suspension to 1×10^6 cells/mL (=1,000,000 cells) with medium, add same volume of medium with 20% DMSO. The final concentration of cells calculated is 500,000 cells/mL, which is sufficient for 2 wells of a 6-well plate. The final concentration of DMSO is 10%.
6. Freeze aliquots of 1 mL in cryotube at –20°C for >1 h, then at –80°C for >1 h, then in liquid nitrogen.
7. Store in liquid nitrogen until use.

3.1.4. Restitute Cells from Cryoconservation

1. Remove cells from liquid nitrogen and bring tubes to 37°C prewarmed water bath.
2. As soon as the ice is dissolved, add 500 μL of cell suspension to 1 well of a 6-well plate in 4 mL NB media.
3. Feed cells after 2–3 days, thus last DMSO is removed.

3.2. Two-Dimensional Gel Electrophoresis

3.2.1. Sample Preparation

Sample Preparation for Cultured Cells

1. Remove cells from medium, for attached cells, scrap or trypsinize, centrifuge 1 min at 800× *g*.
2. Wash pellet 3 times in D-PBS, 1 min each, 800× *g*, alternatively wash in 300 mosmol/L Tris-HCl sucrose, pH 7.4, if osmolarity matters.
3. Dissolve pellet in detergent lysis buffer for 1 h at 18°C in an orbital shaker.
4. Centrifuge at 21,000× *g* for 30 min at room temperature.

Sample Preparation for Animal Tissue

1. Suspended approx. 2 g of animal tissue in 2 mL tissue sample buffer.
2. Homogenize in a Potter-Elvehjem glass homogenisator.
3. Centrifuge for 60 min at room temperature at maximum speed in a table centrifuge (alternatively at 100,000× *g* ultracentrifugation to remove also smaller particles).
4. Store the supernatant which constitutes the cytoplasmic protein extract at –80°C until use.

3.2.2. First Dimension: Isoelectric Focusing

1. Determine the protein concentration of the sample using the Bradford assay (see Note 4). Do not use the BCA assay, as it is incompatible with high CHAPS concentrations.
2. Apply 100–500 μg (should be about 1–10 μL) of the sample lysate to the IEF buffer, resulting in a final volume of 350 μL.
3. Apply samples to the pH 3–10 nonlinear gradient isoelectric focussing (IEF) gel strips for IEF in the IPGphor apparatus.
4. The IEF program should contain the following steps: 12 h of reswelling at 30 V, then 200 V, 500 V, and 1,000 V for 1 h each. Voltage was increased to 8,000 V in 30 min and kept constant at 8,000 V for 12 h, resulting in a total of 100,300 Vh.
5. Continue immediately with the second dimension or store the gel strips at −80°C.

3.2.3. Second Dimension: Polyacrylamide Gel Electrophoresis

1. Equilibrate the gel strips after the first dimension to preserve the reduced state of proteins for 20 min at room temperature in DTT equilibration buffer. Soak the gels in the buffer; you don't need a special strip holder.
2. Equilibrate the gel strips to alkylate proteins to prevent reoxidation for 20 min in IAA equilibration buffer. Soak the gels in the buffer; you do not need a special strip holder.
3. Place the gel strip on top of the acrylamid gel, press the gel strip carefully on top of the acrylamide gel using a spacer, and seal the gel strip tightly with hand-warm (30–40°C) agarose sealing solution. Wait for 5 min for the agarose to harden.
4. First fill the inner (upper) electrophoresis chamber with running buffer to make sure the apparatus is tightly assembled. Then fill the outer (lower) chamber. Do not exceed filling lines, since this will cause errors in the power supply.
5. Run gels at 30 mA for 30 min, which allows the proteins to enter the gel, and then at 100 mA for about 4 h in a 20 cm × 20 cm water-cooled vertical electrophoresis apparatus. Make sure that the temperature of the cooling system is not below 10°C, since this will cause precipitation of the SDS. You can stop the run when the bromophenol blue reaches the bottom of the glass plate.
6. Always turn off and disconnect the power supply when handling the apparatus for your own safety. Follow the manufacturer's instructions.

3.2.4. Protein Visualization and Image Digitization

The following staining protocol allows the detection of 1 ng of protein.

Silver Staining

1. Fix gel in 40% (v/v) methanol, 10% (v/v) acetone. Incubation should be at least 1 h, best is overnight.
2. Wash gel in 30% (v/v) ethanol, three times, 20 min each, at room temperature.

3. Incubate in 0.02% (w/v) $Na_2S_2O_3 \cdot 5H_2O$ (0.2 g/L) for 1 min.
4. Wash briefly three times in ddH_2O, about 20 s each.
5. Stain gel in 0.2% (w/v) $AgNO_3$ (2 g/L); 0.02% (v/v) HCOH 37% (200 μL) for 20 min.
6. Wash briefly three times in ddH_2O, about 20 s each.
7. Develop color in 3% (w/v) Na_2CO_3 (30 g/l); 0.05% (v/v) HCOH 37% (500 μl/L); 0.0005% (w/v) $Na_2S_2O_3 \cdot 5H_2O$ (50 μL of a 10% solution), 3–5 min.
8. Wash briefly three times in ddH_2O, about 20 s each.
9. Stop reaction in 0.5% (w/v) glycine (5 g/L), 5 min.
10. Seal gel in a plastic bag.
11. Scan gel in a through-light desktop scanner, at least 300 dpi, TIFF.

Destaining of Silver-Stained Gels

Most silver stains are incompatible with consecutive mass spectrometry. Thus, Coomassie stains have to be applied. Here, we describe the destaining of silver stained gels, which then can be used to restain with a mass spectrometry compatible stain.

1. Prepare a fresh solution of Farmer's reagent by mixing components (A) and (B), ratio 1:1.
2. Soak gel in Farmer's reagent and destain until the gel gets clear. Typically, a 20×20 cm^2 2D gel will need 3–5 min to destain, darker spots may be incubated for 20 min to 1 h.
3. Stop reaction by washing in H_2O. If necessary, increase speed by soaking in stop solution.
4. Repeat washing steps until the (yellow) background turns clear. This step may take more than 3 h. Replace water from time to time.

"Blue Silver" Sensitive Coomassie Gel Stain

This staining is compatible with most mass spectrometry protocols. Make sure to wear gloves to prevent the contamination of the gels with human skin proteins.

1. Fix gel in 40% (v/v) methanol, 10% (v/v) acetone. Incubation should be at least 1 h, best is over-night.
2. Soak gels overnight in the "Blue silver" staining solution.
3. The next day, wash gels in H_2O until desired background is reached.
4. Seal gel in a plastic bag.
5. Scan gel in a through-light desktop scanner, at least 300 dpi, TIFF.

3.3. Software Analysis

The protocol for the software analysis depends strongly on the software used. Here, general steps are provided (10, 11) and demonstrated for the Phoretix/Progenesis software.

1. Create a new experiment. Copy all gel images into a single folder. The images should have a resolution of at least 300 dpi, black and white, and be stored in TIF format. In the software, create a new experiment by "File > New." This will open the "Analysis Wizard," where the main analysis parameters will be defined and an automated analysis will be started. Once you have created a new experiment, you can redefine individual parameters by hand.
2. Import the gels to be analyzed in this experiment: "File > Add/ Remove Gels."
3. Detect spots by using a defined spot detection algorithm, e.g., "2005 Detection." This will number and annotate all spots in the gel: "Analysis > Detect spots." The program will calculate geometric characteristics of the protein spots such as area, optical density, or spot volume. You can select the parameters in the "Measurements window."
4. Next, you will have to filter individual protein spots by "Analysis > Spot Filtering." This will exclude artifacts. Of note, it is not advisable to delete gel artifacts detected as spots, since this will add to the background. The background is defined as the whole area outside of a spot (see next step). Thus, a filtered spot is not regarded as background, but it is also not a spot—it is not visible any more, and does not add to calculations. You can also exclude spots of a certain size, for example, spots smaller than 100 px^2.
5. Subtract gel backgrounds using the settings "Mode-of-non-spot" and a margin of 45 pixels.
6. (optional:) Average multiple sample gels, "Edit > Averaged Gels." This will create a virtual gel averaged of all gels in the experimental group.
7. Choose a reference gel, "Edit > Choose Reference Gel" (see Note 4).
8. Warp gel images, "Analysis > Warping." This will overlay the gels to the reference gel.
9. Match spots in the different sample gels to the corresponding spots in the reference gel. "Analysis > Matching."
10. Modify the reference gel by adding new spots, "Analysis > Add to Reference Gel." After this step, you must rematch all gels to the modified reference gel to update the current experiment.
11. Synchronize spot number, "Edit > Synchronize Spot Numbers." Thus all (matched) spots will have the identical number in all gels. You should repeat this step whenever matching is modified.

12. Normalize spots for intergel comparison, “Analysis > Normalization.”
13. (optional:) Calibrate the 2D image for the isoelectric point and molecular weight, “Analysis > 2D Calibration.”
14. Analyze the results statistically. You can use the three built-in statistical tests, which are *t*-test, paired *t*-test, and ANOVA. Choose the setting under “Tools > Options > Comparisons.”
15. (optional:) Direct automated spot picking devices, “Edit > Spot picking.”
16. (optional:) Include additional data such as mass spectrograms. This feature is not included in the current version.
17. (optional:) Create a web-based federated database for the publication of the data on the internet, “Tools > Web Page Builder.”
18. Export the data, “Edit > Export to file” or “Edit > Export to Excel.”
19. Make sense out of the data. This is the most challenging step of the whole analysis, but there is no general protocol. The analysis depends on what your original question and study rationale was. The reader is referred to Table 2, and the literature (12).

Table 2
Selected internet tools for the analysis of proteomic data

Name	Description	URL
Pathway analysis		
BioCarta	Interactive graphical display of pathways	http://www.biocarta.com/genes/index.asp
Pathway Interaction Database	Biomolecular interactions and cellular processes assembled into authoritative human signaling pathways	http://pid.nci.nih.gov
Human Protein Atlas	Protein expression atlas in humans	http://www.proteinatlas.org
GeneCards	Hyperlinked integrated database of human genes	http://www.genecards.org
PathGuide	Compilation of pathway resources	http://www.pathguide.org
GeneOntology Tree Machine	Compares gene lists to GO terms	http://bioinfo.vanderbilt.edu/gotm/
Literature mining		
iHOP—Information Hyperlinked over Proteins	Literature-based information tool	http://www.ihop-net.org
PubGene	Search for relationship between two (or many) genes/proteins, or lists thereof	http://www.Pubgene.org
ChiliBot		http://www.Chilibot.net
FABLE	Mines the biomedical literature for information about human genes and proteins	http://www.Fable.chop.edu
LitInspector		http://www.Litinspector.org
PLIPS—Protein Lists Identified in Proteomics Studies	Protein/gene identifiers extracted from published tables	http://mips.helmholtz-muenchen.de/proj/plips/

(continued)

Table 2 (continued)

Name	Description	URL
Software		
BaCelLo	Prediction of subcellular localization	http://gpcr.biocomp.unibo.it/bacello/
The E-Cell Project	Computational biology modeling of complex cellular systems	http://www.e-cell.org/ecell/
Cytoscape	Visualizing and integrating molecular interaction networks and gene expression profiles	http://www.cytoscape.org
DAVID—Database for Annotation, Visualization and Integrated Discovery (DAVID) 2008	Integrates functional genomic annotations with graphical summaries. Annotation and summary of lists of identifiers for Gene Ontology, protein domain, and biochemical pathway membership	http://david.abcc.ncifcrf.gov/home.jsp
MeV—MultiExperiment Viewer	Analysis, visualization and data-mining of large-scale genomic data, and microarray tool	http://www.tm4.org/mev
Gene list Venn diagrams	Creates "Venn diagrams" (overlapping circles) of gene lists	http://www.bioinformatics.org/gvenn/
L2L Microarray Analysis Tool	Database of published microarray gene expression data, and comparing own data to published microarray results	http://depts.washington.edu/l2l/
GOMiner	Biological interpretation of "omic" data	http://discover.nci.nih.gov/gominer/index.jsp

This list assembles some important freely available internet tools which I found useful for further proteomics analysis. The list is neither comprehensive, nor does it contain commercially available links or software (all links were last accessed on 2010-02-07)

4. Notes

1. Percoll separation:
   (a) In a 50-mL tube, resuspend pellet in 18 mL medium, add 9 mL Percoll/D-PBS solution consisting of 8.1 mL Percoll and 0.9 mL 10× D-PBS (use within several days, store at 4°C). This will give a 30% gradient. For a 50% gradient, mix equal volumes of medium and Percoll/D-PBS solution.
   (b) Centrifuge at 20,000× *g*, 30 min at room temperature.
   (c) Watch out for 3 layers: the top layer contains myelin (ochre), the medium layer contains cell culture media (rose), and the bottom layer contains neurospheres and red blood cells (red). The cells are in a sharp band right next to the bottom in the 30% gradient, or in the same band with myelin in the 50% gradient.
   (d) Place the cells in a 10-mL tube, add 2–3 mL DMEM Ham's F12, and wash 3 times in DMEM Ham's F12, at least in 10 isovolumes, to remove the Percoll. Mix well.

(e) Centrifuge for 10 min, 800× *g*, 4°C.

2. Accutase is used for gentle lysis of adherent cells. It is well suited for sensitive cells and for retaining the cell surface structure. Incubation takes longer, but the cell surface epitopes will stay intact. Growth in the new medium will speed up. Moreover, no inactivation is necessary. AccuMax is used for cells in suspension. It dissolves cell aggregates smoothly and reliably within minutes. Rate of protein synthesis is increased in AccuMax treated cells.

   For Accutase treatment, resuspend the pellet in 500 μL Accutase (preheated to 37°C), triturate with a blue tip, and incubate 10–30 min, 37°C. For AccuMax treatment, resuspend the pellet in 1 mL AccuMax (preheated to 37°C) and add 1 mL DPBS, and incubate 5 min, 37°C.

3. Trypan blue exclusion test:

   (a) Mix 45 μL trypan blue solution (Sigma) with 5 μL cell suspension (= dilute 1:10).

   (b) Calculate the cell number: Count cells in a Bürkle hematocytometer, 1 big quadrant, multiply number by 10,000 = cells/mL.

   (c) Multiply by 10 to correct for trypan blue 1:10 dilution.

4. Selection of the reference gel: It is a crucial step of the whole analysis to find a good reference gel. The "perfect" reference gel shows a maximum number of well separated spots (Fig. 2). In our hands, it is advisable to do 3–5 technical replicates of each of the gels in the "control" group. After spot detection, choose the gel with the most spots. If the gels shows a good resolution (some programs offer data on spot overlap, singularity, shape, or eccentricity etc., but you can do this also by visual control), this gel is your reference gel.

   By adding spots to this reference gel from other gels after gel matching, the software allows creating "virtual" gels which contain all spots of the experiment. Then, the spots of all gels in the experiment are matched to this virtual gel. With this approach, it actually does not seem very important which gel you choose first. Despite that, I still suggest choosing a "good" gel in the first instance.

Acknowledgments

The author was supported by the European Union (EU) within the Seventh Framework Programme (FP7), the German Ministry of Education and Research (BMBF) within the National Genome Research Network (NGFN-2), the German Research Foundation (DFG), the Intramural Research Grant of the Medical Faculty,

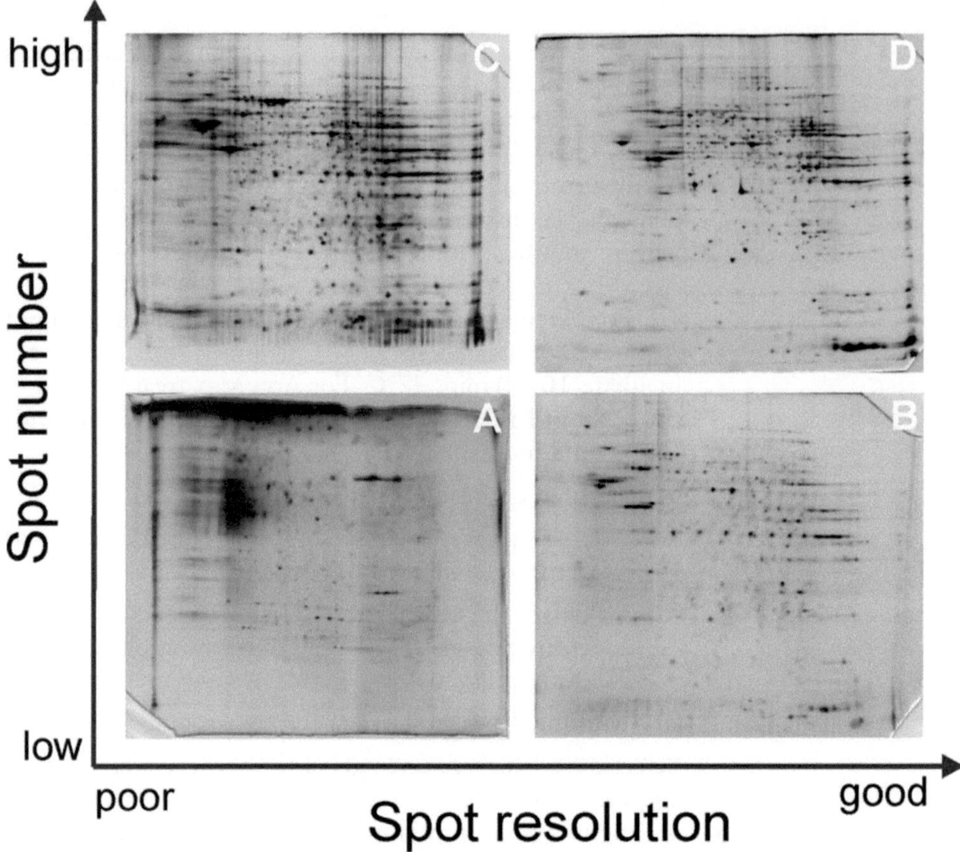

Fig. 2. How to choose the reference gel. The optimal reference gel combines a maximum number of spots with a good spot separation. For example, in (**a**), neither number nor resolution is good. In (**b**), the spot resolution is acceptable, but there are only few spots in the gel. In (**c**), there are numerous spots, but they are overlapping and blurry. Only in (**d**), spot resolution and number are good. This is the gel I would recommend to choose as reference gel in the first instance. Afterwards, the reference gel is modified by the software.

University of Heidelberg, the Steuben-Schurz Society, and the Estate of Friedrich Fischer.

The author thanks Prof. em. Dr. Wolfgang Kuschinsky for his sustained support for many years, Mrs. Maria Harlacher, Mrs. Tilly Lorenz, and Mrs. Inge Keller for excellent technical support.

References

1. Brüstle, O., and McKay, R. D. (1996) Neuronal progenitors as tools for cell replacement in the nervous system, *Curr. Opin. Neurobiol.* 6, 688–695.
2. Gage, F. H. (2000) Mammalian neural stem cells, *Science* 287, 1433–1438.
3. Okano, H. (2002) Stem cell biology of the central nervous system, *J. Neurosci. Res.* 69, 698–707.
4. Maurer, M. H., Feldmann, R. E., Jr., Fütterer, C. D., and Kuschinsky, W. (2003) The proteome of neural stem cells from adult rat hippocampus, *Proteome Sci.* 1, 4.
5. Link, A., (Ed.) (1999) *2-D proteome analysis protocols*, Humana Press, Totowa, N.J.
6. Maurer, M. H., and Kuschinsky, W. (2007) Proteomics, in *Handbook of Neurochemistry and*

Molecular Neurobiology, vol. 5: Brain Energetics. Integration of Cellular and Molecular Processes. (Lajtha, A., Gibson, G. E., and Dienel, G. A., Eds.) 3 rd ed., pp 737–769, Springer, New York.

7. Görg, A., Obermaier, C., Boguth, G., Harder, A., Scheibe, B., Wildgruber, R., and Weiss, W. (2000) The current state of two-dimensional electrophoresis with immobilized pH gradients, *Electrophoresis* 21, 1037–1053.
8. Rabilloud, T., Vaezzadeh, A. R., Potier, N., Lelong, C., Leize-Wagner, E., and Chevallet, M. (2009) Power and limitations of electrophoretic separations in proteomics strategies, *Mass Spectrom. Rev.* 28, 816–843.
9. Berkelman, T., and Stenstedt, T. (2002) *2-D Electrophoresis using immobilized pH gradients. Principles and Methods*, 2nd ed., Amersham, Uppsala.
10. Maurer, M. H. (2006) Software analysis of two-dimensional electrophoretic gels in proteomic experiments, *Curr. Bioinformatics.* 1, 255–262.
11. Sheehan, D., and Tyther, R., (Eds.) (2009) *Two-Dimensional Electrophoresis Protocols*, Vol. 519, Humana Press, New York.
12. Maurer, M. H. (2004) The path to enlightenment: making sense of genomic and proteomic information, *Genomics Proteomics Bioinformatics* 2, 123–131.

Chapter 8

iTRAQ Proteomics Profiling of Regulatory Proteins During Oligodendrocyte Differentiation

Mohit Raja Jain, Tong Liu, Teresa L. Wood, and Hong Li

Abstract

Recent evolution in proteomics approaches from two-dimensional gel electrophoresis to peptide-based "shotgun proteomics" methods has greatly enhanced the abilities of scientists to uncover expression changes among "low abundant" proteins. Shotgun proteomics methods typically employ stable isotope labeling techniques to distinguish peptides from the various sources that are compared. Recently, a new shotgun quantitative proteomics technology called *i*sobaric *t*ags for *r*elative and *a*bsolute *q*uantification (iTRAQ) has been developed for protein expression analysis. The major strength of the iTRAQ technology is its ability to compare the proteomic changes among multiple samples in a single experiment. Here we present a protocol on using the 8-plex iTRAQ approach for the discovery of molecular targets in oligodendrocyte progenitor cells during rapamycin-induced inhibition of differentiation. We provide the technical details on peptide labeling, chromatography, mass spectrometry, database search, and bioinformatics procedures for the identification of differentially expressed proteins.

Key words: Mass spectrometry, iTRAQ, Neuroproteomics, Expression proteomics, mTOR, Oligodendrocyte progenitor cell, Rapamycin

1. Introduction

Systems biological approaches are increasingly used as unbiased discovery tools to gain insights into the functional molecular alterations during neurological diseases (1). Currently microarray analysis has been a method of choice for "system-wide" gene expression analysis; however, the application of microarray approaches has limitations in select biological scenarios, e.g., in biological fluids that are mostly devoid of mRNAs. Furthermore, mRNA changes do not always correlate directly to changes at either protein levels or activities as the result of varying regulatory steps during translation, posttranslational modifications, subcellular localization, and

Yannis Karamanos (ed.), *Expression Profiling in Neuroscience*, Neuromethods, vol. 64,
DOI 10.1007/978-1-61779-448-3_8, © Springer Science+Business Media, LLC 2012

degradation (2). An evolving approach toward comprehensive understanding of the molecular mechanisms underlying neurological diseases is shotgun neuroproteomics, i.e., the large-scale quantification of peptides and proteins within the contexts of stem cell differentiation, neurodegeneration and neuroregeneration, etc (3, 4). Accurate quantification of peptides relies on the availability of high resolution mass spectrometers that are increasingly robust and are equipped with sophisticated bioinformatic tools for routine high-throughput analysis. As an example of earlier shotgun methods, isotope-coded affinity tags (ICAT) reagents are used to label proteins (5). They are thiol-reactive chemical "tags" that are designed to contain either a light [^{12}C] or heavy [^{13}C] ICAT reagent with a mass difference of 9 Da. Relative protein abundance is determined from the relative MS ion abundance of the corresponding ICAT-labeled peptides. Since the ICAT method selectively quantifies only cysteine-containing peptides, it is able to quantify both protein expression and oxidative modification changes, but unable to quantify proteins that do not contain cysteines (6, 7). Alternatively, in *s*table *i*sotope *l*abeling by *a*mino acids in *c*ell culture (SILAC) approach (8), the proteomes from two or more cell populations are compared which are metabolically labeled during protein synthesis with either light or heavy stable isotope-incorporated amino acids during cell culture. Quantification of proteins is achieved by comparing the relative abundance of peptides with identical sequences, yet distinguishable in MS by their relative mass due to "heavy" amino acid incorporation. Although SILAC has been proven effective for large-scale quantification of proteins and their phosphorylation changes, it cannot be readily used for studying terminally differentiated cells including neurons since they are not rapidly dividing to incorporate SILAC-specific amino acids into proteins. In addition, this method is not suitable for routine analysis of tissues. By comparison, the iTRAQ approach has been successfully used for multiplexed protein expression analysis from both animal tissues and neuronal cells (9–15). This method utilizes the covalent labeling of peptides with isobaric mass tags for simultaneous identification and quantification of peptides derived from up to eight categories using tandem mass spectrometry (MS/MS) methods. iTRAQ reagents are identical in mass, therefore isobaric, and are consisted of a variety of isotope-incorporated reporter groups, corresponding stable isotope mass balance groups, and a peptide reactive group. These reagents are covalently linked with peptides via primary amines at lysine side chains and N-termini. The iTRAQ-labeled peptides from up to eight samples can be combined, fractionated using two-dimensional liquid chormatography (LC) and identified and quantified using MS/MS. iTRAQ-labeled peptides labeled with any one of the eight iTRAQ tags are eluted at the same retention time during the LC steps and they display identical mass during MS analysis because they are isobaric. During MS/MS peptide fragmentation, the reporter ions that contain different combinations of stable isotopes (*m/z* 113, 114, 115, 116, 117, 118, 119, and 121) are released from the iTRAQ-tagged peptides. The peak areas of the reporter ions are used to determine the relative abundance of each peptide and of the corresponding proteins in each sample. A series of peptide y- and b-ions fragments is also generated by MS/MS analysis for protein identification through protein database matching (16, 17). Since iTRAQ reagents are efficient at labeling nearly all peptides, this method is effective at providing high protein identification sequence coverages, allowing sensitive quantification of low-abundant proteins such as signal transducers, transcription regulators, and membrane receptors (10–12, 14).

There are currently two versions of iTRAQ reagents that are available from ABSciex (Foster City, CA) for the comparison of either four or eight sample categories simultaneously. We will present here an 8-plex iTRAQ-based shotgun neuroproteomics method that we have used to elucidate the target pathways downstream from mammalian target of rapamycin (mTOR) in oligodendrocyte progenitor cells (OPCs), following rapamycin-induced inhibition of OPC differentiation (4). mTOR is a member of the phosphatidylinositol 3-kinase-like family of serine–threonine kinases that integrates signals from growth factor stimulation and nutrient sensing to modulate a number of biological processes including cell growth, proliferation, protein translation, differentiation, and autophagy (18). mTOR forms two intracellular signaling complexes known as mTOR Complex 1 (mTORC1) and mTOR Complex 2 (mTORC2), defined respectively by the association of mTOR with the adaptor proteins raptor or rictor that direct mTOR's kinase activity toward distinct downstream signaling effectors (19, 20). Inhibition of mTORC2 during OPC differentiation causes a reduction in the mRNA levels of several key myelin genes, whereas inhibiting mTORC1 results in decreased myelin protein levels most likely by interfering with the translation of these transcripts (21). By comparison, recent reports have shown that the rate of protein translation increases during OPC differentiation in an mTOR-dependent fashion (22). However, the targets of the mTOR pathway that regulate oligodendrocyte differentiation are unknown. To identify novel targets regulated by the mTOR pathway during oligodendrocyte differentiation, proteins from four independent control and rapamycin-treated OPCs were processed sequentially with disulfide reduction, alkylation, and trypsin digestion (Fig. 1). The resulting peptides were individually labeled with the iTRAQ reagents. The four iTRAQ reagents (113, 114, 115, and 116) were utilized to label the peptides derived from the four control samples, and another four iTRAQ reagents (117, 118, 119, and 121) were used for labeling the peptides derived from rapamycin-treated samples. Equal amounts of the labeled peptides were combined and quantified using two-dimensional liquid chormatography coupled with tandem mass spectrometry.

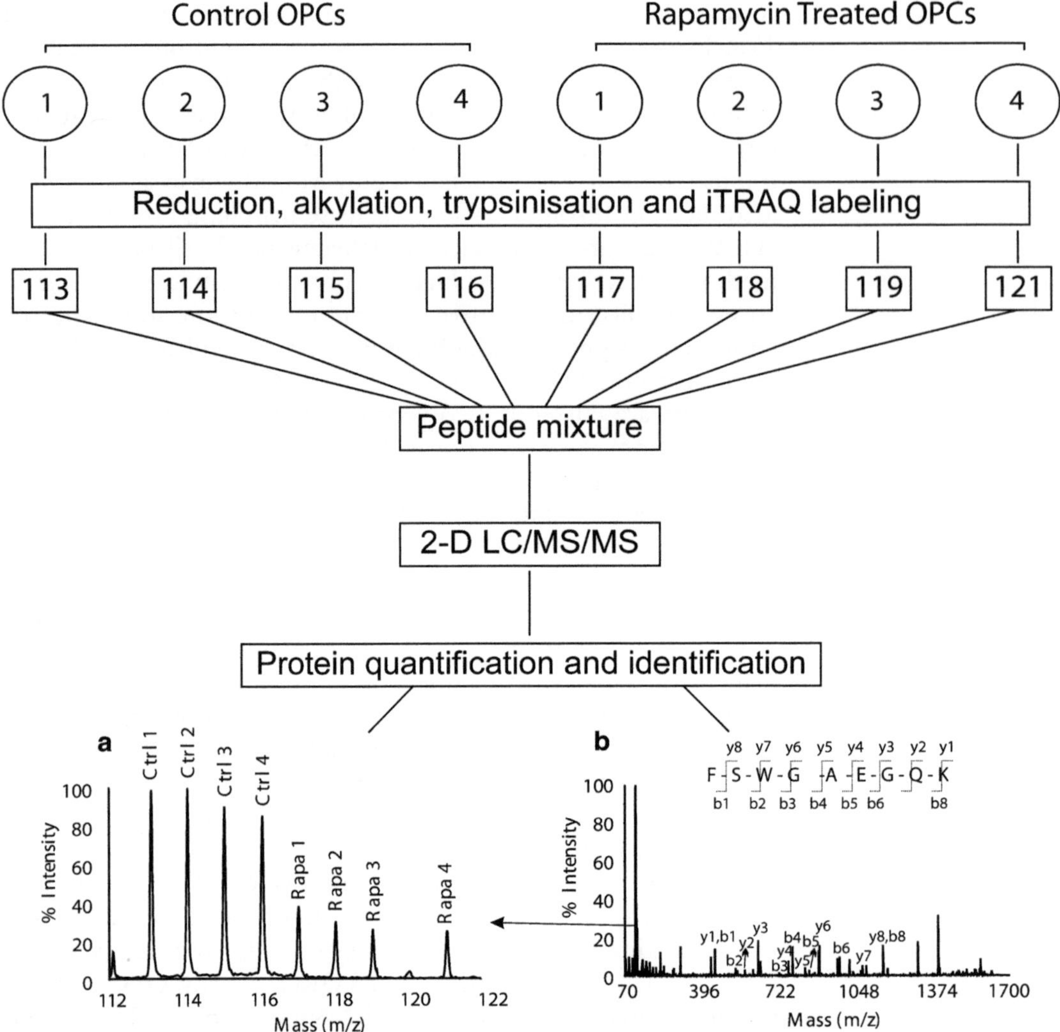

Fig. 1. iTRAQ-based expression proteomics work flow for identification of proteins regulated by the mTOR pathway following a rapamycin inhibition of oligodendrocyte differentiation. Proteins extracted from four control and four rapamycin-treated OPCs groups were sequentially reduced, alkylated, and then digested by trypsin. The resulting peptides were individually labeled with the iTRAQ reagents 113, 114, 115, 116, 117, 118, 119, and 121 as indicated. The labeled peptides were combined and analyzed using strong cation exchange liquid chromatography (SCXLC) and reversed phase liquid chromatography (RPLC) 2-dimensional liquid chromatography coupled with tandem mass spectrometry (2D LC/MS/MS). Database search and bioinformatics procedures were used for protein (**a**) quantification and (**b**) identification reprinted with permission from Tyler et. al. (4).

2. Materials

2.1. Protein Extraction

1. Control and rapamycin-treated OPCs
2. Phosphate buffer saline (PBS): 1.54 mM Potassium phosphate monobasic (KH_2PO_4), 155.17 mM Sodium chloride (NaCl), 2.71 mM Sodium phosphate dibasic ($Na_2HPO_4{\cdot}2H_2O$)

3. Lysis buffer: 500 mM Triethylammonium bicarbonate (TEAB), 1.0% Igepal CA630 (NP-40), 1.0% Triton X-100, 0.1% v/v each of protease inhibitor cocktail (Sigma, St. Louis, MO), phosphatase inhibitor cocktail 1 (Sigma, St. Louis, MO) and phosphatase inhibitor cocktail 2 (Sigma, St. Louis, MO), pH 8.5
4. Ultrasonic Homogenizer with a 5/32″ Micro-Tip (Omni International, Kennesaw, GA)
5. Ice bath
6. Micro centrifuge 5415 R (Eppendorf, Hauppauge, NY)

2.2. Protein Estimation

1. BCA protein assay kit (Thermo Scientific, Rockford, IL)
2. Protein standard (2.0 mg/ml bovine serum albumin)
3. SpectraMax 190 micro plate reader (Molecular Devices, Sunnyvale, CA)

2.3. Eight-plex iTRAQ Labeling

1. Reducing reagent: 50 mM tris-(2-carboxyethyl) phosphine (TCEP) (AB SCIEX, Foster City, CA)
2. Cysteine alkylation reagent: 200 mM methyl methanethiosulfonate (MMTS) (AB SCIEX, Foster City, CA)
3. HPLC grade isopropanol
4. HPLC grade water
5. Trypsin (20 μg/vial, Promega, Madison, WI)
6. iTRAQ reagents: 113, 114, 115, 116, 117, 118, 119, and 121 (Part Number 4390811, AB SCIEX, Foster City, CA)
7. Vacuum concentrator 5301 (Eppendorf, Hauppauge, NY)

2.4. Strong Cation Exchange Liquid Chromatography

1. PerSeptive BioCAD SPRINT Perfusion chromatography system (PerSeptive Biosystems, Cambridge, MA)
2. Mobile phase A: 10 mM KH_2PO_4 and 25% acetonitrile (ACN), pH 3.0
3. Mobile phase B: 500 mM KCl, 10 mM KH_2PO_4, and 25% ACN, pH 3.0
4. Mobile Phase C: 600 mM KCl, 10 mM KH_2PO_4 and 25% ACN, pH 6.0
5. Column: Polysulfoethyl-A column (4.6 mm × 200 mm, particle size 5.0 μm, 300 Å) (Poly LC Inc., Columbia, MD)
6. pH paper (Whatman Inc., Piscataway, NJ)
7. 2-ml fraction collection tube

2.5. Peptide Desalting

1. PepClean C_{18} spin columns (Thermo Scientific, Rockford, IL)
2. Activation solution: 50% ACN
3. Equilibration solution: 5% ACN containing 0.5% trifluoroacetic acid (TFA)

4. Elution solution: 70% ACN containing 0.1% TFA

2.6. Reversed-Phase Liquid Chromatography

1. Solvent A: 2% ACN containing 0.1% TFA
2. Solvent B: 85% ACN containing 0.1% TFA
3. Matrix-assisted laser desorption ionization (MALDI) matrix solution: 6 mg/ml α-cyano-4-hydroxycinnamic acid (Sigma, St Louis, MO) in 50% ACN, 5 mM monobasic ammonium phosphate and internal calibrants (50 fmol/ml each of [Glu^1]-Fibrinopeptide B (GluFib), *m/z* 1,570.677 (Sigma, St Louis, MO) and adrenocorticotropic hormone 18–39 (ACTH 18–39), *m/z* 2,465.199 (Sigma, St Louis, MO))
4. LC-Packings Ultimate chromatography system equipped with a Probot MALDI spotting device (Dionex, Sunnyvale, CA, USA)
5. C_{18} PepMap trapping column (0.3 mm i.d. × 5 mm length, 5 μm, 100 Å, Dionex, Sunnyvale, CA, USA)
6. C_{18} PepMap capillary column (0.1 mm i.d. × 150 mm length, 3 μm, 100 Å, Dionex, Sunnyvale, CA, USA)

2.7. Mass Spectrometry

1. 4800 Plus MALDI TOF/TOF Analyzer (AB SCIEX, Foster City, CA)
2. MALDI plates (AB SCIEX, Foster City, CA)
3. Mass standards kit (AB SCIEX, Foster City, CA)
4. 4000 Series Explorer (AB SCIEX, Foster City, CA)

2.8. Data Analysis Software

1. ProteinPilot (AB SCIEX, Foster City, CA, http://www.absciex.com)
2. Scaffold (Proteome Software Inc., Portland, OR, http://www.proteomesoftware.com)
3. TS2Mascot (Matrix Science Inc., Boston, MA, http://www.matrixscience.com)
4. Mascot (Matrix Science Inc., Boston, MA, http://www.matrixscience.com)
5. Excel (Microsoft corporation, Redmond, WA, http://office.microsoft.com)

3. Methods

3.1. Protein Extraction

1. For each sample category, wash 5×10^6 cells with ice cold PBS twice to remove any residual media (see Note 1).
2. Carefully harvest the cells in 1 ml of PBS with a corning cell scraper.
3. Pellet the cells in a micro centrifuge at 2,500× *g* at 4°C and remove the PBS completely (see Note 1).

4. Resuspend each cell pellet in 250 μl of cold iTRAQ lysis buffer by vortexing for 15 s at high speed (see Note 2).
5. Lyse the cells by a 10 s sonication pulse followed by a 30 s incubation in the ice bath. Repeat three times. Incubate the cell lysate on the ice bath for 10 min.
6. Clarify the cell lysate by centrifugation at 16,100× *g* for 30 min at 4°C in a micro centrifuge.
7. Carefully transfer the supernatant to a fresh Eppendorf tube.
8. Estimate the protein concentrations using the BCA protein assay kit with bovine serum albumin protein standard diluted in the iTRAQ lysis buffer as the standards. Adjust the protein concentrations of all eight samples to same level by diluting the samples in the iTRAQ lysis buffer (see Note 3).

3.2. Protein Digestion with Trypsin

1. Transfer 75 μg of proteins from each of the eight samples into fresh tubes (see Note 4).
2. Add 2.0 μl of the reducing reagent to each sample. Mix well by vortexing for 15 s and centrifuge briefly to bring down the solution (see Note 4).
3. Incubate the mixture at 60°C for 1 h with mixing (see Note 4).
4. Add 1 μl of the cysteine alkylating solution. Mix well by vortexing for 15 s and centrifuge briefly to bring down the solution. Incubate the solution for 10 min at room temperature (see Note 4).
5. Reconstitute two vials (20 μg/vial) of sequencing grade trypsin with 80 μl each in HPLC grade water by slowly pipetting the solution up and down a few times. Vortex for 30 s and centrifuge briefly to bring down the solution (see Note 5).
6. Add 20 μl of the trypsin solution to each sample tube. Vortex for 1 min and centrifuge briefly to bring the solution down. Incubate the digestion reaction vials at 37°C for 16 h in a water bath. Centrifuge briefly to bring all the solution down at the bottom (see Note 6).

3.3. Peptide Labeling with the 8-plex iTRAQ Reagents

1. Bring the iTRAQ reagents out of the freezer to room temperature. Centrifuge briefly to bring the solution to the bottom of the vial (see Note 7).
2. Add 100 μl of HPLC grade isopropanol to each vial of the iTRAQ reagent.
3. Vortex each vial at high speed for 30 s and then centrifuge briefly.
4. Transfer the entire contents of each freshly prepared iTRAQ reagent to their respective tryptic peptide sample tube. Vortex all the tubes at high speed for 30 s and then centrifuge briefly (see Note 8).

5. Test the pH of each sample by placing 0.5 μl of the solution onto a pH paper. If necessary, add up to 5 μl of 0.5 M TEAB to adjust the pH of the final solution to between 7.5 and 8.5 (see Note 9).
6. Incubate the iTRAQ labeling reaction tubes at room temperature for 2 h. Vortex and spin down briefly to bring solution to the bottom of the tubes (see Note 10).
7. Combine the contents of all eight iTRAQ labeled samples. Vortex to mix, then spin down the solution (see Note 11).
8. Completely dry the combined sample in a speed vac at room temperature (see Note 12).

3.4. Two-Dimension LC Separation of Peptides

The combined iTRAQ-labeled peptides are fractionated first with strong cation exchange chromatography and then with reversed phase chromatography.

3.4.1. Strong Cation Exchange Liquid Chromatography

Various reagents (e.g., TEAB, isopropanol, TCEP, detergents and excess iTRAQ reagents) used during protein extraction, digestion and labeling may interfere with either reversed-phase liquid chromatography (RPLC) steps or MS identification and quantification steps; therefore they must be removed completely beforehand. Peptide mixture is initially fractionated on a PerSeptive BioCAD SPRINT Perfusion chromatography system equipped with a PolySULFOETHYL A strong cation exchange column.

1. Reconstitute the iTRAQ labeled peptide mixture by adding 4 ml of the Strong Cation Exchange Liquid Chromatography (SCXLC) mobile phase A (see Note 13). Test the pH of the peptide solution using a pH paper. If necessary, adjust the pH of the solution to between 2.7 and 3.0 by addition of 1 M phosphoric acid.
2. Centrifuge the sample at 20,000× *g* for 15 min at 25°C to pellet the precipitates and particulates. Carefully transfer the clarified solution into a fresh tube.
3. Equilibrate the column for 15 column volume (~30 ml) with the mobile phase A.
4. Inject the iTRAQ-labeled peptides onto the SCXLC column through a 5-ml sample loading loop (see Note 14).
5. After the injection, wash the column with 15 column volume (~30) with mobile phase A to remove the unbound iTRAQ reagents and detergents (see Note 15).
6. Elute the peptides with a 2-segment linear gradient at a flow rate of 1 ml/min according to table below. Collect 2 min fractions during SCXLC in Eppendorf tubes (see Note 16).

Time (min)	Mobile phase A (%)	Mobile phase B (%)	Mobile phase C (%)
0	100	0	0
45	50	50	0
60	0	0	100
75	0	0	100

7. Dry each fraction completely in a speed vac.

3.4.2. Concentration and Desalting the SCXLC Fractions with C_{18} Spin Columns

1. Resuspend the peptides in each SCXLC fraction in 200 μl of the equilibration solution (see Note 17). Sonicate in a water bath for 15 s, vortex, and then spin briefly.
2. Activate the C_{18} resin by adding 200 μl of activation solution to the C_{18} spin columns. Centrifuge at 1,500× *g* for 1 min. Repeat step 2 once more.
3. Equilibrate the C_{18} resin by adding 200 μl of the equilibration solution. Centrifuge at 1,500× *g* for 1 min. Repeat step 3 again.
4. For each resuspended SCXLC fraction, transfer the solution completely onto an equilibrated C_{18} spin column. Centrifuge at 1,500× *g* for 1 min. Collect the flow through and load the flow through again onto the spin column. Centrifuge at 1,500× g for 1 min. Repeat step 4 once more.
5. Wash off the unbound salts by adding 200 μl of the equilibra tion solution onto the column. Centrifuge at 1,500× *g* for 1 min. Repeat step 5 twice.
6. To elute the peptides, add 30 μl of the elution solution onto the column. Collect the eluted peptides by centrifugation at 1,500× *g* for 1 min in a fresh Eppendorf tube. Repeat step 6 twice and collect all the eluted peptides from each SCXLC fraction in same tube.
7. Dry the desalted peptides in speed-vac for further fractionation using reversed-phase liquid chromatography.

3.4.3. Reversed-Phase Liquid Chromatography

1. Reconstitute the peptides in each SCXLC fraction in 20 μl of RPLC solvent A. Vortex at high speed for 1 min and then spin briefly. Sonicate the sample in a water bath for 15 s and vortex. Centrifuge the tubes at 16,100× *g* for 5 min. Transfer each sample solution into the bottom of an auto-sampler vial and place all the vials in cooled auto sampler tray.
2. Equilibrate the RPLC column for at least 20 min with 2% solvent B at 0.300 μl/min.
3. For each SCXLC fraction, load 5 μl of the reconstituted peptides onto a C_{18} trapping column using a μl pickup injection

method at a flow rate of 20 μl/min (see Note 18). Subsequently, the bound peptides are resolved in a high resolution C_{18} PepMap column at a flow rate of 0.3 μl/min with the following gradient.

Time (min)	Solvent A (%)	Solvent B (%)
0	98	2
6	98	2
7	90	10
51	77	23
77	54	46
89	5	95
99	5	95
100	98	2
121	98	2

4. The eluted peptides are mixed with the MALDI matrix solution in a 1:1 ratio through a 30-nl mixing tee and directly spotted onto a MALDI plates in a 33 × 10 spot array format using Probot, which produces a spot every 12.5 s (see Note 19).
5. Repeat the RPLC steps for each of the SCXLC fractions.

3.5. Mass Spectrometry

Peptides spotted on the MALDI plates are analyzed on a 4800 Plus MALDI TOF/TOF Analyzer using 4000 Series Explorer Software.

1. Tune and optimize the sensitivity and resolution of the mass spectrometer using the mass standard mixture kit. Check and optimize both metastable ion suppressor and the timed-ion-selector for specific precursor ion selection at the maximum resolution of 400, corresponding to ±2.5 Da at *m/z* of 1,000. Optimal performing instrument is very important for accurate iTRAQ quantification outcome (13).
2. Using the 6 peptide masses within the mass standard mixture kit, update all three MS calibration parameters (Detector Offset, TOF Offset, B-Factor) of the instrument using update default calibration function to ensure maximum mass accuracy. Update the MS/MS calibration parameters (Detector Offset, TOF Offset, B-Factor) using the MS/MS ion spectra of the GluFib (*m/z* 1,570.677) (see Note 20).
3. For each quantitative project, create a new project on the 4000 Series Explorer Software. Then create a new spot set using a predefined spot set template. Load the sample MALDI plate into the mass spectrometer using the newly established spot set.

4. Align the plate using the alignment MALDI spots specified in the corresponding spot set template. Confirm that the laser target crosshair in the video viewer is aligned with the laser spot.
5. Create an acquisition, a processing, and a job-wide interpretation method for both MS and MS/MS analyses.
6. For the MS acquisition method, use positive ion and reflector mode as the operating mode. Specify the mass range of interest as *m/z* 850–3,000 and the focus mass as 1,950 *m/z*. Set the laser intensity to 3,000 and the detector voltage multiplier at 0.90. Each MS spectrum is averaged over 1,000 laser shots. In the processing method, GluFib (*m/z* 1,570.677) and ACTH 18–39 (*m/z* 2,465.199) masses are used as the internal calibrants. For the interpretation method, set following criteria for precursor selection for the subsequent MS/MS analysis; 15 most abundant precursors per spot, minimum S/N filter at 50, spot to spot precursor mass tolerance at 200 ppm and from the rarest to the most abundant MS/MS ion as the data acquisition order.
7. For MS/MS acquisition method, use a 2 KV positive ion MS–MS method as the operating mode. Set the laser intensity to 4,000 and detector voltage multiplier at 0.90. Specify the metastable suppression as "on", CID as "on" and the precursor mass window at relative 400 resolution (FWHM). Each MS/MS spectrum is accumulated over 2,000 laser shots. In the MS/MS processing method, each spectrum is smoothed using the Savitsky–Golay algorithm with points across peak set at 3 and polynomial order set at 4.
8. Set the medium CID gas recharge pressure to medium with a threshold of 5.0×10^7 torr.

3.6. Bioinformatics

Peptide identification and quantification is determined by ProteinPilot software (v. 2.0.1) against the rat IPI database (v3.55, Release date Feb 12, 2009, 39,874 sequence entries) using the Paragon algorithm (23) in the search engine.

3.6.1. Protein Identification

1. The following default parameters are used for peptide identification by the ProteinPilot: identification focused on biological modifications, "thorough" search is engaged, iTRAQ 8-plex as sample type, MMTS as Cys alkylation reagent, trypsin as the digestion enzyme (see Note 21) and instrument type of 4800 MALDI TOF-TOF are selected. Protein detection threshold is set at an Unused ProtScore of 1.3, corresponding to 95.0% confidence interval (C.I.). The Paragon algorithm set instrument-appropriate mass error tolerance automatically. The iTRAQ reagent isotopic carryover bias correction is set as automatic.
2. In ProteinPilot software, in order to eliminate protein identification redundancy, raw peptide identification results are first

processed by the Pro Group algorithm to create nonredundant protein groups from the peptides identified. A minimal set of proteins is produced for a given protein confidence interval threshold. In each group, one protein is designated as the winner protein for having the highest Unused ProtScore, meaning with the most number of peptides matched within the homologous group.

3. To minimize the probability of false identification, consider only the winner proteins with an Unused ProtScore of at least 1.3 and having at least two distinct peptides with a minimum C.I. of 95% as identified.

4. To estimate the protein false discovery rate (FDR), all spectra are also searched against a decoy IPI Rat database containing all the same proteins with reversed sequences using the same search parameters as described above. The FDR is calculated as

$$\text{FDR} = 2\times\left(N_{\text{decoy}}\right)/\left(N_{\text{decoy}} + N_{\text{forward}}\right)$$

where N_{decoy} is the number of proteins identified using the decoy database, N_{forward} is the number of proteins identified using the regular protein database (in this case Rat IPI protein database) (13, 24).

5. If the FDR is higher than the generally accepted 1.0%, then repeat the search using higher Unused ProtScore until the FDR reach less than 1.0%.

3.6.2. Protein Quantification

1. For relative protein quantification, the Pro Group algorithm calculates the relative protein expression ratios using only the iTRAQ ratios obtained from the peptide(s) that are distinct to each protein. The following peptides are excluded from quantification calculation (a) iTRAQ ion S/N ratio ≤6. (b) C.I. % <1.0%, (c) peptides that are identified without an iTRAQ modification tag information, (d) shared peptides – the spectrum matched to a peptide sequence that is claimed by more than one protein.

2. The relative protein expression for each iTRAQ labeled sample is calculated as the weighted average for all the corresponding peptides.

3. Mean protein ratios are normalized for correcting the experimental bias introduced during labeling. Protein ratios for all iTRAQ-labeled proteins are presented as the relative ratios compared to iTRAQ 113-labeled proteins (see Note 22).

4. To calculate the average protein expression ratios between rapamycin-treated and the control samples, all eight quantification ratios for each protein is exported to Microsoft Excel and is calculated as follows: (average of the four rapamycin treated)/(average of the four control) values.

5. To identify differently expressed proteins, *p*-values are calculated from the 2-tailed Student's *T*-test for each protein by comparing the four rapamycin treated OPCs ratios with the four control OPCs ratio values. Proteins exhibiting a greater than 20% changes ($p < 0.05$) are considered significant (see Notes 23 and 24). These changes are beyond the 10% analytical coefficient of variance for iTRAQ analysis on our MS system (13).

3.6.3. Alternate Protein Identification and Quantification Procedures

Different database search algorithms identify only a fraction of the large number of spectra generally acquired during a shotgun proteomics experiment. They may also produce different identification and quantification results based on the identification cutoff criteria (25). It has been suggested that by using multiple search engines, a higher proportion of the proteome can be quantified (26, 27). Scaffold software is designed to analyze MS/MS-based protein identifications by comparing tandem mass spectrometry data that have been analyzed by the more than one search algorithms (e.g., Mascot, X! Tandem etc). To compare and validate the search results from multiple search engines; Scaffold uses PeptideProphet and ProteinProphet algorithms to assign statistical threshold for confidant protein identifications (28, 29). Quantitative analysis of both peptide and protein changes can be further evaluated with the aid of the Scaffold Q+ 2.0 software module. The use of Scaffold analysis of Mascot and X! Tandem search results is described below.

1. Generate a peak list using TS2Mascot as a Mascot Generic File (MGF) from the tandem MS spectra using following parameters: mass range from 20 to 60 Da below precursor, S/N ratio of at least 10.
2. Submit the peak list for automated protein sequence database search using a local Mascot server (version 2.3) against the rat IPI database (v3.55, Release date Feb 12, 2009, 39,874 sequence entries). Set following search parameters, iTRAQ 8-plex (K), iTRAQ 8-plex (N-terminal) and methylthio (C) as fixed modifications; iTRAQ 8-plex (Y) and Oxidation (M) as variable modifications; trypsin as the cleaving enzyme with maximum of one missed cleavage allowed; monoisotopic, peptide tolerance 50 ppm; MS/MS mass tolerance 0.3 Da. Reverse sequence database search was engaged. Once the search is completed, mascot generate a search result file as *.dat file for further analysis.
3. Open Scaffold, set up a new file, select quantitative technique as iTRAQ (8-plex), and insert the iTRAQ reagent purity correction parameters provided by AB SCIEX.
4. Load the *.dat file generated by Mascot and select the IPI protein database to be used for additional analysis by X! Tandem, using the same search parameters as used for the Mascot search (see Note 25).

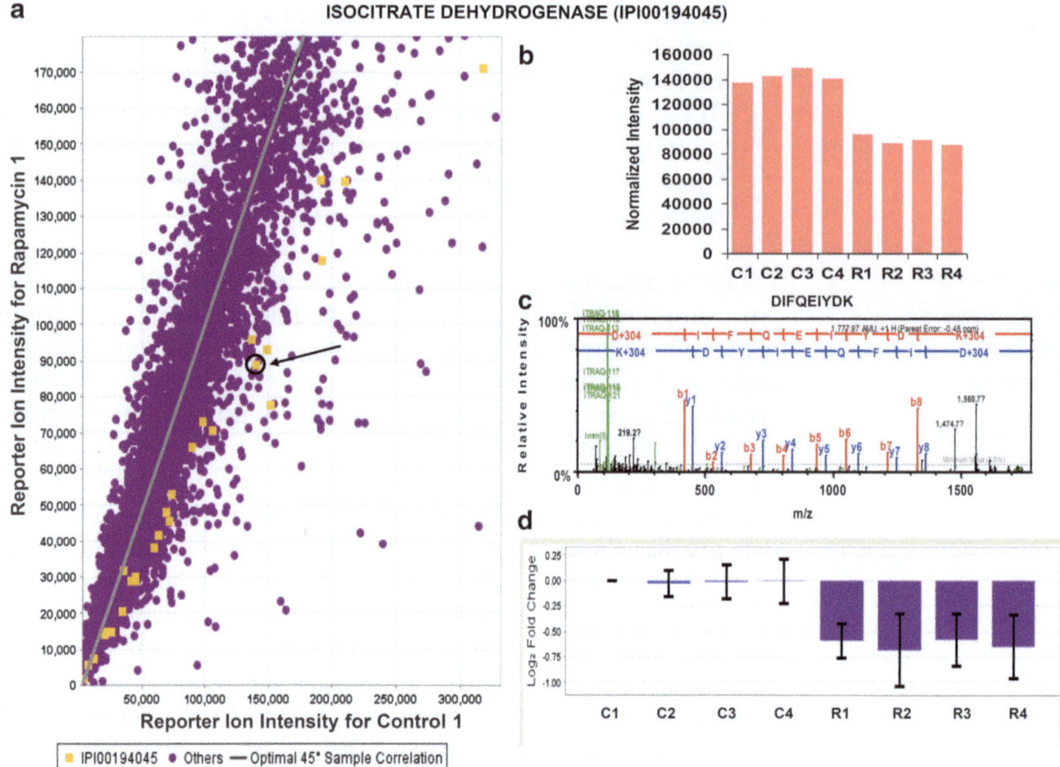

Fig. 2. Downregulation of isocitrate dehydrogenase is ascertained from iTRAQ analysis by Scaffold. (**a**) A representative Scaffold graphical plot of normalized reporter ion intensity values of iTRAQ reporter ion 113 (see Fig. 1, Control 1) versus those of iTRAQ reporter ion 117 (see Fig. 1, Rapamycin treatment 1) for all the identified peptides is shown (*purple dots*). The iTRAQ plot dots fall near the center straight line represent the peptides that are not differentially regulated by ramamycin treatment. Each *yellow dot* represents the iTRAQ signals derived from a peptide matched to isocitrate dehydrogenase. iTRAQ ion plots of various intensities derived from the tryptic peptides obtained from this protein revealed similar down-regulated iTRAQ expression trend of 117/113 ratios of ~0.7. (**b**) A *bar graph* of the normalized intensities of the iTRAQ reporter ions for a representative peptide (DIFQEIYDK) ion (*circled* and *pointed* with *arrow* in **a**) of isocitrate dehydrogenase indicates consistent downregulation of this peptide (R/C of ~0.7) across all the replica of the experiment (C1–C4 represents the control samples and R1–R4 represents rapamycin treated samples) (see Fig. 1). (**c**) Continuous series of the b and y ions allows the identification of the peptide as DIFQEIYDK (*circled* in **a**). (**d**) A *bar graph* of the means and standard deviations of the protein level iTRAQ ratios in relation control signal obtained from all the peptides derived from isocitrate dehydrogenase. Quantification variability among the peptides is shown by the *error bars*. Notably the means for each independent experiment were very consistent among the replicates.

5. After the search is completed, view the list of all the matched proteins in the sample pane of Scaffold. Set the minimum protein identification probability score to 95%, minimum number of peptides to 2, and minimum peptide identification probability score to 95% (see Note 26).
6. Invoke the statistics pane in the Q+ module of Scaffold to evaluate the consistency of expression changes among different peptides belonging to the same protein (Fig. 2). If most of the

light dots (each representing a peptide) fall near a center reporter ion correlation line drawn over the scatter plot of all the peptides identified from a control and rapamycin-treated sample, such observation represents an 1:1 ratio between the selected samples. Therefore, the protein is not differentially expressed. If the selected protein is upregulated, then the light spots representing the peptides within this protein on the scatter plot would be placed above the center line. If the protein is downregulated, the peptide spots would be below the center line (Fig. 2).

7. To obtain the relative quantification of proteins in each sample, invoke the Q+ module in Scaffold. Organize sample comparison by assigning the comparison categories: the iTRAQ-labeled samples 113, 114, 115, and 116 in the control category and iTRAQ-labeled samples 117, 118, 119, and 121 in the rapamycin treated category.
8. Scaffold transforms quantitative ratios into Log_2 format and normalizes the values.
9. All eight quantification ratios for each protein can be exported to Microsoft Excel and converted back to anti-log fold changes. The relative protein expression between rapamycin-treated and the control samples is calculated as follows: (average of the four rapamycin treated)/(average of the four control) values.
10. To identify differently expressed proteins, *p*-values are calculated from the 2-tailed Student's *T*-test for each protein by comparing the four rapamycin treated OPCs ratios with the four control OPCs ratio values.

4. Notes

1. Removal of the residual media is very important. Otherwise, the components in cell culture media may interfere with subsequent digestion and iTRAQ labeling, resulting in external variability in the samples to be analyzed. We use PBS without any ammonium salt to completely wash the cells and to remove remaining media components. Traces of PBS should be removed by using a fine pipette tip, preferably with the gel loading tip without disturbing the cell pellets.
2. The lysis buffer described here has been successfully used to extract both soluble and membrane proteins. Other alternate buffers can also be used. One such buffer contains 20 mM TEAB, 25 mM Na_2CO_3, and 0.1% (v/v) of protease inhibitor cocktail, pH 8.5. Buffer based on urea can be also used. However,

care must be taken to avoid buffers that contain primary amines (e.g., tris) which interfere with subsequent iTRAQ labeling.

3. It is very important to estimate the protein concentration of all the samples accurately. A protein concentration assay that is fully compatible with the lysis buffer should be employed. Here, we use the detergent-compatible BCA protein assay. For urea containing lysis buffer, the Bradford protein assay should be used.
4. As per iTRAQ manufacturer instruction, each protein sample should be between 5 and 100 μg for each iTRAQ labeling reaction. To ensure maximum labeling efficiency, sample volumes should not be more than 50 μl each. If the sample volume is larger than 50 μl, a speed vac or lyophilizer can be used to reduce the sample volume before iTRAQ labeling. To achieve effective reduction of protein disulfides, final concentration of TCEP should be maintained at 4 mM. If urea is used in lysis buffer, samples can be incubated at 37°C during the reduction step to avoid any carbamylation of amino acids (30). Final concentration of MMTS should be maintained at 8 mM for effective alkylation of reduced cysteines.
5. For effective trypsin digestion, dilute the sample so that the final concentrations of the detergents and other chaotropic reagents do not inhibit trypsin activity.
6. Check the protein digestion efficiency before proceeding to iTRAQ labeling. Take 1 μl from each of the digested samples. Cleanup the peptides with a SCX ZipTip® (Millipore, Billerica, Massachusetts). Mix the eluted peptides with the MALDI matrix solution in a 1:1 ratio and spot them onto a MALDI plate. Acquire the MS spectra and check if the peptide ion signals are comparable across all the eight samples. If the digestion is not complete, repeat the trypsin digestion step with additional enzyme.
7. Each iTRAQ tag may be provided in different volumes by vendor. It is normal if iTRAQ reagent volumes are not same in all the vials.
8. Make sure that the organic solvent concentration is at least 60% by adding HPLC grade isopropanol to the digested peptides before adding the iTRAQ solution for labeling. Lower organic concentration would result in rapid hydrolysis of iTRAQ reagents and poor labeling efficiency or no labeling.
9. For optimal labeling efficiency, the pH of the peptide/iTRAQ mixture solution must be between 7.5 and 8.5. If the pH is less than 7.5, the labeling efficiency would be significantly reduced.
10. Labeling with the 8-plex iTRAQ reagent requires a reaction time of 2 h as compared to 1 h with the 4-plex version of the iTRAQ reagents. It is important to incubate the samples for at

least 2 h before combining in the next step. Failure to do so may result in incomplete labeling and artificial variations via cross-sample labeling.

11. Before combining the iTRAQ-labeled peptides, analyze an aliquot of each sample by MS/MS to verify the presence of each of the eight iTRAQ reporter ion peak in the spectra. First, remove the organic solvent by speed vac, and then clean the peptides using a SCX ZipTip. In MS/MS spectra, verify the presence of peaks at the *m/z* of the appropriate iTRAQ reporter groups. If not, relabel the protein digests.
12. To remove the TEAB completely, resuspend the dried sample in 250 μl of HPLC grade water and dry again completely using a speed vac. Repeat the process twice. Dried peptides can be stored at –80°C till the next step.
13. After resuspension of the peptides, make sure that concentration of the buffer salts is less than 10 mM by diluting the sample mixture with 4.0 ml of SCXLC mobile phase A.
14. To maximize the binding of the peptides to the column, samples should be loaded with relatively slow flow rate (e.g., 0.25 ml/min). Alternatively the flow through can be reloaded.
15. It is very important to wash the bound peptides with large volumes of mobile phase A to remove the detergents and residual iTRAQ reagents completely from the column prior to the start of elution gradient. Typically, wait until the UV absorbance reading reaches baseline before the initiation of the salt gradient.
16. At pH 3.0, basic residues in peptides (His, Arg, Lys) are positively charged, as are the N-termini; acidic residues (Asp, Glu) are uncharged and the C-termini are mainly uncharged. Thus, most peptides with free N-termini will have net charges of at least +1 and should bind to PolySULFOETHYL A resin and can then be eluted with a salt gradient. Generally 2/3 of all the peptides has +1 or +2 charge and is eluted with a gradient up to 0.25 M KCl. The others typically have charges of +3 or +4. It may be necessary to use a gradient to 0.6 M salt to elute these peptides. In order to distribute tryptic peptides as uniformly as possible in the collected fractions we prefer to use a linear gradient with two segments; one segment goes to 0.25 M salt which involves ~75% of the gradient and a second segment from 0.25 to 0.6 M salt with pH 4.5–6.0 over the remaining 25% of the gradient.
17. For high salt concentration buffer fractions, 200 μl of the loading solution may not be enough to dissolve the salts and peptides completely. Salts and peptides should be solubilized completely by adding more of the loading solution. Entire contents of the tube should be dissolved and loaded onto the spin column for desalting.

18. Volume of the peptide solutions to be loaded depends upon the concentrations of the peptides. In order to maintain peptide separation and resolution, saturating the reversed phase trap with excessive peptides should be avoided. The trapping column capacity for this protocol is ~20 pmoles. Absorbance at 214 nm during SCXLC fraction collections can be used as a guide to guage the concentration of peptides in each fraction. To further determine the complexity of the peptides in each fraction, mix an aliquot (1.0 μl) of the desalted peptides with 1.0 μl of the MALDI matrix. Acquire the MS spectra. SCXLC fractions whose MS spectra contain less than 100 peaks with S/N ≥5 between *m/z* 1,000 and 2,000 and with a relatively low absorbance at 214 nm during SCXLC (generally at the beginning and end of the SCXLC gradient) can be combined prior to subsequent RPLC separations.
19. On each MALDI plate, eluted peptides from five different SCXLC fractions can be spotted, resulting in a total of 1,650 MALDI spots. To avoid carry over between RPLC runs, a blank run should be performed between two RPLC runs to clean the chromatography system.
20. Turn on the high voltage for the MALDI source at least 30 min prior to update the default calibration. The warm-up period ensures the MS to deliver highest mass-accuracy during the calibration step by reducing the variability in accelerating voltages and yielding more reproducible and accurate calibration results.
21. ProteinPilot software allows the search for tryptic, nontryptic, and semitryptic cleavages simultaneously. In conjunction with the iTRAQ technique, this feature has been successfully used to uncover regulated proteolytic events in spinal cords of experimental autoimmune encephalomyelitis animals by analyzing semitryptic and nontryptic peptides independent from the tryptic peptides (11). This approach does not interfere with typical protein expression studies and may produce very useful information that is relevant to the underlying biology. This additional information may complement differential expression data commonly sought after by those performing shotgun proteomics studies. It can be obtained with relatively little additional effort and cost, thus improving proteomic research productivity.
22. In ProteinPilot, Tag 113 is set as denominator for peptide ratio calculation by default. If it is required to set any other tag as the denominator, it can be selected under the quant function. Similarly, auto bias correction is set by default. However, in some cases, manual bias parameters can be set to compensate for extreme sample variability as a result of sample preparation artifacts and iTRAQ reagent heterogeneity.

23. Using the method described here, we acquired 54,485 MS/MS spectra. Total of 622 proteins were indentified with an estimated FDR of 0.1%. Two hundred twenty one (221) proteins exhibited a greater than 20% change ($p < 0.05$) in OPCs by rapamycin treatment.
24. It is very important to validate the iTRAQ quantification results using alternate methods including Western blotting, immunohistochemistry or other biological methods.
25. Scaffold needs to access the same protein sequence database that is used by Mascot search. Upload the same database (Rat IPI protein database in this case) into Scaffold beforehand.
26. Minimum protein identification probability set the threshold requirement for Scaffold to calculate the probability of correct protein identification. Minimum number of peptides set the number of unique peptides that must be found for a given protein in order to consider the protein to be identified. Minimum peptide identification probability set how certain (C. I. %) a peptide identification must be before it can be counted toward the minimum number of peptides.

Acknowledgments

This work is supported in part by National Institutes of Health: Grant no. NS37560, Grant no. NS056097 and Grant no. NS046593; National Multiple Sclerosis Society: Grant no. RG4015A2/2.

References

1. Noorbakhsh F, Overall CM, Power C (2009) Deciphering complex mechanisms in neurodegenerative diseases: the advent of systems biology. *Trends Neurosci* **32**, 88–100
2. Gygi SP, Rochon Y, Franza BR et al (1999) Correlation between protein and mRNA abundance in yeast. *Mol Cell Biol* **19**, 1720–1730
3. Bayes A, Grant SG (2009) Neuroproteomics: understanding the molecular organization and complexity of the brain. *Nat Rev Neurosci* **10**, 635–646
4. Tyler WA, Jain MR, Cifelli SE et al (2011) Proteomic identification of novel targets regulated by the mammalian target of rapamycin pathway during oligodendrocyte differentiation. *Glia* **59**, 1754–1769
5. Gygi SP, Rist B, Gerber SA et al (1999) Quantitative analysis of complex protein mixtures using isotope-coded affinity tags. *Nat Biotechnol* **17**, 994–999
6. Wu C, Parrott AM, Liu T et al (2011) Distinction of thioredoxin transnitrosylation and denitrosylation target proteins by the ICAT quantitative approach. *J Proteomics* **74**, 2498–2509
7. Fu C, Wu C, Liu T et al (2009) Elucidation of thioredoxin target protein networks in mouse. *Mol Cell Proteomics* **8**, 1674–1687
8. Ong SE, Blagoev B, Kratchmarova I et al (2002) Stable isotope labeling by amino acids in cell culture, SILAC, as a simple and accurate approach to expression proteomics. *Mol Cell Proteomics* **1**, 376–386
9. Ross PL, Huang YN, Marchese JN et al (2004) Multiplexed protein quantitation in Saccharomyces cerevisiae using amine-reactive isobaric tagging reagents. *Mol Cell Proteomics* **3**,1154-1169

10. Liu T, Donahue KC, Hu J et al (2007) Identification of differentially expressed proteins in experimental autoimmune encephalomyelitis (EAE) by proteomic analysis of the spinal cord. *J Proteome Res* **6**, 2565–2575
11. Jain MR, Bian S, Liu T et al (2009) Altered proteolytic events in experimental autoimmune encephalomyelitis discovered by iTRAQ shotgun proteomics analysis of spinal cord. *Proteome Sci* **7**, 25
12. Grant JE, Hu J, Liu T et al (2007) Post-translational modifications in the rat lumbar spinal cord in experimental autoimmune encephalomyelitis. *J Proteome Res* **6**, 2786–2791
13. Hu J, Qian J, Borisov O et al (2006) Optimized proteomic analysis of a mouse model of cerebellar dysfunction using amine-specific isobaric tags. *Proteomics* **6**, 4321–4334
14. Liu T, Hu J, Li H (2009) iTRAQ-based shotgun neuroproteomics. *Methods Mol Biol* **566**, 201–216
15. Jain MR, Liu T, Hu J et al (2008) Quantitative Proteomic Analysis of Formalin Fixed Paraffin Embedded Oral HPV Lesions from HIV Patients. *Open Proteomics J* **1**, 40–45
16. Hunt DF, Yates JR, 3rd, Shabanowitz J et al (1986) Protein sequencing by tandem mass spectrometry. *Proc Natl Acad Sci* U S A **83**, 6233–6237
17. Steen H, Mann M (2004) The ABC's (and XYZ's) of peptide sequencing. *Nat Rev Mol* Cell Biol **5**, 699–711
18. Sarbassov DD, Ali SM, Sabatini DM (2005) Growing roles for the mTOR pathway. *Curr Opin Cell Biol* **17**, 596–603
19. Kim DH, Sarbassov DD, Ali SM et al (2002) mTOR interacts with raptor to form a nutrient-sensitive complex that signals to the cell growth machinery. *Cell* **110**, 163–175
20. Sarbassov DD, Ali SM, Kim DH et al (2004) Rictor, a novel binding partner of mTOR, defines a rapamycin-insensitive and raptor-independent pathway that regulates the cytoskeleton. *Curr Biol* **14**, 1296–1302
21. Tyler WA, Gangoli N, Gokina P et al (2009) Activation of the mammalian target of rapamycin (mTOR) is essential for oligodendrocyte differentiation. *J Neurosci* **29**, 6367–6378
22. Bibollet-Bahena O, Almazan G (2009) IGF-1-stimulated protein synthesis in oligodendrocyte progenitors requires PI3K/mTOR/Akt and MEK/ERK pathways. *J Neurochem* **109**, 1440–1451
23. Shilov IV, Seymour SL, Patel AA et al (2007) The Paragon Algorithm, a next generation search engine that uses sequence temperature values and feature probabilities to identify peptides from tandem mass spectra. *Mol Cell Proteomics* **6**, 1638–1655
24. Peng J, Elias JE, Thoreen CC et al (2003) Evaluation of multidimensional chromatography coupled with tandem mass spectrometry (LC/LC-MS/MS) for large-scale protein analysis: the yeast proteome. *J Proteome Res* **2**, 43–50
25. Kapp EA, Schutz F, Connolly LM et al (2005) An evaluation, comparison, and accurate benchmarking of several publicly available MS/MS search algorithms: sensitivity and specificity analysis. *Proteomics* **5**, 3475–3490
26. Resing KA, Meyer-Arendt K, Mendoza AM et al (2004) Improving reproducibility and sensitivity in identifying human proteins by shotgun proteomics. *Anal Chem* **76**, 3556–3568
27. Elias JE, Haas W, Faherty BK et al (2005) Comparative evaluation of mass spectrometry platforms used in large-scale proteomics investigations. *Nat Methods* **2**, 667–675
28. Keller A, Nesvizhskii AI, Kolker E et al (2002) Empirical statistical model to estimate the accuracy of peptide identifications made by MS/MS and database search. *Anal Chem* **74**, 5383–5392
29. Nesvizhskii AI, Keller A, Kolker E et al (2003) A statistical model for identifying proteins by tandem mass spectrometry. *Anal Chem* **75**, 4646–4658
30. Stark GR, Stein WH, Moore S (1960) Reactions of the Cyanate Present in Aqueous Urea with Amino Acids and Proteins. *J Biochem Mol Biol* **235**, 3177–3181

Chapter 9

Protein Profiling of the Brain: Proteomics of Isolated Tissues and Cells

Nicole Haverland and Pawel Ciborowski

Abstract

A relatively new proteomic technology platform known as stable isotope labeling of amino acids in cell culture (SILAC) has been developed to record quantitative changes in in vitro experimental systems. This method allows for monitoring of changes due to various experimental conditions and also dynamics of changes in time or, e.g., during administration of drugs, etc. Furthermore, SILAC technology has advanced in such a way to include the analyses of protein turnover in both dividing and nondividing cells, a technology now known as pulse SILAC (pSILAC). The ability to profile proteomes of isolated cells or tissues of the brain opens new experimental opportunities in neuroscience. The focus of this protocol is to provide a step-by-step outline including brain tissue collection and isolation, sample preparation for iTRAQ and pSILAC labeling of primary cells, fractionation of proteins and in-gel tryptic digest, mass spectrometry-based identification and quantitation of peptides, and validation of differentially expressed proteins using Western blot.

Key words: Proteomics, Neuroproteomics, Biomarker, Validation, Brain, SILAC, iTRAQ

1. Introduction

The ability for global profiling using proteomic-based technologies has elevated the expectations of new discoveries across all fields of study, including a wide range of neuroscience disciplines (1). These expectations are limited not only to elucidation of molecular mechanisms underlying neurodegenerative diseases or finding new biomarkers and drug targets, but also includes the understanding of how a normal, healthy brain develops and functions (2).

Early neuroproteomic experiments used either the whole brain (3, 4) or organelle(s) isolated from the whole brain (5); however, it has been quickly realized that global analysis of proteins is both

Yannis Karamanos (ed.), *Expression Profiling in Neuroscience*, Neuromethods, vol. 64,
DOI 10.1007/978-1-61779-448-3_9,

too complex and not specific enough to draw explicit conclusions (6). In comparison, isolated brain regions can provide useful information about the function of specific structures in the central nervous system (7), yet such an approach is usually limited to changes in one region of interest and does not take into account other regions with concurrent pathological changes occurring related to the same disease.

The limitations mentioned above helped drive the development of new isolation methods for biological materials. These technologies include laser capture microdissection, manual isolation of specific brain regions, isolation of specific organelles, and protein enrichment procedures (8), each with its own distinguishing set of pros and cons. For instance, manual dissection provides a very crude spatial map, yet it allows for the analysis of much protein via mass spectrometry, whereas laser capture microdissection provides a much more refined spatial map but limits the amount of protein available for mass spectrometric analysis. Subsequently, a tissue MALDI profiling approach has been developed to use mass spectrometry in combination with histopathology and/or imaging to get insights into protein/peptide and drug distribution (9). One advantage of this approach is that it eliminates operator's error, which is inevitable when brain regions are excised manually.

Despite the advances in technological development, neuroproteomics is a branch of proteomics that faces additional experimental challenges due to two main factors: (1) human brain tissue can only be obtained postmortem in which the tissue quality is limited due to rapid changes occurring after death and (2) only a limited amount of cerebrospinal fluid (CSF) can be drawn from any one individual at a time point. Despite these serious drawbacks, the postmortem brain and the limited availability of CSF are the sources that are and will continue to be used for in vivo human proteomic profiling.

A solution to these problems may be met through the use of well-defined experimental models including both primary cell-based in vitro experiments and a wide range of animal models. Cell-based in vitro experiments remain a very attractive model to study various aspects of neurodegenerative disorders. Such a model includes several types of cells in which their interactions and communication can be tested (Note 1). Although this does not fully reflect the complexity of the functional brain, many features can be tested in detail. The nonhuman primate model is also a favorable model due to its close resemblance of pathological changes in the human brain. Additionally, a wide range of animal models are available for studying neuroproteomics including MPTP injection in squirrel monkeys (10), rotenone treatment in rats (11) to study Parkinson's disease, or transgenic mice expressing human APP or Aβ (12, 13) to study Alzheimer's disease, just to name a few (Note 2).

Regardless of the sample source (animal vs. human, tissue vs. cells), protocols for proteomic profiling of samples share many similarities. One exception is that only cells can be cultured in the presence of stable isotope-labeled amino acids, a technique known as stable isotope labeling with amino acids in cell culture (SILAC) (Note 3, Fig. 1). SILAC is a quantitative proteomics technology developed for identifying and quantifying proteins in dividing cells (14, 15), but has expanded to include protein turn-over (16–18), post-translational modifications including methylation (19, 20), and phosphoproteomics (16, 21–23).

SILAC technology was originally based on the integration of heavy labeled amino acids into synthesized proteins of dividing cells (14); however, recently applications of SILAC for nondividing cells—like neurons—have been developed (16). Additionally, a pulsed SILAC (pSILAC) technique has been developed to measure protein turn-over *de novo* between different experimental conditions (17, 18, 24, 25) (Note 4). In this technique, cells are first plated in non-SILAC labeled media. Next, the media is exchanged for media supplemented with SILAC amino acids, whose composition varies between treatment conditions. This media is applied for only a limited period of time—typically ranging from several minutes to 96 h—thereby allowing for measurement of protein turn-over between conditions (17). pSILAC will be the proteomic technique focused on in this chapter and the overall scheme is presented in Fig. 1.

2. Materials

General Laboratory Materials

10-mL Regular tip serological pipet, sterile and disposable (BD, #357551)

15-mL BD Falcon™ Conical Tube (BD, #352916)

25-mL Regular tip serological pipet, sterile and disposable (BD, #357515)

50-mL BD Falcon™ Conical Tube (BD, #352770)

5-mL Regular tip serological pipet, sterile and disposable (BD, #356543)

Aluminum Foil

Cell lysis buffer: 50 mM Tris–HCl (pH 7.4), 5 mM EGTA, 1% Triton X-100, and 1 mM DTT; prepare shortly before use and keep on ice or at 4°C

Dry ice

Liquid nitrogen

Milli-Q ultrapure water

Non-stick RNase- and DNase-free microfuge tubes (2.0 mL; Invitrogen, #AM12475)

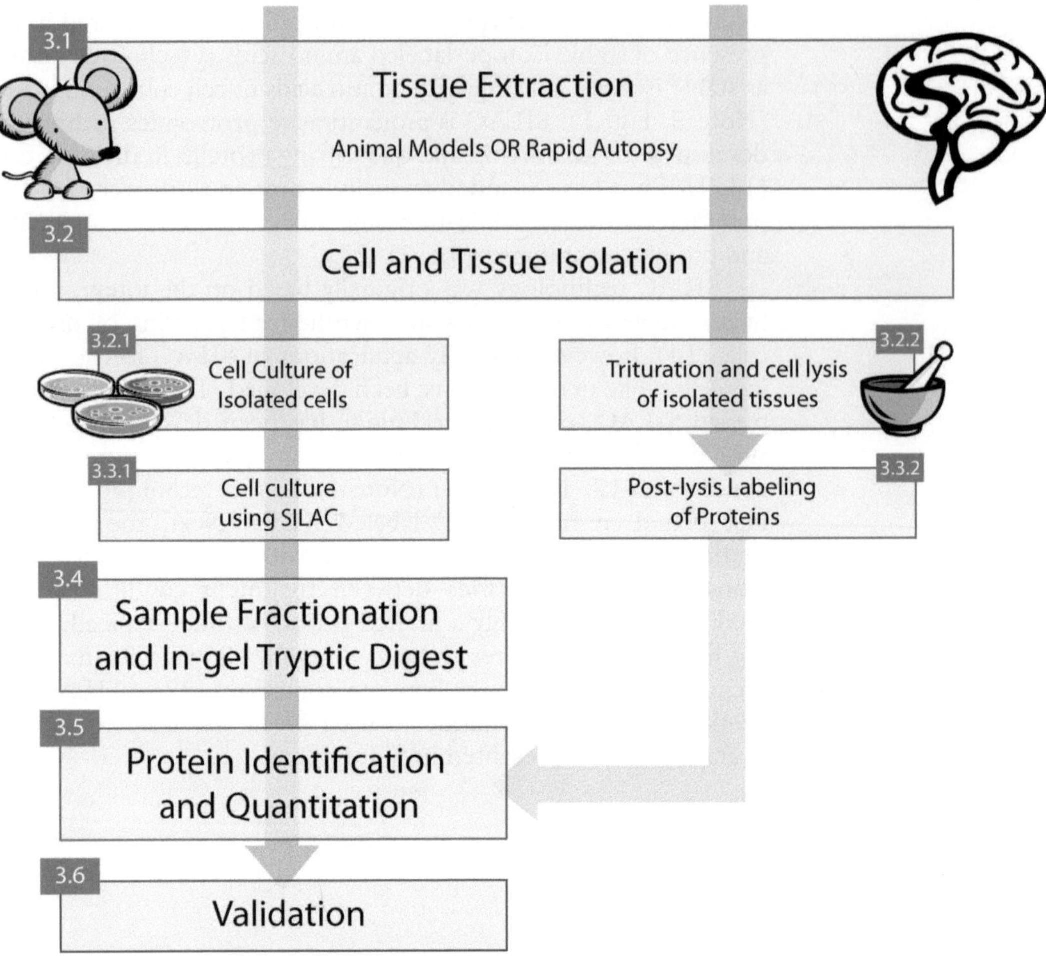

Fig. 1. SILAC-based neuroproteomics workflow beginning with tissue extraction (Sect. 3.1). Following tissue collection via rapid autopsy or from animal models, cell (Sect. 3.2.1) and tissue (Sect. 3.2.2) isolation occurs. This protocol describes in detail how isolated cells are cultured in SILAC (Sect. 3.3.1) and then lysed for sample fractionation and in-gel trypitic digestion (Sect. 3.4). This protocol also briefly describes the isolation of tissues, trituration, lysing (Sect. 3.3.2), and preparations for post-lysis labeling of proteins (Sect. 3.3.2). Following labeling, peptides are identified and ratios between experimental conditions are quantitated using liquid chromatography coupled to tandem mass spectrometry (LC-MS/MS; Sect. 3.5). Differentially expressed proteins are then validated using Western blot (Sect. 3.6).

Phosphate-buffered saline (PBS): 137 mM sodium chloride, 2.7 mM potassium chloride, and 11.9 mM phosphate buffer (Fisher Scientific, #BP399)

Protease inhibitor: 1 mL (1×) Reconstituted protease inhibitor cocktail per 1 g of tissue; add to lysis buffer immediately before lysis (Sigma, #P2714). This specific inhibitor contains AEBSF, aprotinin, bestatin hydrochloride, E-64, EDTA, and leupeptin hemisulfate salt.

Protein quantification assay kit

Razor blade

Equipment

Light box

LTQ Orbitrap (Thermo-Fisher Scientific) coupled with an LC system or LCQ Deca (ThermoElectron) coupled with HPLC

Protein and peptide identification software (i.e.,: Bioworks 3.3.1, Mascot, etc.)

Protein and peptide quantitation software (i.e.,: MSQuant, MaxQuant, or other software of choice that is able to handle SILAC data)

Speed-Vac (Thermo Scientific)

XCell SureLock® Mini-Cell and XCell II™ Blot Module Kit CE Mark (Invitrogen, #EI0002)

X-ray film cassette (BioExpress, #SB810) with intensifying screen (BioExpress, #IS810)

2.1. Tissue Extraction

60-mm BD Falcon™ non-treated culture dish (BD, #351007)

Hank's balanced salt solution, 10× (HBSS; Sigma, #H4641)

Micro-dissection instruments (sterilized): small dissecting scissors; Dumont forceps, straight and angled; curved micro-dissecting scissors

2.2. Cell and Tissue Isolation

2.2.1. Cells

10-mL wide tip serological pipet, sterile and disposable (BD, #357504)

B27 Supplement (serum free; Invitrogen, #17504-044)

BD Biocoat™ Poly-D-lysine 6-well multi-well plate (BD, #354413)

Deoxyribonuclease I (DNaseI; Sigma, #DN25)

Dulbecco's modified eagle media (DMEM, high glucose; Invitrogen, #11960)

Fetal bovine serum (FBS, heat inactivated; Invitrogen, #10082)

GlutaMAX supplement (Invitrogen, #35050-061)

Neurobasal media with phenol red (Invitrogen, #21103-049 for fetal tissues or 10888-022 for post-natal and adult tissues)

Neurobasal media *without* phenol red (Invitrogen, #12348-017 for fetal tissues or 12349-015 for post-natal and adult tissues)

Treatment solution suspended in the appropriate media (see Sect. 3 for details)

Trypsin solution, 10× (Sigma, #T4549)

2.2.2. Tissues

Small mortar and pestle set (Fisher Scientific)

2.3. Cell Culture Using SILAC and Tissue Preparation for Post-Lysis Labeling of Proteins

Cell scraper (1.8 cm blade; BD, #353085)

L-Arginine (Sigma-Aldrich, #A6969)

L-Lysine (Sigma-Aldrich, #L8662)

Modified neurobasal media without L-arginine and L-lysine (Millipore Custom Media Services)

Protein quantification kit of choice: NanoDrop, BCA assay, or 2D Quant

SILAC stable isotopic $^{13}C_6$, $^{15}N_2$-L-lysine (Sigma-Aldrich)

SILAC stable isotopic $^{13}C_6$, $^{15}N_4$-L-arginine (Sigma-Aldrich)

2.4. Sample Fractionation and In-Gel Tryptic Digest

2.4.1. Fractionation

Destain solution A: 25% (v/v) methanol and 10% (v/v) acetic acid in Milli-Q water

Destain solution B: 25% (v/v) methanol in Milli-Q water

Fixing solution: 40% (v/v) methanol and 7% (v/v) acetic acid in Milli-Q water

NuPAGE® antioxidant (Invitrogen, #NP0005)

NuPAGE® LDS sample buffer (4×; Invitrogen, #NP0007)

NuPAGE® MES SDS running buffer (for Bis-Tris Gels only, 20×; Invitrogen, #NP-0002)

NuPAGE® Novex® 4–12% Bis-Tris gel (Invitrogen, #NP0322BOX)

NuPAGE® reducing agent (10×; Invitrogen, #NP0004)

SeeBlue® Plus2 pre-stained standard (Invitrogen, #LC5925)

Staining solution: 4 parts 1× Brilliant Blue G-Colloidal stock solution (prepared as directed; Sigma, #B2025) and 1 part methanol

2.4.2. Excision of Protein Bands and In-Gel Tryptic Digest

Sequencing grade modified trypsin (Promega, #V5111)

0.5% TFA in Milli-Q water

50% (v/v) acetonitrile (ACN) in Milli-Q water

60% (v/v) ACN and 0.1% (v/v) TFA in Milli-Q water

10 mM ammonium bicarbonate (NH_4HCO_3) in Milli-Q water

50% (v/v) ACN and 10 mM NH_4HCO_3 in Milli-Q water

50% (v/v) ACN and 50 mM NH_4HCO_3 in Milli-Q water

2.5. Protein Identification and Quantitation

Thermo-Scientific LTQ Orbitrap XL or LTQ Orbitrap Velos or equivalent

Manufacturer's software for searching of obtained spectra (i.e. Proteome Discoverer by Thermo-Scientific)

Software suitable for the analysis of SILAC data (i.e. Proteome Discoverer by Thermo-Scientific or MSQuant [open source] by CEBI research group at the University of Southern Denmark)

2.6. Validation

PBST: 1× PBS solution supplemented with 0.2% Tween-20

Blocking buffer: 10% (w/v) dry skim milk in PBST

GeneMate Blue Lite Autoradiography Film (ISC BioExpress, #F-9024)

Immun-Blot PVDF/Filter Paper Sandwiches (BioRad, #162-0218) or compatible nitrocellulose or nylon membrane

PerfectWestern Container, large (Fisher, #NC9471025)
Primary for protein of choice and compatible secondary antibody

Sodium azide

SuperSignal® West Pico chemiluminescent substrate solution (Pierce, #34077)

Transfer buffer: NuPAGE® transfer buffer (20×; Invitrogen, #NP0006) diluted to 1× in Milli-Q water with 10% (v/v) methanol

SILAC Media Preparations

Heavy media: Modified neurobasal media without phenol red with 2.0% (v/v) B27 supplement and 0.25% (v/v) GlutaMAX supplement; add 2 mg/L (0.0225 mM) heavy isotope ($^{13}C_6$,$^{15}N_4$-L-arginine) arginine and 146 mg/L (0.798 mM) heavy isotope ($^{13}C_6$,$^{15}N_2$-L-lysine) lysine.

Medium media: Modified neurobasal media without phenol red with 2.0% (v/v) B27 supplement and 0.25% (v/v) GlutaMAX supplement; add 2 mg/L (0.0225 mM) normal isotope arginine and 146 mg/L (0.798 mM) heavy isotope ($^{13}C_6$,$^{15}N_2$-L-lysine) lysine.

Control/light media: Modified neurobasal media without phenol red with 2.0% (v/v) B27 supplement and 0.25% (v/v) GlutaMAX supplement; add 2 mg/L (0.0225 mM) unlabeled arginine and 146 mg/L (0.798 mM) unlabeled isotope lysine.

3. Methods

3.1. Tissue Extraction

1. Harvest the whole brain or brain tissue from specific region of interest. Rinse tissue with ice-cold PBS to remove excess blood and debris. Isolate area of interest using Dumont forceps and fine scissors (Note 5).
2. Transfer brain tissue to a 60-mm culture dish containing enough 1× HBSS (diluted from 10× stock to 1× using Milli-Q water) to cover the bottom of the plate.

3.2. Cell and Tissue Isolation

3.2.1. Cells

Isolation of cells resulting in a homogenous culture varies from cell type to cell type (Note 6). Protocols typically follow the same general methods: (1) dissection of a specific tissue (i.e., the hippocampus or substantia nigra), (2) further sample prep and manual dissection, (3) incubation with trypsin and DNAseI to promote a single cell dispersion, (4) dissociation of any remaining cell clumps via gentle trituration using a pipette tip, and (5) plating (on poly-D lysine, purchased with coating already applied or prepared the day before). An example protocol is provided for hippocampal cells from an embryonic day 16 mouse following tissue extraction (26):

1. Manually dissect meninges from the cortex brain tissue to expose the hippocampus. Remove the hippocampus from the cortex using fine scissors. Clean-up of the hippocampus involves the removal of the fimbrie using forceps and fine scissors. Transfer the hippocampal tissue to sterile 15-mL conical tube containing 1 mL of HBSS.
2. Adjust HBSS level to 1.8 mL by adding additional HBSS. Add 200 μL 10× trypsin stock (thereby making it a 1× trypsin in HBSS solution). Incubate at 37°C for 30 min with gentle rocking.
3. Add 50 μg DNaseI and incubate at room temperature for 30 s with gentle rocking (Note 7).
4. Remove the HBSS and briefly wash the tissue by gentle inversion using 10 mL DMEM/10% (v/v) FBS media. Before aspirating the wash, be sure to let the tissue settle completely (2–10 min).
5. Repeat the wash twice and after the final wash, suspend the tissue in 10 mL DMEM/10% (v/v) FBS media.
6. Triturate the tissue gently using a wide-tipped pipette. This will reduce the size of any clumps that may still exist. Next, triturate the sample gently using a regular-tipped pipette. This will promote dissociation of remaining clumps of tissue to a homogenous cell suspension. Each trituration step should consist of 10 slow, trituration series and should be done producing minimal bubbles.
7. Incubate for 2 min undisturbed at room temperature. This allows any non-triturated tissue to settle at the bottom of the tube.
8. Transfer only the supernate to a new 15-mL conical tube. The supernate contains suspended individual cells, whereas the particulate contains non-triturated cell clumps.
9. Determine total cell number and viability.
10. Adjust cells to the required concentration for plating using the DMEM/10% (v/v) FBS media. When using a 6-well plate, 3 million cells per well should be plated using a concentration of 1 million cells/mL for a total of 3 mL media per well.
11. Plate cells on poly-D-lysine-coated plates (purchased pre-coated or prepared on the previous day) and incubate at 37°C for 2 h. After 2 h, check cell adherence. Cells should be adhered

and media can be exchanged to 3 mL/well neurobasal media with phenol red containing 2.0% (v/v) B27 supplement and 0.25% (v/v) GlutaMAX supplement.

12. Incubate at 37°C for up to 10 days changing media when necessary.

3.2.2. Tissues

Proceed with manual dissection until tissue(s) are cleaned of all debris/unwanted materials. When dissection is complete, transfer tissue to pre-chilled mortar. Using a pre-chilled pestle, slowly grind frozen tissue to a fine powder, adding liquid nitrogen dropwise as necessary. Transfer ground tissue to a sterile tube that holds at least four times the volume of brain tissue obtained (i.e., 50-mL conical tube).

3.3. Cell Culture Using SILAC and Tissue Preparation for Post-Lysis Labeling of Proteins

3.3.1. SILAC for Cells: Pulse SILAC

Media Exchange and Treatment

1. Cell media is exchanged at $t=0$ (less than 10 days after tissue isolation and cell culture) for either "medium" or "heavy" media containing labeled amino acids (Note 8).
2. Aspirate neurobasal media at $t=0$ and apply modified neurobasal media: 3 mL/well "medium" neurobasal media is applied to control cells, and 3 mL/well "heavy" neurobasal media with treatment conditions is applied to treatment cells. "Light" neurobasal media should be applied to another set of cells for an additional treatment control.
3. Treat cells for a predetermined duration based on the experimental design, but for no longer than 12 h. This predetermined duration can also include a treatment curve (e.g., samples are collected at 2, 4, 6, and 12 h).

Collection and Lysis

1. Collect the supernatant, aliquot, and transfer to a sterile 2.0-mL microfuge tube. Snap-freeze using dry ice and store at −80°C.
2. Apply lysis buffer to cell culture, incubate at room temperature for 30 s, and scrape cells from plate surface. Collect cell lysate and transfer to a sterile 2.0-mL microfuge tube.
3. Centrifuge lysate at 15,000 RPM for 10 min at 4°C to sediment non-dissolved materials.
4. Transfer the supernatant to a clean microfuge tube making sure not to disturb the pellet. Snap-freeze and store at −80°C. If multiple freeze–thaw cycles are anticipated, it is best to aliquot samples to prevent protein degradation. Save a small portion of each sample for protein quantification.

Protein Quantification: Each Method of Protein Quantification Has Its Own Set of Positives Along with Drawbacks or Inabilities (Note 9)

1. NanoDrop protein concentration is a common method due to its ease, quick speed, and requiring no additional reagents. Additionally, NanoDrop is preferred for low-volume samples as only 1–2 μL of each sample is needed for accurate protein quantitation. However, if the sample is complex or if nonprotein factors are present, considerable error may occur due to

changes in the reading for absorption of UV wavelengths (A230, A260, A280).

2. The BCA assay (Pierce, #23227) is another common method; it is compatible with the detergents, reducing agents, chelating agents, and buffers that were used in cell lysis. The BCA assay is a colorimetric assay.
3. The 2D Quant kit (GE Life Sciences, #80-6483-56) is another common method as it too is compatible with the detergents, reducing agents, and buffers that were used in cell lysis. The 2D Quant kit works by having copper selectively bind proteins, and the absorption recorded at 480 nm is inversely proportional to protein concentration. The 2D Quant kit assay is comparable in time to the BCA assay.

3.3.2. Post-Lysis Labeling of Proteins from tissues

1. Cell lysis: To the tube containing the ground brain tissue, add a volume of cell lysis buffer (with protease inhibitor added) equal to at least four times the volume of brain powder obtained. Lyse tissue via rocking for 90 min at 4°C. Vortex tube every 15 min.
2. Centrifugation: Following lysis, centrifuge lysate at 15,000 RPM for 10 min at 4°C to sediment non-dissolved materials (Note 10).
3. Transfer the supernatant to clean conical tube, making sure not to disturb the pellet. Snap-freeze and store at –80°C until needed. If multiple freeze–thaw cycles are anticipated, it is best to aliquot samples into microfuge tubes to prevent protein degradation. Save an aliquot of each sample for protein quantification.
4. Protein quantification: For details, see Sect. 3.3.1.3 (SILAC for cells: protein quantification).
5. Sample preparation for post-lysis labeling of proteins typically requires ethanol precipitation of proteins and followed by labeling of samples (Note 11). For details concerning labeling using an iTRAQ 4-plex system, refer to Chap. 13.

3.4. Sample Fractionation and In-Gel Tryptic Digest

There are many means for protein fractionation and is dependent on the type of labeling used. Fractionation of SILAC labeled proteins can occur by polyacrylamide gel electrophoresis (PAGE) or isoelectric focusing (IEF). For more information on IEF (including sample clean-up), refer to Chap. 13. The methods below describe fractionation by sodium dodecyl sulfate (SDS)–PAGE:

3.4.1. Fractionation

1. Measure no more than 12.5 μg of each sample (control and experimental) and mix at a 1:1 ratio in a 2.0-mL microfuge tube.
2. Desiccate the sample to a pellet using a Speed-Vac.

3. Rehydrate the sample in 20 μL sample buffer with reducing agent. Heat at 95°C for 5 min to denature proteins (alternatively, the sample can be heated at 70°C for 10 min according to NuPAGE®). During this time, assemble the Mini-gel apparatus per the manufacturer's protocols. Fill inner chamber with 1× SDS running buffer, making sure to cover the top of the wells with running buffer. Ensure that there are no leaks from the inner chamber.
4. Briefly vortex samples and tap spin to bring the entire sample to the bottom of the microfuge tube. Load the entire 20 μL of each sample to its designated well in the NuPAGE® Novex® 4–12% Bis-Tris gel. Load 7 μL of SeeBlue® Plus2 pre-stained standard to its designated well. Fill outer chamber with 1× SDS running buffer so that the outer chamber is roughly 2/3 full.
5. Run the gel: 100 V constant voltage with an expected current start of 125 mA/gel and an end expected current of 160 mA/gel for 90 min.
6. Immediately after running the gel, separate the two sides of the gel cassette using the gel knife. Remove one side of the plates and leave the gel on the other. Strip away the wells and bottom of the gel using a gel knife. Rinse the gel with Milli-Q water and transfer the gel to fixing solution (Note 12). Allow the gel to fix for 1 h.
7. While the gel is fixing, prepare staining solution. Decant and dispose of fixing solution in an appropriate container. Cover the gel with staining solution and incubate at room temperature with rocking in the dark for 1–2 h. If bands are not apparent, incubate at 4°C overnight with rocking.
8. Decant staining solution into appropriate container and cover the gel in destaining solution A. Incubate at room temperature for 60 s with rocking.
9. Decant destaining solution A into appropriate container and cover the gel in destaining solution B. Incubate at 4°C for up to 24 h with rocking. When the gel background appears clear, destaining is complete.
10. Decant staining solution B into appropriate container and cover the gel with Milli-Q water. Incubate at 4°C for up to 24 h with rocking (Note 13).

3.4.2. Excision of Protein Band and In-Gel Tryptic Digest

1. Excision of protein bands: Using a light box, remove selected bands using a clean razor blade. Ensure complete excision of the band but avoid removing excess, unstained gel. Any large bands should be further cut into smaller sections, but transferred to the same centrifuge tube. Transfer each band to its own 2.0-mL microfuge tube.

2. Washes: Add 200 μL of 50% ACN to each sample and wash at room temperature by rocking for 5 min. Aspirate wash. Add 200 μL 50% ACN and 50 mM NH_4HCO_3 to each samples and wash for 30 min at room temperature by rocking for 30 min. Aspirate wash. Add 200 μL 50% ACN and 10 mM NH_4HCO_3 to all samples and wash for 30 min at room temperature by rocking for 30 min. Aspirate wash.
3. Speed-Vac gel pieces to complete dryness. At this point, the gel should appear translucent and the band should appear white.
4. Tryptic digest: Add 10 μL (0.1 μg/μL) modified trypsin to each sample (use more volume if gel piece is not saturated, should be completely transparent). Incubate at room temperature for 5–10 min. Add 50 μL 10 mM NH_4HCO_3 to each sample. Incubate at 37°C for at least 16 h (overnight).
5. Peptide extraction: Add 200 μL 60% ACN and 0.1% TFA, and incubate at room temperature for 60 min with rocking. Remove buffer containing peptides and transfer to a new centrifuge tube. Repeat peptide extraction, transferring peptides in buffer to previous extract. Dry the combined washes by Speed-Vac.
6. Re-suspend peptides in 0.5% TFA. For methods of clean-up, please refer to the Chap. 13.

3.5. Protein Identification and Quantitation

1. Extracted peptides should be analyzed by LC-MS/MS using a high performance mass spectrometer utilizing electrospray ionization (i.e., LTQ Orbitrap or similar). For the best outcome, spectra obtained should be searched using multiple algorithms such as SEQUEST, Mascot, and X!Tandem. We routinely use BioWorks 3.3.1, part of the Xcalibur software, using the SEQUEST algorithm with the modified masses of SILAC amino acids listed in possible modifications. For details concerning protein identification, database searching, and statistical analysis, please refer to the Chap. 13.
2. Specific peptide quantitation is determined using information retrieved from peptide identification along with retention time information. Next, the peptide extracted ion chromatogram (XIC) is determined for heavy, medium-heavy, and light labeled peptides. These XIC graphs are then used to determine the ratio of signal intensities and thereby the relative abundance for that specific peptide using heavy to light, medium-heavy to light, and heavy to medium-heavy ratios (based on experimental design) of that specific peptide (Note 14).

3.6. Validation

Validation of proteins can be accomplished through several means (Note 15). Western blot remains a popular method; however, results can vary based on the type of antibody used and the region of the target protein to which the antibody was raised (Note 16). An example of Western blot variation between antibodies is shown

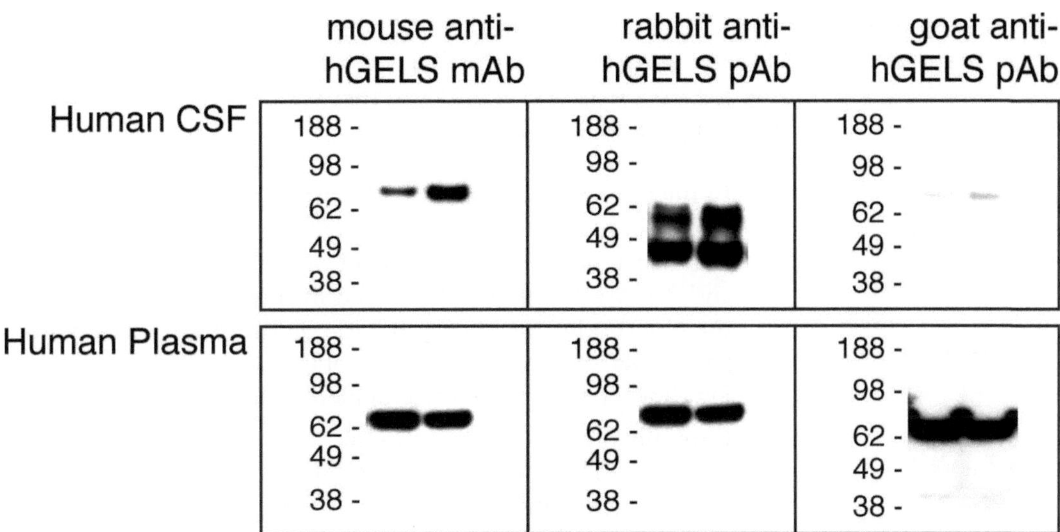

Fig. 2. Antibody variation can occur between different types of antibodies and different protein sources. In this figure, there is much variation apparent between mouse monoclonal antibodies raised against human gelsolin (mouse anti-hGELS mAb), rabbit polyclonal antibodies raised against human gelsolin (rabbit anti-hGELS pAb), and goat polyclonal antibodies raised against human gelsolin (goat anti-hGELS pAb) in human cerebrospinal fluid (CSF) samples. Furthermore, variation is also noticable for each antibody between human CSF and human plasma samples.

in Fig. 2 for human cerebrospinal fluid and human plasma using antibodies raised against human gelsolin in mice, goat, and rabbit. We recommend using multiple antibodies for validation due to the variation that can occur between different hosts. The Western blot protocol is straightforward: gel electrophoresis, transfer, blocking, incubation with primary antibody, incubation with secondary antibody, and detection.

3.6.1. Gel Electrophoresis

Please refer to Sect. 3.4.1. After beginning the 1DE, prepare blot pads, transfer membrane, and filter paper.

1. Soak the blotting pads in transfer buffer to saturate. Remove bubbles by squeezing the blotting pads while they are submerged in the buffer. Bubbles can block the transfer of biomolecules.
2. Cut selected transfer membrane and filter paper to the dimension of the gel or use pre-cut sandwiches (1 per gel).
3. Pre-wet the membrane:
   - PVDF: Pre-wet by dipping briefly in methanol. Rinse in deionized water and place in a shallow dish containing 30 mL of transfer buffer for several minutes (up to 90 min).
   - Nitrocellulose/nylon membrane: Place directly in tray containing transfer buffer for several minutes.
4. Soak filter paper in transfer buffer immediately before preparing transfer sandwich (see Sect. 3.6.3).

3.6.2. Removing the Gel After Electrophoresis

Immediately after running the gel, separate the two sides of the gel cassette using the gel knife. Remove one side of the plate and leave the gel on the other. Strip away the wells and bottom of the gel using a gel knife.

3.6.3. Preparing the Transfer Apparatus (Sandwich)

1. In the blot module, place (from bottom to top) 3 blotting pads soaked in transfer buffer, pre-wetted filter paper, the gel (rinsed in transfer buffer), transfer membrane, pre-wetted filter paper, and 3 more blotting pads soaked in transfer buffer. Complete the blot module by placing the top on the sandwich of blotting pads, filter papers, transfer membrane, and gel. Please refer to Fig. 3 for a diagram of the completed blot module.
2. Carefully pick up the blot module and place inside the transfer apparatus. The completed blot sandwich should fit horizontally across the bottom of the module, leaving a 1-cm gap at the top of the electrodes.
3. Fill the blot sandwich with transfer buffer such that the gel/membrane sandwich is covered in transfer buffer. DO NOT overfill (fill to the top), as this will generate extra conductivity and heat. Remove any excess buffer via aspiration by pipette.

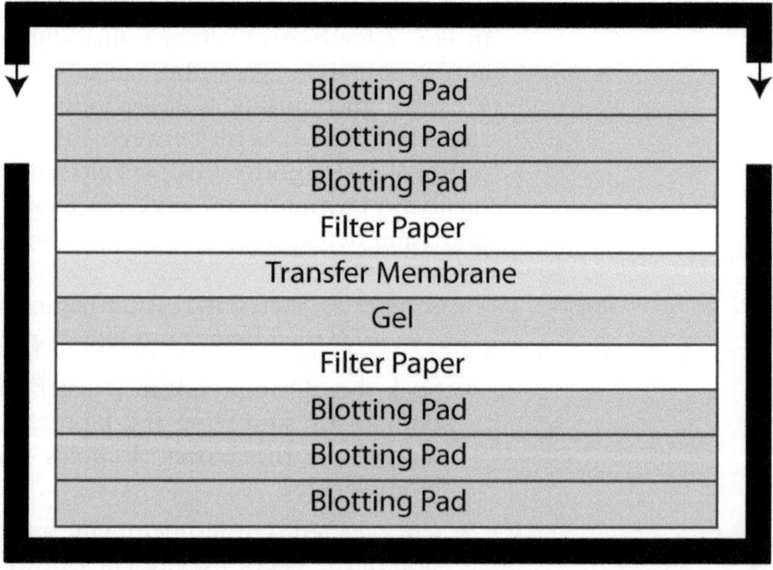

Fig. 3. The blot module as described in text (see step 1 in Sect. 3.6.3). In the bottom of the blot module, first place three blotting pads soaked in transfer buffer. Ensure that there are no bubbles or air pockets in the blotting pads as this will interfere with transfer. Next, layer a filter paper briefly dipped in transfer solution on top of the blotting pads. Gently place the gel on the filter paper. Placement of the gel can be helped by placing the entire apparatus in a plastic container filled with transfer buffer; in this way, the gel can float and soak until proper placement is achieved. This will also prevent tears in the gel which may occur if the gel is not properly placed on the first attempt. The transfer membrane is placed atop the gel. At this point, using a gel roller, any residual bubbles can be removed. However, take care in preventing stretching of the gel by using minimal force. Atop the transfer membrane, place another pre-wetted filter paper and three more blotting pads.

4. Ensure that there are no leaks from the inner chamber. Fill the outer buffer chamber with Milli-Q water or transfer buffer. Place lid on gel assembly unit. With power off, plug into the power supply.

3.6.4. Transfer Proteins

30 V constant voltage for 90 min with an expected current start of 170 mA and an end expected current of 110 mA.

3.6.5. Blocking the Membrane

1. Immediately following completion of the transfer, remove the blot and place in blocking buffer (Note 17). Incubate for at least 1 h at room temperature while rocking.
2. If this fails to block properly, incubate/block overnight at 4°C.

3.6.6. Incubation with Primary Antibody

1. Dilute antibody following the manufacturer's or literature recommendations to appropriate dilution factor (i.e., 1:1,000 or 1:500) in appropriate volume of blocking buffer.
2. Decant blocking buffer and apply primary antibody to membrane. Incubate for at least 1 h at room temperature while rocking.
3. Primary antibody can be applied with shaking for 1–4 h at room temperature or overnight at 4°C (Note 18).
4. Following incubation with primary antibody, decant antibody solution. Perform three washes using PBST for 5 min per wash at room temperature with rocking.

3.6.7. Incubation with Secondary Antibody

1. Dilute antibody following the manufacturer's or literature recommendations to appropriate dilution factor (i.e., 1:20,000) in 10 mL blocking buffer.
2. Decant final wash and apply secondary antibody to membrane. Incubate for 30 min at room temperature with rocking.
3. Following incubation with secondary antibody, decant antibody solution. Perform five washes using PBST for 3 min per wash at room temperature with rocking.

3.6.8. Detection of Target Protein Using Chemiluminescence

1. Prepare chemiluminescent substrate solution. Only 5 mL/membrane is needed, so add 2.5 mL of each solution to small conical tub and keep in the dark. Try to use immediately.
2. Following the final PBST wash, add chemiluminescent working solution and cover with foil. Rock for 5 min.
3. Remove membrane and excess chemiluminescent substrate (drag across a KimWipe delicate task tissue while holding the membrane with a forceps). Place the membrane between a sheet-protector in a prepared Western blot cassette. Although this step does not need to take place in the dark room, try to keep the membrane in the dark as much as possible. With that said, if this step is done at the bench rather than in the dark room, these steps should be done as quickly as possible.
4. Expose to film (in a dark room) and process.

3.6.9. Membrane Preparation for Subsequent Western Blot Analyses and/or Storage

Following detection, transfer membrane from sheet-protector to Western blot container containing PBST and approximately 50 mg of sodium azide (this is a spatula head's amount); sodium azide will kill the chemiluminescent signal. Rock the membrane for no less than 30 min to ensure that the signal has been completely eliminated. Store at 4°C (Note 19).

4. Notes

1. Many brain-inspired co-culture systems exist, including blood–brain barrier model focusing on microvascular endothelial cell–macrophage interactions (27), neuron–microglia interactions (28), cortico-hippocampal support cultures, nigro-striatal support cultures (26), and neuron–astrocyte interactions (29) to name a few.
2. For a fairly extensive list, including protocols, of animal models used in a wide range of neurological and psychiatric disorders, see Current Protocols in Neuroscience (30). Topics include migraine pain, eating disorders, encephalopathy, cerebral ischemia, amyotrophic lateral sclerosis, and Alzheimer's and Parkinson's disease.
3. Stable isotope labeling of mammals (SILAM) is an additional technology that has been developed. SILAM focuses on metabolically labeling the entire organism in an attempt to provide an internal standard for complex tissue samples being analyzed by quantitative proteomics (31). Additionally, stable isotope labeling in the chicken has also been developed for measuring protein turnover (32). In both situations, experimental animals are fed a special diet including the isotopically labeled amino acid of choice, resulting in the integration of the modified amino acid into the whole organism, not just a subset of cells.
4. Although the term "pulsed SILAC" was not coined until 2008 by Selbach et al. (25), pulse-chase style methods using heavy labeled amino acids to measure protein turnover was performed earlier in 2006 by Milner et al. (24). In this protocol, cells were first plated in "heavy" media and then swapped to normal media for collection. This is in comparison to Selbach, in which the cells are first plated in normal media and swapped to "heavy" and "medium-heavy" SILAC media for treatment and collection.
5. Several online brain atlases are available online for use.
   - Brainmaps.org by Jones at UC Davis includes brain atlases for several primates, cat, dog, rat, mouse, and several other models.

- The Laboratory of Neuro Imaging (LONI) by Toga at UCLA includes brain atlases for human, monkey, mouse, and rat, among several other models and variations (www.loni.ucla.edu).

6. A brief reference to protocols involving the isolation of specific CNS cells:
   - Booher and Sensenbrenner (33): Neurons and glial cells from embryonic chick, rat, or human brain
   - Brewer and Torricelli (34): Rat or mouse adult neurons and neurospheres
   - Chen et al. (35): Oligodendrocyte precursor cells
   - Dvorak et al. (36): Chick embryo sympathetic ganglia neurons
   - Fath et al. (26): Rat or mouse hippocampal and substantia nigra neurons
   - McCarthy and de Vellis (37): Rat cerebral astroglia and oligodendrocytes
   - Taylor et al. (38): Chick lumbar spinal cord astrocytes and motoneuron cells
   - Uchida et al. (39): Human CNS stem cells
7. DNaseI treatment will break down any DNA that may have been released via rupture or lysis of cells occurring from treatment with trypsin and will promote continued disaggregation of cells, therefore preventing cell clumping.
8. There are several assumptions made by exchanging the media in both the control and experimental conditions: (1) integration of labeled amino acids into proteins occurs at the same rate in both conditions and (2) proteins remaining in the light form can be disregarded as they are not actively undergoing synthesis.
9. The Bradford Assay (Pierce, #23200) is not a good choice in this scenario; there is an increased probability for reagent precipitation due to the presence of surfactants in the cell lysis buffer. Additionally, although the Bradford assay is quick and easy, there is substantial variation compared to other methods.
10. Following centrifugation, a pellet should be visible; this contains non-lysed cells and lysed cell debris. The supernate contains both membrane and cytosolic proteins.
11. Alternatively, depending on experimental design and number of samples, an 8-plex iTRAQ® labeling of samples can take place; however, with additional labeling of samples, there is an increased risk of added variability.
12. Gels that are not immediately removed and fixed will have proteins diffuse, resulting in broad and indistinguishable bands,

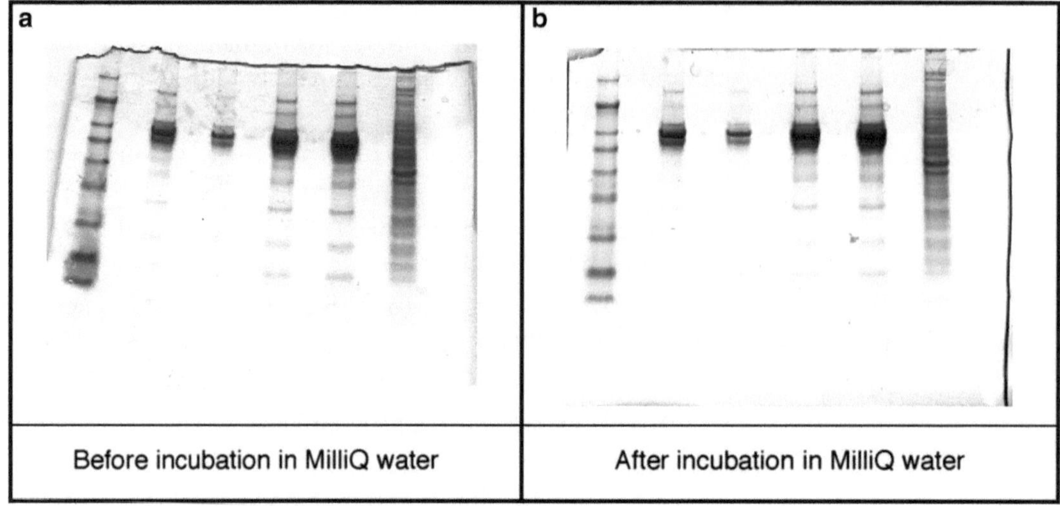

Fig. 4. Stained gel before (**a**) and after (**b**) a 24-h incubation in Milli-Q water. Incubation with Milli-Q water will return the gel to its original shape prior to exposure to organic compounds; the acetic acid and methanol used throughout fixing, staining and destaining of the gel results in shrinking of the gel. Additionally, incubation of the gel in Milli-Q water for up to 24 h after staining will continue to destain the gel, resulting in a cleaner gel background.

making it difficult, if not impossible, to proceed to gel cutting and tryptic digestion.

13. Allowing the gel to soak in Milli-Q is not necessary, but will allow the gel to return to its original shape (Fig. 4) and continue to destain any residual background. If necessary, following treatment with Milli-Q water, store the gel in 25% ammonium sulfate at room temperature for up to 3 weeks.
14. Ong (2006) (15) provides a complete, in-depth protocol for the identification and quantitation of peptides from SILAC samples along with troubleshooting.
15. Several additional means for validation are available including Stable Isotope Standards and Capture by Anti-Peptide Antibodies (SISCAPA) (40), Statistical Analysis (41, 42), and MRM.
16. Polyclonal antibodies are made in vivo by direct injection into a host and the antibody suspension is composed of a heterogeneous assortment of antibodies with different affinities to various epitopes of the same protein. Monoclonal antibodies are made in vitro through the fusion of a myeloma cell to an antibody-secreting plasma cell, and the antibody suspension consists of a homogenous solution of antibodies with the same affinity to a single epitope of the protein of interest. Additional variation may occur depending on the region of the protein used to raise the antibody (i.e., C-terminal end, N-terminal end, somewhere in the middle).

17. Following transfer, the 1DE can be saved and re-stained to verify efficacy of transfer.
18. For results with high noise, the primary antibody incubation can be decreased or the dilution factor increased so that the antibody is more dilute (i.e., 1:500 1:1,000).
19. Alternatively, the membrane can be dried on a KimWipe and stored in a standard sheet protector. To rehydrate the membrane, briefly dip in methanol, block in blocking buffer, and begin the protocol at primary antibody incubation.

Acknowledgments

We would like to thank Dr. Gwënaël Pottiez and Ms. Jayme Wiederin for help in preparation of this manuscript. This work was partially supported by the National Institutes of Health 1 P20DA026146-01 and 2 P01 NS043985-05.

References

1. Taurines, R., Dudley, E., Grassl, J., Warnke, A., Gerlach, M., Coogan, A. N., and Thome, J. (2010) Proteomic research in psychiatry, *J Psychopharmacol.* **25**, 151–96
2. Reckow, S., Gormanns, P., Holsboer, F., and Turck, C. W. (2008) Psychiatric disorders biomarker identification: from proteomics to systems biology, *Pharmacopsychiatry* **41** Suppl 1, S70–77.
3. Klose, J., and Feller, M. (1981) Genetic variability of proteins from plasma membranes and cytosols of mouse organs, *Biochem Genet* **19**, 859–870.
4. Fountoulakis, M., Hardmaier, R., Schuller, E., and Lubec, G. (2000) Differences in protein level between neonatal and adult brain, *Electrophoresis* **21**, 673–678.
5. Krapfenbauer, K., Berger, M., Lubec, G., and Fountoulakis, M. (2001) Changes in the brain protein levels following administration of kainic acid, *Electrophoresis* **22**, 2086–2091.
6. Hirano, M., Rakwal, R., Shibato, J., Agrawal, G. K., Jwa, N. S., Iwahashi, H., and Masuo, Y. (2006) New protein extraction/solubilization protocol for gel-based proteomics of rat (female) whole brain and brain regions, *Mol Cells* **22**, 119–125.
7. Ishihama, Y., Sato, T., Tabata, T., Miyamoto, N., Sagane, K., Nagasu, T., and Oda, Y. (2005) Quantitative mouse brain proteomics using culture-derived isotope tags as internal standards, *Nat Biotechnol* **23**, 617–621.
8. Rohlff, C. (2000) Proteomics in molecular medicine: applications in central nervous systems disorders, *Electrophoresis* **21**, 1227–1234.
9. Stoeckli, M., Chaurand, P., Hallahan, D. E., and Caprioli, R. M. (2001) Imaging mass spectrometry: a new technology for the analysis of protein expression in mammalian tissues, *Nat Med* **7**, 493–496.
10. Langston, J. W., Forno, L. S., Rebert, C. S., and Irwin, I. (1984) Selective nigral toxicity after systemic administration of 1-methyl-4-phenyl-1,2,5,6-tetrahydropyrine (MPTP) in the squirrel monkey, *Brain Res* **292**, 390–394.
11. Betarbet, R., Sherer, T. B., MacKenzie, G., Garcia-Osuna, M., Panov, A. V., and Greenamyre, J. T. (2000) Chronic systemic pesticide exposure reproduces features of Parkinson's disease, *Nat Neurosci* **3**, 1301–1306.
12. Quon, D., Wang, Y., Catalano, R., Scardina, J. M., Murakami, K., and Cordell, B. (1991) Formation of beta-amyloid protein deposits in brains of transgenic mice, *Nature* **352**, 239–241.
13. LaFerla, F. M., Tinkle, B. T., Bieberich, C. J., Haudenschild, C. C., and Jay, G. (1995) The Alzheimer's A beta peptide induces neurodegeneration and apoptotic cell death in transgenic mice, *Nat Genet* **9**, 21–30.

14. Ong, S. E., Blagoev, B., Kratchmarova, I., Kristensen, D. B., Steen, H., Pandey, A., and Mann, M. (2002) Stable isotope labeling by amino acids in cell culture, SILAC, as a simple and accurate approach to expression proteomics, *Mol Cell Proteomics* **1**, 376–386.
15. Ong, S. E., and Mann, M. (2006) A practical recipe for stable isotope labeling by amino acids in cell culture (SILAC), *Nat Protoc* **1**, 2650–2660.
16. Spellman, D. S., Deinhardt, K., Darie, C. C., Chao, M. V., and Neubert, T. A. (2008) Stable isotopic labeling by amino acids in cultured primary neurons: application to brain-derived neurotrophic factor-dependent phosphotyrosine-associated signaling, *Mol Cell Proteomics* **7**, 1067–1076.
17. Schwanhausser, B., Gossen, M., Dittmar, G., and Selbach, M. (2009) Global analysis of cellular protein translation by pulsed SILAC, *Proteomics* **9**, 205–209.
18. Kelly, I., and Foster, L. J. (2010) Reducing artifacts and defining the linear quantitative range of pSILAC, in Poster session presented at the 58th ASMS Conference on Mass Spectrometry, Salt Lake City.
19. Ong, S. E., Mittler, G., and Mann, M. (2004) Identifying and quantifying in vivo methylation sites by heavy methyl SILAC, *Nat Methods* **1**, 119–126.
20. Zee, B. M., Levin, R. S., Xu, B., LeRoy, G., Wingreen, N. S., and Garcia, B. A. (2010) In vivo residue-specific histone methylation dynamics, *J Biol Chem* **285**, 3341–3350.
21. Blagoev, B., Kratchmarova, I., Ong, S. E., Nielsen, M., Foster, L. J., and Mann, M. (2003) A proteomics strategy to elucidate functional protein-protein interactions applied to EGF signaling, *Nat Biotechnol* **21**, 315–318.
22. Gruhler, A., Olsen, J. V., Mohammed, S., Mortensen, P., Faergeman, N. J., Mann, M., and Jensen, O. N. (2005) Quantitative phosphoproteomics applied to the yeast pheromone signaling pathway, *Mol Cell Proteomics* **4**, 310–327.
23. Zhang, G., Spellman, D. S., Skolnik, E. Y., and Neubert, T. A. (2006) Quantitative phosphotyrosine proteomics of EphB2 signaling by stable isotope labeling with amino acids in cell culture (SILAC), *J Proteome Res* **5**, 581–588.
24. Milner, E., Barnea, E., Beer, I., and Admon, A. (2006) The turnover kinetics of major histocompatibility complex peptides of human cancer cells, *Mol Cell Proteomics* **5**, 357–365.
25. Selbach, M., Schwanhausser, B., Thierfelder, N., Fang, Z., Khanin, R., and Rajewsky, N. (2008) Widespread changes in protein synthesis induced by microRNAs, *Nature* **455**, 58–63.
26. Fath, T., Ke, Y. D., Gunning, P., Gotz, J., and Ittner, L. M. (2009) Primary support cultures of hippocampal and substantia nigra neurons, *Nat Protoc* **4**, 78–85.
27. Ricardo-Dukelow, M., Kadiu, I., Rozek, W., Schlautman, J., Persidsky, Y., Ciborowski, P., Kanmogne, G. D., and Gendelman, H. E. (2007) HIV-1 infected monocyte-derived macrophages affect the human brain microvascular endothelial cell proteome: new insights into blood–brain barrier dysfunction for HIV-1-associated dementia, *J Neuroimmunol* **185**, 37–46.
28. Zhou, Y., Wang, Y., Kovacs, M., Jin, J., and Zhang, J. (2005) Microglial activation induced by neurodegeneration: a proteomic analysis, *Mol Cell Proteomics* **4**, 1471–1479.
29. Sergent-Tanguy, S., Chagneau, C., Neveu, I., and Naveilhan, P. (2003) Fluorescent activated cell sorting (FACS): a rapid and reliable method to estimate the number of neurons in a mixed population, *J Neurosci Methods* **129**, 73–79.
30. Gerfen, C. R., Holmes, A., Sibley, D., Skolnick, P., and Wray, S. (Eds.) (2010) Preclinical Models of Neurologic and Psychiatric Disorders, in Current Protocols in Neuroscience. Wiley InterScience.
31. Wu, C. C., MacCoss, M. J., Howell, K. E., Matthews, D. E., and Yates, J. R., 3 rd. (2004) Metabolic labeling of mammalian organisms with stable isotopes for quantitative proteomic analysis, *Anal Chem* **76**, 4951–4959.
32. Doherty, M. K., Whitehead, C., McCormack, H., Gaskell, S. J., and Beynon, R. J. (2005) Proteome dynamics in complex organisms: using stable isotopes to monitor individual protein turnover rates, *Proteomics* **5**, 522–533.
33. Booher, J., and Sensenbrenner, M. (1972) Growth and cultivation of dissociated neurons and glial cells from embryonic chick, rat and human brain in flask cultures, *Neurobiology* **2**, 97–105.
34. Brewer, G. J., and Torricelli, J. R. (2007) Isolation and culture of adult neurons and neurospheres, *Nat Protoc* **2**, 1490–1498.
35. Chen, Y., Balasubramaniyan, V., Peng, J., Hurlock, E. C., Tallquist, M., Li, J., and Lu, Q. R. (2007) Isolation and culture of rat and mouse oligodendrocyte precursor cells, *Nat Protoc* **2**, 1044–1051.
36. Dvorak, D. J., Gipps, E., and Kidson, C. (1978) Isolation of specific neurones by affinity methods, *Nature* **271**, 564–566.
37. McCarthy, K. D., and de Vellis, J. (1980) Preparation of separate astroglial and oligodendroglial cell cultures from rat cerebral tissue, *J Cell Biol* **85**, 890–902.

38. Taylor, A. R., Robinson, M. B., and Milligan, C. E. (2007) In vitro methods to prepare astrocyte and motoneuron cultures for the investigation of potential in vivo interactions, *Nat Protoc* **2**, 1499–1507.
39. Uchida, N., Buck, D. W., He, D., Reitsma, M. J., Masek, M., Phan, T. V., Tsukamoto, A. S., Gage, F. H., and Weissman, I. L. (2000) Direct isolation of human central nervous system stem cells, *Proc Natl Acad Sci* U S A **97**, 14720–14725.
40. Anderson, N. L., Anderson, N. G., Haines, L. R., Hardie, D. B., Olafson, R. W., and Pearson, T. W. (2004) Mass spectrometric quantitation of peptides and proteins using Stable Isotope Standards and Capture by Anti-Peptide Antibodies (SISCAPA), *J Proteome Res* **3**, 235–244.
41. Nesvizhskii, A. I., and Aebersold, R. (2004) Analysis, statistical validation and dissemination of large-scale proteomics datasets generated by tandem MS, *Drug Discov Today* **9**, 173–181.
42. Nesvizhskii, A. I., Vitek, O., and Aebersold, R. (2007) Analysis and validation of proteomic data generated by tandem mass spectrometry, *Nat Methods* **4**, 787–797.

Chapter 10

The Proteome of Brain Capillary Endothelial Cells: Toward a Molecular Characterization of an In Vitro Blood–Brain Barrier Model

Sophie Duban-Deweer, Christophe Flahaut, and Yannis Karamanos

Abstract

The blood–brain barrier (BBB) represents 95% of the molecular exchange area between the blood and the cerebral compartment. The BBB plays a major role in the brain homeostasis, essentially by regulating the passage of endogenous and exogenous circulating compounds. There is direct and indirect evidence that proteins such as membrane transporters, enzymes, and tight junction proteins interact to establish and maintain this metabolic and physical cellular barrier. First, we describe a method for harvesting of brain capillary endothelial cells (BCECs) from our in vitro BBB model. Then, we develop the protein preparation (cell lysis and protein isolation) and the two-dimensional gel electrophoresis steps, including the isoelectric focusing and SDS-PAGE conditions, the gel staining, the image acquisition, and the comparative study. Finally, we present examples of protein identification by peptide mass fingerprint (PMF) measured by MALDI-TOF-MS analysis, and complementary data issued from peptide fragmentation fingerprints (PFF) that allow successful protein identifications. PMF and PFF analyses provide complementary datasets and thus, a more comprehensive sequence coverage of the BCEC proteome, especially when they are combined together (PMF/PFF).

Key words: Proteomics, Blood–brain barrier, Cerebral endothelial cells, Mass spectrometry, 2D electrophoresis

1. Introduction

The Blood–brain Barrier (BBB) is the vascular endothelial interface that isolates the brain from the bloodstream. The cerebral capillaries (referred to as the BBB), comprising approximately 95% of the total area of barriers between blood and brain, are the main entry route for molecules into the central nervous system (CNS) (1).

Yannis Karamanos (ed.), *Expression Profiling in Neuroscience*, Neuromethods, vol. 64, DOI 10.1007/978-1-61779-448-3_10,

The main role of the BBB is to create the ionic homeostasis required for the neuronal functions (2), and this dynamic interface also provides the CNS nutrients and protects it from toxic elements by several elaborated systems of transport (3).

This barrier is supported by the brain capillaries and more specifically, the brain capillary endothelial cells (BCECs). These capillaries are surrounded by a continuous tubular sheath of astrocytic end-feet, which induces the BBB properties to BCECs (1). Moreover, closely associated to perivascular neurons, pericytes, and astrocytes, the BCECs constitute a functional neurovascular unit (4). The main physiological characteristics of BCECs displaying BBB properties are a lack of fenestrations and pinocytic vesicles, associated with a reinforcement of tight junctions that lead to the setting up of a physical barrier (5). Moreover, the BBB also forms a metabolic barrier. Indeed, the increased activities of intracellular enzymes, such as monoamine oxidase or γ-glutamyl-transpeptidase (6), suggest that the BCEC metabolic activities are also modified. Together, these fence features give a high trans-endothelial electrical resistance (TEER) due to a low paracellular and transcellular permeability of various compounds (7).

This BBB permeability is frequently a rate-limiting factor for penetration of drugs or peptides into the brain (8, 9). This transport barrier prevents the permeation for drugs that are hydrophilic and have a molecular mass higher than 400 Da, and the presence of efflux transporters at the luminal and basal membranes of BCECs limits the brain penetration of lipophilic xenobiotics and drugs. Thereby, the BBB prevents 98% of potential therapeutic drugs or peptides from reaching their targets in the CNS (10). For these reasons, the treatment of CNS diseases such as stroke, brain tumors, or Alzheimer's disease remains still limited, showing that progress in drug delivery to the brain is essential for the success of therapies for neurological disorders (11).

Based on web-based literature mining, we have made the choice of a proteomics approach to better define and decipher the astrocyte-induced phenotype of the BCECs from our BBB model. The aim of this approach is to obtain a large-scale protein characterization using the technique of two-dimensional gel electrophoresis and to determine, by differential proteomic analysis, the observable changes occurring in functional BBB cells: BCECs in co-culture with astrocytes compared to BCECs having lost BBB functions: cultured without astrocytes. Then the proteins of interest are identified by mass spectrometry as well as by peptide mass fingerprint (PMF), peptide fragmentation fingerprint (PFF), or both.

2. Materials

For all water-based solutions or buffers, ultrapure water (MQ water) is needed.

2.1. Cell Processing

2.1.1. Cell Harvesting Procedure

1. 100-mm Petri dishes (Corning).
2. 15-mL conical tubes (BD Falcon).
3. 10-mL aspirating pipettes (BD Falcon).
4. Pipet-aid.
5. 37°C incubator.
6. Vortexer.
7. Refrigerated bench-top centrifuge capable of 500 × *g*.
8. Pasteur pipettes.
9. Vacuum faucet aspirator pump.
10. Calcium magnesium-free phosphate-buffered saline (CMF–PBS), pre-warmed at 37°C. see Note 1.
11. Calcium, magnesium-containing phosphate-buffered saline (CaMg–PBS), pre-warmed at 37°C. see Note 1.
12. 0.1% (w/v) *Clostridium histolycum* collagenase, type XI (Sigma). see Note 2.

2.1.2. Cell Lysis and Acetone-Based Protein Precipitation

1. 1.5-mL microtubes (Eppendorf).
2. Sonicator probe for microtubes.
3. Polypropylene pestle microtubes (Sigma).
4. Refrigerated bench-top centrifuge capable of 18,000× *g*.
5. Vortexer.
6. Lysis buffer: 10 mM Tris/HCl, pH 6.8, containing 1% (v/v) Triton X-100, 1 mM ethylendiamine tetracetic acid (EDTA), 0.1% (v/v) 2-mercaptoethanol (2-ME), and 1 tablet of protease inhibitors (Roche Molecular Biochemicals) per 10 mL.
7. Acetone, HPLC grade, stored at −20°C.
8. 80% Acetone solution, stored at −20°C.
9. Protein solubilization solution: 7 M urea, 2 M thiourea, 2% (w/v) CHAPS, and 2% (v/v) Triton X-100. see Note 3.
10. Protein assay kit (e.g., 2D Quant kit 80-6483-56, GE Healthcare Life Sciences).

2.2. Two-Dimensional Electrophoresis

2.2.1. Isoelectric Focusing

1. Isoelectricfocusing (IEF) cell apparatus (Bio-Rad).
2. Focusing tray, for 24-cm strips (Bio-Rad).
3. Rehydration/equilibration tray, 24 cm, strips (Bio-Rad).
4. DeStreak rehydration solution (GE Healthcare Life Sciences).

5. Biolyte pH 3/10:4/7:5/8 (Bio-Rad).
6. Immobiline DryStrip, pH 3–10, 24 cm (GE Healthcare Life Sciences).
7. Mineral oil (Bio-Rad).
8. Wicks (Bio-Rad).

2.2.2. SDS-PAGE

1. Electric field generator, capable of 1,000 V.
2. Ettan DALTsix electrophoresis unit (GE Healthcare Life Sciences).
3. Orbital shaker.
4. Microwave oven.
5. Equilibration buffer with dithiothreitol (DTT): 6 M urea, 20% v/v glycerol, and 2% w/v SDS, in 93 mM Tris-HCl buffer, pH 8.8, supplemented with 20 mM DTT and traces of bromophenol blue; see Note 4.
6. Equilibration buffer with iodoacetamide: same buffer as above, except that DTT is replaced by 100 mM iodoacetamide (light sensitive); see Note 4.
7. SDS-PAGE buffer: 25 mM Trizma base, 192 mM glycine, and 0.2% (w/v) SDS, pH 8.8.
8. Electrophoresis gel mixture: 10% Duracryl™ (Proteomic Solutions, France) T30%, C2.6%, in SDS-PAGE buffer, 0.05% ammonium persulfate, and 0.05% *N,N,N′,N′*-tetramethylethylenediamine TEMED (harmful); see Note 5.
9. 0.5% (w/v) agarose low-melting point solubilized in SDS-PAGE buffer (Note 6).

2.2.3. Silver Nitrate Staining

Prepare 250 mL of following solutions per gel, except for revelation solution (500 mL is required).

1. Orbital shaker.
2. Sensitization solution: 0.2% (v/v) sodium thiosulfate ($Na_2S_2O_3$) from a concentrated solution of 10% (w/v) of sodium thiosulfate in MQ water.
3. Fixing solution: 30% ethanol and 5% acetic acid.
4. Impregnation solution: 1 g/L silver nitrate ($AgNO_3$) and 0.028% (v/v) 37% formalin (harmful), see Note 7.
5. Revelation solution: 24 g/L sodium carbonate (Na_2CO_3), 12.5% (v/v) sodium thiosulfate from a concentrated solution of 10% (w/v) of sodium thiosulfate, and 0.028% (v/v) of 37% formalin; see Note 8.
6. Stop solution: 0.33 M Trizma base and 2% (v/v) acetic acid.
7. 1% acetic acid solution.

2.2.4. Image Acquisition and Comparative Study

1. Computer-assisted densitometer, freshly calibrated (such as the Umax Scanner, GE Healthcare Life Sciences).
2. 2D gel analysis software (e.g., Progenesis SameSpots, Nonlinear Dynamics).

2.3. Mass Spectrometry

2.3.1. Trypsin Digest

1. Speed-Vac® vacuum centrifuge (Thermo Fisher Inc).
2. Thermostatic bath.
3. Vortexer.
4. 37°C incubator.
5. Destaining solution: 1.6% (w/v) sodium thiosulfate and 1% (w/v) potassium ferricyanide in MQ water.
6. DTT solution: 10 mM DTT in 20 mM ammonium bicarbonate buffer, pH 8.0.
7. Iodoacetamide solution: 20 mM ammonium bicarbonate, pH 8.0, and 55 mM iodoacetamide.
8. Acetonitrile HPLC grade.
9. Trypsin stock solution: 20 μg trypsin sequencing grade (Promega) is solubilized in 0.01% trifluoroacetic acid (TFA) and stored at –20°C.
10. 40 mM ammonium bicarbonate, pH 8.0. It will also be diluted to obtain 20 mM ammonium bicarbonate, pH 8.0 buffer.
11. Solution of 0.1% TFA.

2.3.2. Sample Preparation for MALDI-TOF Analysis

1. Acetone HPLC grade.
2. Ethanol HPLC grade.
3. Matrix stock solution: 5 mg/mL α-cyano-4-hydroxycinnamic acid (HCCA) solubilized with acetone, and stored at –20°C.
4. AnchorChip™600 target (Bruker Daltonics).
5. Peptide calibration standard II (Bruker Daltonics).

2.3.3. Database Search and Protein/Peptide Identification

1. We use a matrix-assisted laser desorption/ionization time-of-flight mass spectrometer (MALDI-TOF/TOF-MS, UltraFlex II instrument from Bruker Daltonics) for the mass analyses.

3. Methods

Bovine BCECs were isolated and characterized as described in a previous study (12). Petri dishes (100 mm) were coated with rat tail collagen (2 mg/mL) containing tenfold concentrated Dulbecco's modified Eagle's medium (DMEM) and 0.4 M NaOH. The BCECs (4×10^5 cells/mL) were seeded and cultured in DMEM supplemented with 10% (v/v) heat-inactivated calf serum, 10% (v/v)

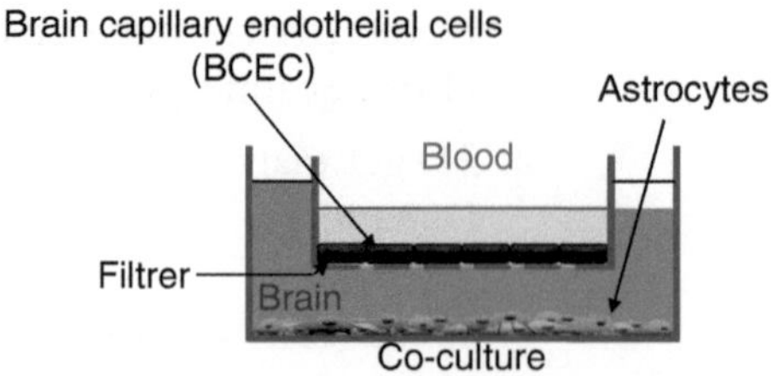

Fig. 1. Schematic drawing of our in vitro blood–brain barrier (BBB) co-culture model. For this in vitro BBB model, endothelial cells are grown on filter inserts together with glial cells (astrocytes) at the bottom of 100-mm Petri dishes. Glial soluble factors secreted in the culture medium re-induce the BBB phenotype in the capillary endothelium.

heat-inactivated horse serum (Hyclone, laboratories), 2 mM glutamine, 50 mg/mL gentamicin, and 1 ng/mL basic fibroblast growth factor. The culture medium was refreshed every 2 days until confluency (around 6 days). Co-cultures were set in Transwell™ cell culture inserts—0.4 mm pore size, 100-mm diameter (Corning), and coated on the upper side with home-made rat tail collagen (13). Endothelial cells were then seeded in the inserts (Fig. 1) and transferred to a 100-mm Petri dish containing glial cells prepared according to Booher and Sensenbrenner (14). The re-induction of BBB properties (12 days of co-culture) was typically checked by measurement of the paracellular permeability coefficient and by immunostaining of the main tight junction proteins (occludin, claudin-5, and zona occludens 1). This experimental design can not only be used for compound screening in the drug discovery process in the pharmaceutical industry, but is also well suited for studying mechanistic aspects of BBB transport as well as other biological and pathological processes (1). Endothelial cell harvesting was performed after 12 days of co-culture.

3.1. Cell Processing

3.1.1. Cell Harvesting Procedure

In order to obtain reproducible and representative proteome of our in vivo BBB model, we have evaluated different BCEC harvesting procedures (15). Collagenase treatment resulted in a soft and substrate-targeted enzymatic digestion, unlike the trypsin action which released the cells by unspecific cleavage of the membrane anchoring proteins, as shown in the Fig. 2. This particular harvesting method is necessary to avoid the destroying of these proteins, and cellular stress.

1. Remove culture medium by aspiration using a Pasteur pipette coupled to vacuum faucet aspirator pump. Be careful not to perforate the filter of the cell culture insert with the pipette. Do not allow the cells to dry.
2. Remove Transwell™ insert from Petri dish and place it in an unused dish to avoid protein contaminations from astrocytes. see Note 9.

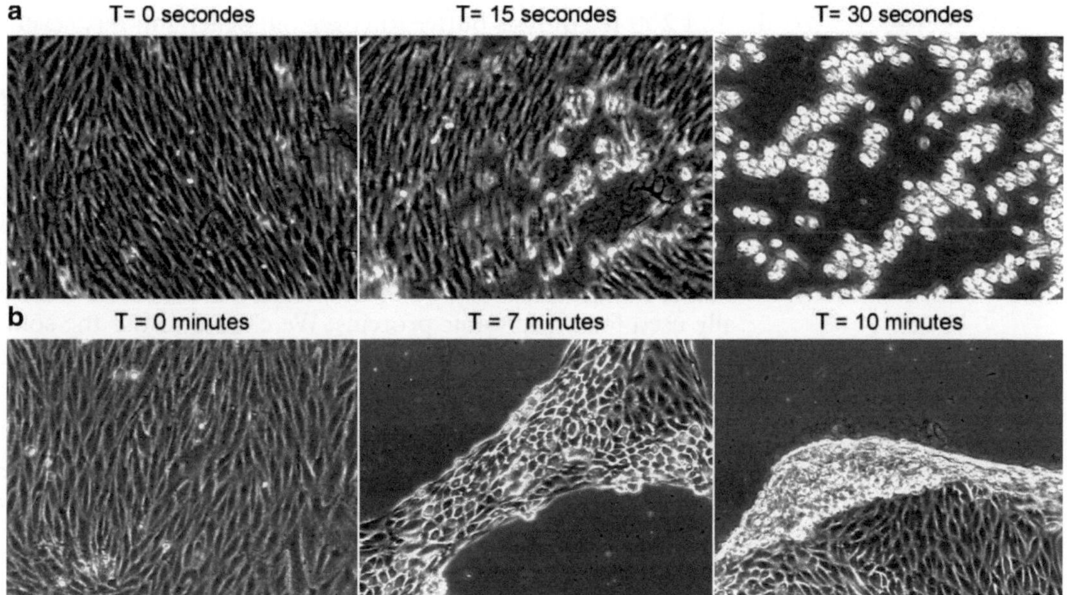

Fig. 2. Phase contrast microscopy follow-up of BCEC detachment following enzymatic treatments. Confluent endothelial cell monolayers were extensively washed with PBS and subjected to either trypsin cell disruption (**a**) or collagenase cell detachment (**b**). Trypsin induces a faster release of the cell monolayer which becomes spherical and refringent. Comparatively, collagenase causes a slower detachment of cells assembled in plates without the loss of their fusiform shape (15). Reproduced with permission of WILEY-VCH Verlag GmbH & Co. KGaA, Weinheim© 2009.

3. Add 10 mL of CaMg–PBS (pre-warmed at 37°C) and wash cells gently two times for 1 min. see Note 10. Remove the buffer with Pasteur pipette coupled to vacuum faucet aspirator pump.
4. Prepare collagenase solution, as described in Note 2. For cell harvesting, one 100-mm Petri dish requires 1.5 mL of collagenase solution. Incubate BCEC dish with collagenase at 37°C in an incubator for 45 min.

   The collagenase-based cell detachment was monitored by phase contrast microscopy (Fig. 1): unlike trypsin, collagenase causes a slow detachment of cells without loss of their integrity and their fusiform shape (15).
5. Harvest cells using aspirating pipette coupled to a 200-μL tip and generate an "aspiration–expulsion" movement of cell suspension. Repeat this up-and-down movement ten times (the cell detachment is visible to the naked eye).
6. Transfer the cell suspension to a clean 15-mL conical tube and centrifuge at 500× *g*, 4°C for 10 min to pellet cells, and then eliminate supernatant. see Note 11.
7. Wash BCEC pellet with 10 mL CMF–PBS (stored at 4°C), centrifuge as above, and then eliminate supernatant as described in Note 11. Repeat this washing once.
8. The cell pellet is stored at −80°C until cell lysis.

3.1.2. Cell Lysis and Acetone Precipitation of Proteins

1. Add 200 μL of lysis buffer to cover cell pellet and mix several times with a laboratory pipette. Incubate on ice for 5 min and transfer the lysate to a clean 1.5-mL microtube. Then vortex vigorously for 5 min.

   Any formulation of lysis buffer is acceptable (RIPA buffer, plasma membrane buffer, etc.), and the composition of the lysis buffer will depend on the purpose of the analysis. For example, the isolation and analysis of membrane proteins will demand a lysis buffer different from a standard lysis buffer usually used for cytoplasmic proteins. We choose to use the solubilization properties of Triton X-100 to obtain a sub-proteome mainly composed of cytosolic and cytoskeleton proteins from BCECs in our in vitro BBB model.
2. Homogenize cells to fully lyse the sample on ice with a disposable pellet pestle (ten rotations) and allow to stay on ice for 5 min.
3. To enhance cell disruption, sonicate on ice for 10 s, three times (30–40 pulses), and leave to incubate for 5 min on ice. Vortex until no suspension particles is visible in the microtube.
4. Centrifuge at 18,000× *g* at 4°C for 60 min to pellet cellular debris. Previous to acetone precipitation, transfer supernatant to a clean 1.5-mL microtube (take a 15 μL aliquot for assaying of proteins; we have good success with the 2D Quant kit).

   Acetone precipitation techniques can be used to isolate the proteins of a sample from potentially interfering substances, or to concentrate the sample.
5. Add four volumes of cold acetone to the supernatant, close the tube, and invert the sample 2–4 times to mix the sample and acetone thoroughly. A precipitate should become clearly visible. Then store the tube at −20°C overnight.
6. Centrifuge at 18,000× *g* for 60 min, at 4°C. Eliminate supernatant. Wash the pellet of precipitated proteins by addition of two volumes of cold 80% acetone with vigorous agitation.
7. Centrifuge at 13,500× *g* for 5 min at 4°C, eliminate supernatant, and repeat the washing twice.
8. Air-dry the pellet (do not dry with Speed-Vac®). Protein pellet can be stored at −20°C for a few weeks before solubilization.

3.2. Two-Dimensional Electrophoresis

3.2.1. Isoelectric Focusing

1. Suspend acetone precipitate in lysis buffer (see Sect. 2.1.2) to obtain at least a 4 μg/μL stock solution.
2. Prepare the 1D buffer by mixing 120 μL of DeStreak rehydration solution and biolytes, pH 3/10:4/7:5/8 (1:1.5:1 μL)—volume necessary for one strip.
3. We commonly use 250 μg of proteins for an analytical run and 0.5–1 mg for a preparative run, per 24-cm IPG strip

(pH 3–10). Caution: the maximum volume of reabsorbed solution by one 24-cm IPG strip is around 420 μL.

4. Mix the sample stock with the 1D buffer to obtain the chosen protein concentration and the adequate volume. (Typically, mix the 120 μL of the 1D buffer with 65 μL of resuspended sample and complete up to 400 μL with the lysis buffer). Incubate at room temperature (RT) and vortex for 15 min.
5. Apply the sample mix between the electrodes of an IEF tray. Recover the electrodes of each track by the MQ water-soaked wicks. Add a precast IEF strip (the gel must be in contact with the sample mix) and recover it with 1.5 mL mineral oil.
6. Perform the IEF with the following parameters and steps at 20°C: 50 μA per strip, passive rehydration period, 7 h; linear 50 V for 11 h30; gradient 50–200 V, 8 h30; linear 200 V, 9 h30; gradient 200–500 V, 2 h30; linear 500 V, 5 h; gradient 500–3,500 V, 5 h; linear 3,500 V, 22 h30; after isolelectrofocusing, the strip can be stored for several months at −80°C.

3.2.2. SDS-PAGE

Perform a 10% Duracryl™ gel according to the manufacturer's instructions. For a 24-cm IPG strip, the dimensions of the gel should be at least 26×20 cm (typically, 1.0-mm thick). see Note 12.

1. Incubate the strip for 15 min at RT under gentle agitation with 5 mL of equilibration solution containing DTT and then remove the excess solution.
2. Incubate the strip for 15 min at RT under gentle agitation with 5 mL of equilibration solution with iodoacetamide.
3. Load the strip on the top of a 10% Duracryl™ gel.
4. Seal the strip with 3–5 mL agarose solution (heated, see Note 6). Gently press on the IPG strip to ensure physical contact between the strip and the top of the gel.
5. Perform the second-dimension run in two steps: (a) 30 min at 16 mA/gel; an initial low-voltage step reduces electroendosmosis (16); (b) 5 h at 45 mA/gel.

3.2.3. Silver Nitrate Staining

After resolving proteins by 2D electrophoresis, proteins are visualized by in-gel silver nitrate staining. The detection sensitivity of this procedure is close to 1–2 ng of proteins per spot (17). This silver nitrate staining is a mass spectrometry compatible staining according to Shevchenko et al. (18). The incubation times indicated in this procedure should be accurately observed in order to ensure reproducible image development. All steps are carried out on an orbital shaker at RT.

1. Fix proteins in gel for at least one hour with typically 150 mL of the fixing solution. see Note 13. Wash in water four times for 10 min.

2. Immerse the gel in the sensitization solution for 1 min and rinse two times for 1 min in MQ water.
3. Incubate the gel for 30–45 min with the impregnation solution; the gel should be covered with aluminum foil (silver nitrate is light sensitive).
4. Rinse with MQ water for 5–10 s.
5. Incubate the gel with the revelation solution (150 mL), discard the solution rapidly after a yellow precipitate appears, and incubate the gel with 350 mL of revelation solution (start a chronometer) until appropriate staining is achieved. Typically, the spots should be visible within 0–3 min.
6. Stop the reaction by replacing the revelation solution by the stop solution (also stop chronometer). The gel is ready for image acquisition (Fig. 3); it can be stored in 1% acetic acid at 4°C. However, the silver nitrate-stained gel pieces that should be subjected to identification must be excised within a week.

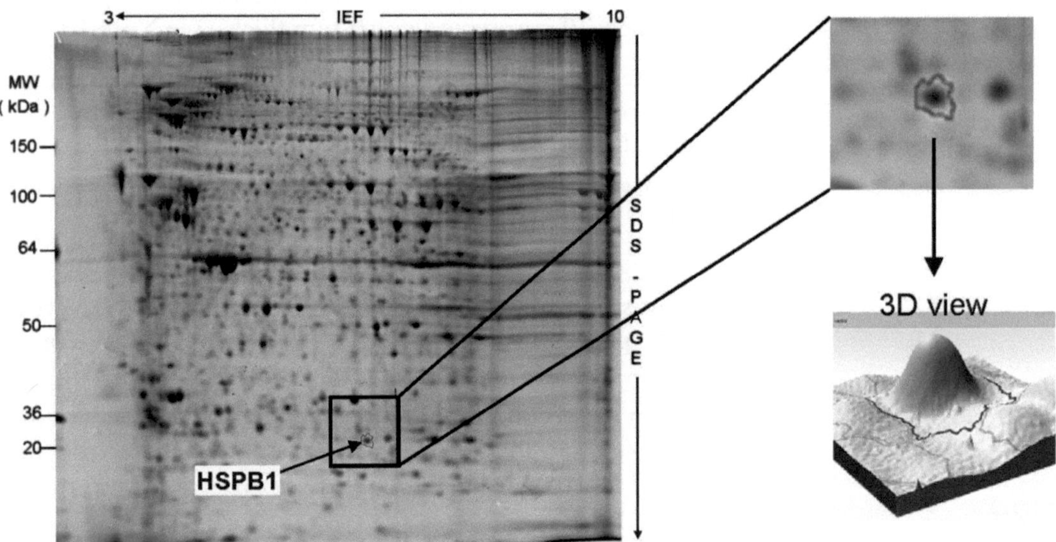

Fig. 3. A typical protein 2D gel obtained with BCEC from our in vitro BBB model. This gel exhibits a large protein spot distribution for all the isoelectric point and molecular weight ranges, especially due to the high sensitivity of silver staining procedure used, which detects proteins in the nanogram range. A differential proteomic study using Samespots™ software has underscored the most significant variations of protein expression in BCECs after co-culture with astrocytes versus BCECs in solo-culture without astrocytes. Among these proteins, several major spots were identified as actin, gelsolin, and filamin-A, and these cytoskeleton-related proteins play a major role in BBB phenotype establishment of BCECs (15). Recently, we identified protein spots of different abundance and these quantitative changes mainly affect proteins involved in cell structure and motility, or in protein metabolism and modification processes. HSPB1 (see spot area of interest and the 3D view adapted from SameSpots™ software) has been chosen to illustrate one of these proteins implicated in the cytoskeleton remodeling of differentiated BCECs.

3.2.4. Image Acquisition and Comparative Study

1. Scan the gels at 300 DPI with a 16-bit densitometer using the Labscan™ software, for example. The densitometer should be freshly calibrated; the intensity level data should be coded over a bit depth greater than 8. No image compression must be used, and the images should not be saturated, and be coded in grayscale. Finally, these recommendations are applicable for this software and most others.
2. Select a reference gel that will be used when aligning each of the gels in your experiment. Choosing a good reference gel will help during the spot alignment step (see Note 14).
3. Select the area of interest in gels: Samespots™ software allows selecting a mask over areas of the image that should be ignored during analysis; the common areas to ignore are the edges, saturated zones, and damaged areas of the gel.
4. Perform the gel alignment and spot detection: process to the gel alignment using match vectors versus the reference image. The spots are detected according to the subsequent criteria: the normalized average volume and the spot area. Total spot volume calculated for each image and to each spot is assigned to a normalized spot volume as a proportion of the total value for adequate comparison.
5. Screen the automatically detected spots: the types of spots to filter out may include spots in damaged areas, on the edge of gel, and outside of the gel (e.g., on the scanner bed).
6. Gather the gel images in groups that you wish to compare (e.g., control gels versus treatment gels) and validate manually spots that will be included in final result.
7. Perform spot quantification: expressed as spot volume relative to either the volume of all spots in each gel or the volume of a single spot found in equal amount in all experimental conditions, to correct for variability resulting from silver staining (spots are ranked by p-value calculated from the one-way ANOVA).
8. Calculate mean values from at least four gels performed from different experiments and apply the appropriate statistical tests [principal components analysis (PCA), correlation analysis, power analysis, and q-values (false discovery rate-adjusted p-values)] included into Samespots™ software to explore the trends in your data (Sect. 3.2.3).

3.3. Mass Spectrometry

In order to optimize protein digestion, the samples are denatured to allow a better accessibility of trypsin and, therefore, to greatly enhance the cleavage rate. If, for particular reasons, the in-gel resolved proteins have not been reduced and alkylated before the second-dimension run, the proteins should be chemically reduced (with DTT solution). Then the reduced cysteine residues should be blocked typically with iodoacetamide solution. Trypsin is the proteomic enzyme of choice for several reasons. Briefly, this is an endoprotease of narrow cleavage specificity (cleaves after Arg and Lys residues) that generates, with a high rate, peptides of size compatible with the mass spectrometry, and ionized positively in acid condition (due to Arg or Lys at the C-terminal end of peptides).

3.3.1. Trypsin Digestion

1. The gel pieces are carefully excised with a disposable gel cutter or a spot picker robot and placed in a 1.5-mL microtube (see Note 15).
2. Destain the gel pieces in 1 mL of the destaining solution containing sodium thiosulfate and potassium ferricyanide, mix vigorously until gel spots become yellow, and then discard the destaining solution.
3. Wash the gel pieces several times in 1 mL of water until they become discolored, and then remove the wash liquid.
4. Transfer the gel spots in a 200-μL microtube and shrink gels by adding 50 μL of 100% acetonitrile. Mix vigorously, after 10 min remove acetonitrile (see Note 16), and dry the pieces of gel in a vacuum centrifuge for 15 min at RT.

   If the protein samples have been reduced and alkylated before the second-dimension run, please go to point #7.
5. Swell the gel pieces with 15 μL of DTT solution, reduce proteins for 45 min at 56°C, and discard the supernatant carefully. To alkylate the proteins, add 15 μL of iodoacetamide solution and incubate for 45 min in the dark at RT.
6. The iodoacetamide surplus is removed. Shrink gels with 50 μL of 100% acetonitrile, vortex for 5 min, and remove acétonitrile. The pieces of gel are dried in a vacuum centrifuge for 15 min at RT.
7. Prepare a fresh digestion solution: 12.5 ng/mL trypsin (from stock solution), 20 mM ammonium bicarbonate, pH 8.0, and 0.5 mM $CaCl_2$, and rehydrate gels on ice for 45 min with 20 μL of this solution.
8. Replace the digestion solution by the same volume of 20 mM ammonium bicarbonate, pH 8.0, and incubate the samples overnight at 37°C in an incubator.
9. After digestion, the remaining supernatant is removed and saved in a 0.2-mL microtube. Extract two times the peptides with 25 μL of 40 mM ammonium bicarbonate, pH 8.0/acetonitrile (v/v) 50:50, mix vigorously for 10 min, and collect the peptide containing supernatant in a clean 0.2-mL microtube.
10. Add 25 μL of 100% acetonitrile to shrink the gels, mix vigorously for 10 min, and pool the supernatant with the previous extraction.

11. Dry down pooled fractions in a vacuum centrifuge and resuspend peptides with 10 μL of 0.1% TFA solution, mix for 30 min at RT, and store peptides at −20°C.

3.3.2. Sample Preparation to MALDI-TOF Analysis and MS Data Processing

1. Prepare a 1 mg/mL matrix solution from stock solution (5 mg/mL) in acetone.
2. Prepare a fresh matrix solution (0.33 mg/mL) in ethanol/MQ water (33:67; v/v).
3. Samples are deposed onto a MALDI target using the dried droplet procedure. We use AnchorChip™ MALDI targets (Bruker Daltonics), see Note 17: briefly, deposit 0.2 μL of peptide mixture on the MALDI target. Do not allow samples to dry totally, quickly add 0.8 μL of 0.33 mg/mL matrix solution, and then air-dry this mix.
4. Apply MS calibrants (peptide calibration standard kit, Bruker Daltonics) with the same method as described for samples, let dry, and place the target in the mass spectrometer.
5. The molecular mass measurements are performed in manual mode using FlexControl™ 2.2 software on an MALDI-TOF/TOF instrument (Ultraflex™ II, Bruker Daltonics) in the reflectron mode for peptide mass fingerprinting, while the LIFT mode is added for peptide fragmentation fingerprinting.
6. An external calibration is performed using the peptide calibration standard kit (Bruker Daltonics) before mass measure of samples. Following MS data acquisition, the MS and MS/MS spectra should be processed to extract the information used for the identification of proteins. The monoisotopic mass lists of peptides or their fragments were generated from MS and MS/MS spectra using Flexanalysis™ 2.4 software (Bruker Daltonics).
7. From calibrated MS spectra of samples, choose a threshold of signal to noise to only select the most intense *m/z* signals (up to 70 compounds) in the mass range 800–3,500 Da (for PMF only).
8. Furthermore, only select the monoisotopic masses of compounds. Remove contaminating peptides (e.g., the auto-proteolysis products of trypsin, peptides found in the control digestion performed on a blank zone of the gel).
9. Export and save the mass lists (also called peak list) as .txt files.

3.3.3. Database Search and Validation of Peptide Identifications

The final step of the analysis consists of identifying the proteins of interest by matching their experimental mass lists issued from MS and MS/MS data against the theoretical mass lists generated from the in silico trypsin digestion or fragmentation of all proteins of one or several sequence databases using a search engine (Figs. 4 and 5).

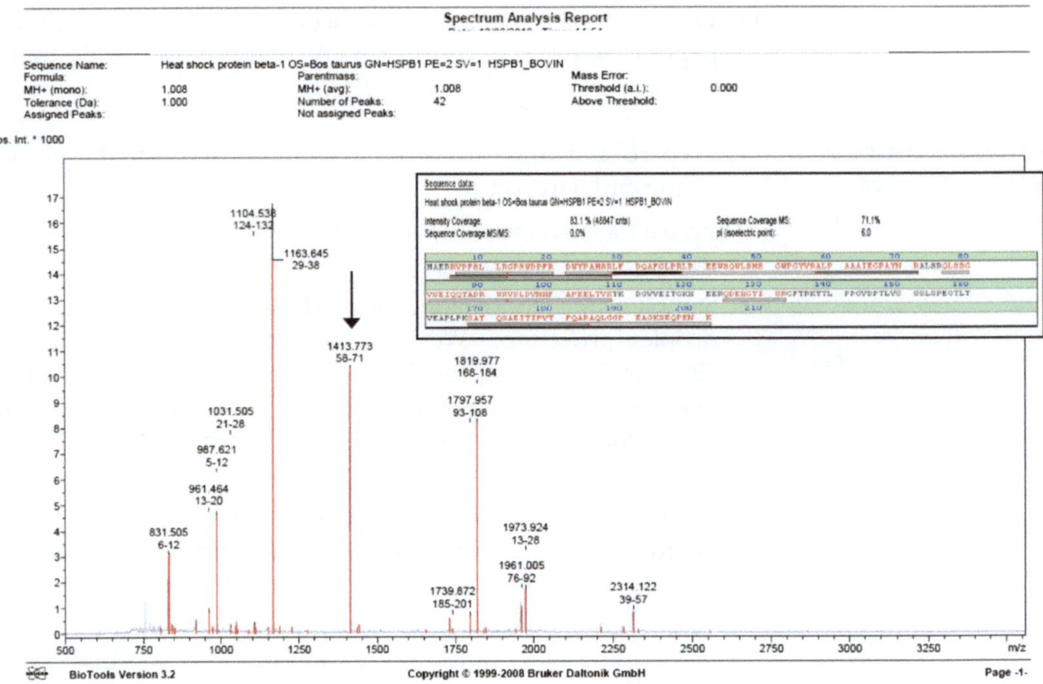

Fig. 4. A screen shot of typical results obtained with Biotools™ software (Bruker Daltonics) after Mascot engine submission. This MALDI-TOF mass spectrum of a digest spot isolated from a silver-stained 2D gel presents experimental mass values that matched with the theoretical masses of protonated tryptic peptides of bovine HSPB1. The peak list generated from PMF data was used to search in sequence databases (e.g., Swiss-Prot: http://www.expasy.org/), and identified the bovine HSPB1. In this example, the protein was successfully identified with a probability-based MOWSE Mascot score (see Note 19) of 170 (data not shown), and the sequence coverage for HSPB1 was 71% (13 out of 42 peptides of the peak list matched with HSPB1). *Black arrow* indicates the *m/z* value (1,413.77) of parent ions subjected to fragmentation.

From PMF, several search engines are web-available (such as Mascot, Profound, MS-Fit, or Aldente) which use their own search algorithms and scoring system. From PFF, the search engines are either web-available but the number of spectra processable is limited (Mascot, MS-Seq, Phenyx, X! Tandem, etc.), or require an on-site installation of hardware and licensed software (Mascot, SEQUEST, etc.) to perform automatic queries and to increase the processable data quantities. In this section, Mascot search engine (version 2.2.1) will be used as an example. Before the search, the peak list files have to be generated. Peak list files are made from the files initially recorded during the MS and MS/MS analysis and contain information about monoisotopic masses of peptides or their fragments from a MS spectrum or precursor selected for MS/MS fragmentation; these files also contain the relative intensities of peptides and fragment ions obtained after fragmentation of a given peptide. Depending on the type of instrument used for the MS and MS/MS analysis, the search parameters have to be adjusted accordingly.

a

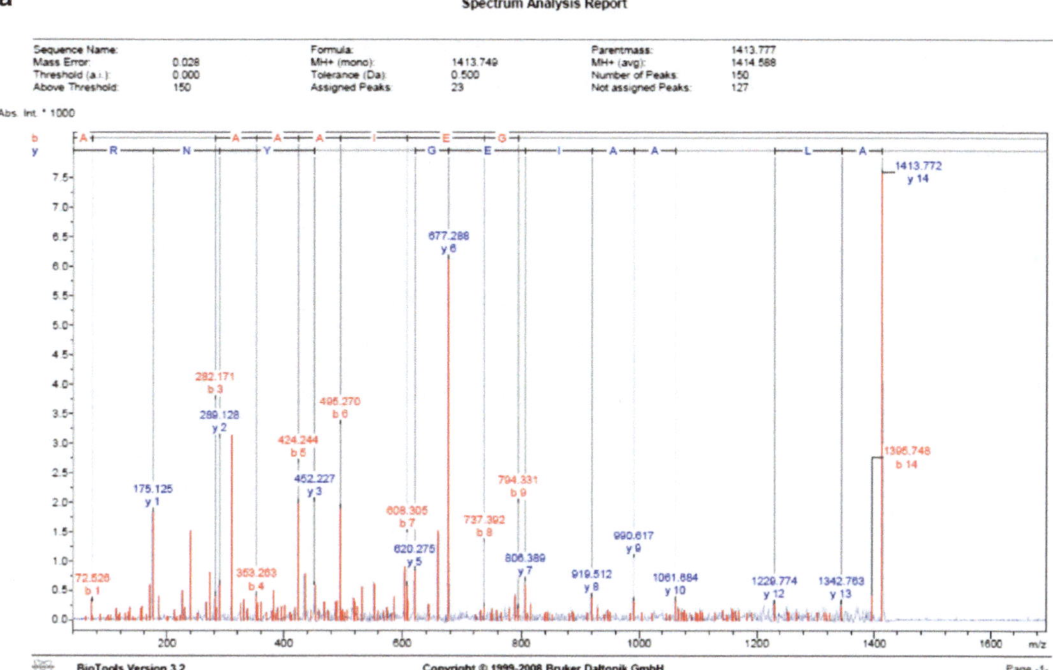

b **Probability Based Mowse Score**

Ions score is -10*Log(P), where P is the probability that the observed match is a random event.
Individual ions scores > 23 indicate peptides with significant homology.
Individual ions scores > 33 indicate identity or extensive homology (p<0.05).
Protein scores are derived from ions scores as a non-probabilistic basis for ranking protein hits.

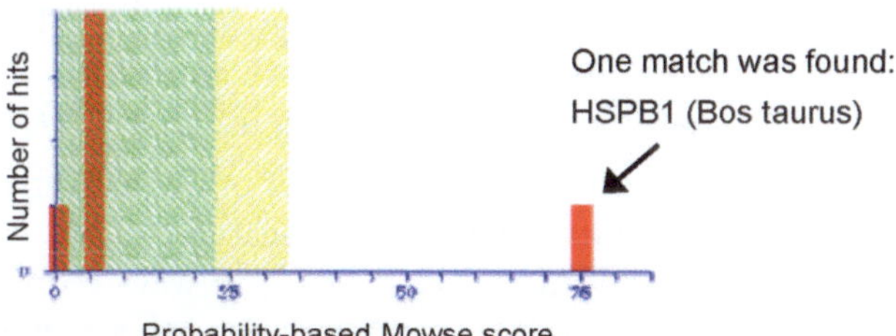

Fig. 5. Example of protein identification. MS/MS data should always confirm the results obtained by PMF analysis. Therefore, MS/MS fragmentation should systematically be performed to confirm the identification of a protein. The fragmentation spectrum of mono-charged ions at *m/z* 1,413.77 was obtained (see **a**), and the monoisotopic mass list (data not shown) corresponding to the fragmentation of selected peptide was submitted in the Mascot engine. The graphical representation in (**b**) (adapted from http://www.matrixscience.com) illustrates the hits obtained by matching the fragmentation spectrum shown in (**a**) and the monoisotopic mass list issued from the theoretical fragmentation spectra of all peptides with a *m/z* of 1,413.77 ± 75 ppm of the Swiss-Prot database. HSPB1 was found with a significant Mascot score equal to 75 (data not shown). Because MS/MS spectra give information about the sequence of peptides rather than their amino acid composition, such MS/MS analysis is more discriminating but requires a minimum of two peptides sequenced.

Table 1
Typical MS (a) and MS/MS (b) search parameters. The following Mascot search parameters can be used for MALDI-TOF/TOF instruments

(a) PMF search parameters

Database searched:	Sprot or/and NCBI
Taxonomy searched:	Mammalia
Peptide mass tolerance:	±75 ppm
Mass value:	Monoisotopic $[M+H^+]$
Variable modifications:	Carbamidomethyl (C) Oxidation (M)
Enzyme:	Trypsin
Max missed cleavages:	1

(b) PFF search parameters

Database searched:	Sprot or/and NCBI
Taxonomy searched:	Mammalia
Peptide mass tolerance:	±75 ppm
Fragment mass tolerance:	±0.5 kDa
Mass value:	Monoisotopic $[M+H^+]$
Variable modifications:	Carbamidomethyl (C) Oxidation (M)
Enzyme:	Trypsin
Max missed cleavages:	1

Adapted from http://www.matrixscience.com

1. Various parameters should be rigorously completed for searching in the sequence databases (see Table 1). The monoisotopic mass lists were submitted in NCBI or Swiss-Prot protein databases using Mascot (http://www.matrixscience.com/), while the database searches, through Mascot (Matrix Science, London, UK), using combined PMF and PFF datasets, were performed via Biotools 3.1 and ProteinScape 1.3 (Bruker Daltonics).
2. One very important aspect of the final analysis is the verification of results obtained from search engines. Indeed, all the protein identifications are far to be always correct; proteins identified by the Mascot search engine with a Mascot score below 60 are usually false positives (nonsignificant) and have to be discarded from the dataset. For the sequenced peptides, a significant score should be greater than 30 (Figs. 4 and 5). see Note 18.

4. Notes

1. Make fresh CMF–PBS (137 mM NaCl, 2.62 mM KCl, and 8.01 mM $Na_2HPO_4 \cdot 12H_2O$, pH 7.4) and CaMg–PBS (PBS–CMF containing 90 mM $CaCl_2$ and 49 mM $MgCl_2$), filter with 0.75-µm filter, and store for one month at 4°C.
2. Collagenase should be re-suspended in CaMg–PBS pre-warmed at 37°C before use.

3. Filtrate solubilization solution (with 0.75 μm filter), and do not heat above 30°C to avoid protein carbamoylation. Indeed, the isocyanate ions generated by thermal decomposition of urea can lead to carbamoylation of proteins and generate artifactual spots on 2D electrophoresis gel (19).
4. Prepare the equilibration solutions freshly, and remember that iodoacetamide is light sensitive (protect with an aluminum foil).
5. Make fresh 10% ammonium persulfate solution. Add 10% ammonium persulfate and TEMED to the gel mixture just before casting it.
6. Low-melting agarose is solubilized by slight boiling using a microwave oven just before use.
7. Add formalin just before using the impregnation solution. Silver nitrate is light sensitive; therefore, the solution must be protected from light.
8. Sodium thiosulfate and formalin are added extemporaneously to carbonate solution.
9. Transwell™ cell culture insert is pressed to be sticked to the bottom of the Petri dish. No air bubbles should be present between the filter and the Petri dish. This procedure avoids perforating of the Transwell™ filter.
10. Add 10 mL (with pipette aid coupled to aspirating pipette) of CaMg–PBS on both sides of the filter and repeat the same procedure for all washes. To keep cell integrity, do not pour the CaMg–PBS out directly on cells, but pour it out along the dish side.
11. Withdraw the supernatant and then invert the conical tube on a bench coat to eliminate the remaining liquid.
12. Cleanliness of glass plates is very important. They should be washed with soap (protein-free soap) to remove all waste of previous gel. They should be rinsed in water and dried with 95% ethanol. All gel-casting systems are handled with clean powder-free gloves, to avoid contamination with other proteins such as keratins.
13. Suggestions: wear powder-free gloves, never touch the gel with your fingers, use fresh formalin solution, and make all solutions freshly for the silver staining procedure.
14. A reference gel is an experiment gel that reaches several criteria. Ideally, a good reference image should show a clear and representative spot pattern, with a minimum of distortion, and show good isoelectrically focused spots for all molecular weight and isoelectric point ranges. For experiments such as kinetic or dose response, choose the middle points that tend to give the best results.
15. Cut a control piece of gel from a blank zone of the 2D gel and process it in parallel with the samples. This control will reveal if the gel is keratin free and will constitute a MS blank spectrum.

16. When dehydrated, the gel pieces have an opaque white color and are significantly smaller in size.
17. Compared with the use of microcolumns or ZipTip tips, the anchorchip™ technology results in some improvements over the original protocol. This target offers sensitivity advantage without loss of sample because it allows sample upconcentration by reducing the surface area of the deposit even when applying rather large sample volumes and a separate desalting step (therefore, the desalting step becomes useless prior the MALDI target loading).
18. In cases where protein identification is based on a single sequenced peptide, special caution has to be taken. The amino acid sequence identified by the Mascot search engine has to be manually checked to ensure that the MS/MS spectrum truly corresponds to the peptide sequence predicted by Mascot. Manual validation is, therefore, essential to avoid a large number of false positives in the list of identified proteins.
19. Probability based MOWSE Mascot score (in our case): the protein score is calculated using the following formula: mascot score = −10*Log(P), where P is the probability that the observed match is a random event. Therefore, typically, protein scores greater than 61 inform the users that the identifications are statistically significant ($p < 0.05$).

References

1. Cecchelli, R., Berezowski, V., Lundquist, S., Culot, M., Renftel, M., Dehouck, M. P., and Fenart, L. (2007) Modelling of the blood–brain barrier in drug discovery and development. Nat Rev Drug Discov 6, 650–61.
2. Abbott, N. J. (2002) Astrocyte-endothelial interactions and blood–brain barrier permeability. J Anat 200, 629–38.
3. Abbott, N. J., Ronnback, L., and Hansson, E. (2006) Astrocyte-endothelial interactions at the blood–brain barrier. Nat Rev Neurosci 7, 41–53.
4. Hawkins, B. T., and Davis, T. P. (2005) The blood–brain barrier/neurovascular unit in health and disease. Pharmacol Rev 57, 173–85.
5. Reese, T. S., and Karnovsky, M. J. (1967) Fine structural localization of a blood–brain barrier to exogenous peroxidase. J Cell Biol 34, 207–17.
6. El-Bacha, R. S., and Minn, A. (1999) Drug metabolizing enzymes in cerebrovascular endothelial cells afford a metabolic protection to the brain. Cell Mol Biol 45, 15–23.
7. Ballabh, P., Braun, A., and Nedergaard, M. (2004) The blood–brain barrier: an overview: structure, regulation, and clinical implications. Neurobiol Dis 16, 1–13.
8. Pardridge, W. M. (1999) Blood–brain barrier biology and methodology. J Neurovirol 5, 556–69.
9. Bickel, U., Yoshikawa, T., and Pardridge, W. M. (2001) Delivery of peptides and proteins through the blood–brain barrier. Adv Drug Deliv Rev 46, 247–79.
10. Pardridge, W. M. (2002) Drug and gene targeting to the brain with molecular Trojan horses. Nat Rev Drug Discov 1, 131–9.
11. Neuwelt, E., Abbott, N. J., Abrey, L., Banks, W. A., Blakley, B., Davis, T., Engelhardt, B., Grammas, P., Nedergaard, M., Nutt, J., Pardridge, W., Rosenberg, G. A., Smith, Q., and Drewes, L. R. (2008) Strategies to advance translational research into brain barriers. Lancet Neurol 7, 84–96.
12. Meresse, S., Delbart, C., Fruchart, J. C., and Cecchelli, R. (1989) Low-density lipoprotein

receptor on endothelium of brain capillaries. J Neurochem 53, 340–5.

13. Cecchelli, R., Dehouck, B., Descamps, L., Fenart, L., Buee-Scherrer, V. V., Duhem, C., Lundquist, S., Rentfel, M., Torpier, G., and Dehouck, M. P. (1999) In vitro model for evaluating drug transport across the blood–brain barrier. Adv Drug Deliv Rev 36, 165–178.
14. Booher, J., and Sensenbrenner, M. (1972) Growth and cultivation of dissociated neurons and glial cells from embryonic chick, rat and human brain in flask cultures. Neurobiology 2, 97–105.
15. Pottiez, G., Sevin, E., Cecchelli, R., Karamanos, Y., and Flahaut, C. (2009) Actin, gelsolin and filamin-A are dynamic actors in the cytoskeleton remodelling contributing to the blood brain barrier phenotype. Proteomics 9, 1207–19.
16. Westermeier R, Naven T, and H-R, H. (2008) Proteomics in Practice (Second Edition), Wiley-VCH, Weinheim.
17. Rabilloud, T., Carpentier, G., and Tarroux, P. (1988) Improvement and simplification of low-background silver staining of proteins by using sodium dithionite. Electrophoresis 9, 288–91.
18. Shevchenko, A., Wilm, M., Vorm, O., and Mann, M. (1996) Mass spectrometric sequencing of proteins silver-stained polyacrylamide gels. Anal Chem 68, 850–8.
19. Katoh, I., Yoshinaka, Y., and Luftig, R. B. (1984) Murine leukaemia virus p30 heterogeneity as revealed by two-dimensional gel electrophoresis is not an artefact of the technique. J Gen Virol 65 (Pt 4), 733–41.

Chapter 11

MALDI Imaging Technology Application in Neurosciences: From History to Perspectives

Michel Salzet, Céline Mériaux, Julien Franck, Maxence Wistorski, and Isabelle Fournier

Abstract

Dynamic properties of the nervous system can now be investigated through mass spectrometry technologies. Generally, the application of these powerful techniques requires the destruction of the specimen under study/examination, but recent technological advances have made it possible to directly analyze tissue sections and perform 2-D or 3-D molecular ions mapping. We review from history to perspective matrix-assisted laser desorption/ionization (MALDI) imaging technology and its application to the analysis of molecular distributions of proteins and peptides in nervous tissues of both invertebrates and vertebrates, focusing in particular on recent studies of neurodegenerative diseases, and early efforts to implement assays of neuronal development.

Key words: Invertebrates, MALDI mass spectrometry imaging, Mammals, Nervous system regeneration, Neurodegenerative diseases, Neurogenesis

1. Introduction

Almost 100 years of constant research and discovery related to progress in electronics and informatics have been instrumental in making mass spectrometry the versatile tool of today. The ability to obtain the measurement of molecular weight of compounds with an increasing level of precision is one of the first steps for identifying molecules. The addition of primary structural information and more recently the secondary structure has made mass spectrometry a very complete technique for identification when analyzing a complex mixture. Biological systems are complex systems with highly heterogeneous compositions, making them inherently difficult to analyze. The road is still long with regard to obtaining reliable

Yannis Karamanos (ed.), *Expression Profiling in Neuroscience*, Neuromethods, vol. 64, DOI 10.1007/978-1-61779-448-3_11, © Springer Science+Business Media, LLC 2012

models of such systems. Thus, it is of primary importance to first identify the constituents of these systems, and then understand how they work in different contexts leading to synergy. Facing such a difficult task, mass spectrometry is now a centrepiece technology, with a more preponderant place than just simply in proteomics. A further understanding of the role of the constituents of biological systems can be aided by the knowledge of where these constituents are distributed. This dimension can now be added by performing mass spectrometry imaging (MSI). However, the imaging dimension is only allowed with a restricted number of sources used for ion production, such as laser desorption ionization (LDI), secondary ion mass spectrometry (SIMS) or matrix-assisted laser desorption ionization (MALDI). The purpose of this review is to define MALDI-MSI, the evolution of this technology as well as the actual and future applications.

2. From Classical Mass Spectrometry to MSI

Since its introduction in the late 1980s, MALDI has become a tool of choice for the analysis of biomolecules, especially large compounds. With a growing need for molecular information on peptides and proteins and because of the ease of MALDI ion sources for the analysis of crude samples, direct tissue analysis was quickly attempted. Various reports of peptides analysis from different invertebrate's cells or organelles have been made since 1994 (1), and many of these concern the study of peptide processing from protein precursors. Taking advantage of the great sensitivity of MALDI mass spectrometry, single cell direct analysis was also successful and allowed the study of specific cell types including neurons. Direct analysis offers the great advantage of studying cells in their original context, while avoiding lengthy steps of purification and separation. In the search of better direct analysis, it was necessary to take advantage of the knowledge of other mass spectrometry techniques such as LDI (2) or SIMS (3), and thus the link to achieve MALDI-MSI was almost complete. MALDI, as with other desorption techniques, was developed from physics where imaging sample surfaces is very common. Indeed, a number of different imaging technologies have emerged on the contrary of other mass spectrometry techniques that find their roots in chemistry which is a field where mostly solutions are studied and for which morphological aspects is generally not of prime important. This missing step was developed at the end of 1990s by the group of Caprioli (4–6) reporting that data acquired on biological samples can be processed to reconstructed ion density curves or molecular images of biomolecules in tissue samples.

3. MALDI Imaging and the Basis of Protein Molecular Imaging

In certain ways, MALDI-MSI is a simple concept. In MALDI, ions are promoted by laser irradiation of a solid sample that is theoretically a homogeneous solid solution constituted of matrix molecules in large excess with analytes. Processes that lead to ion formation in MALDI have progressed since the introduction of the technique; however we are still not able to fully describe the mechanisms that are implicated. This unfortunately means that an important part of our knowledge is empiric and it is therefore difficult to predict the properties of good matrices. What we understand is that transferred energy by short pulses of photon irradiation induces energy transfer to matrix molecules (7, 8) that reach excited states and that relaxation involves dissipation of at least one part of this energy in the sample. Basic studies performed to examine the irradiated area on matrices single crystals demonstrate that the crater formed by the removal of material does not exceed in size the area of the incident laser beam (9, 10). Most of the energy is transferred in the depth of the irradiated sample with very little sideway spread. It follows that if a sample is irradiated at a precise coordinate, the ions will directly come from the area defined by the cross section between the sample surface and the incident laser beam and this can be controlled by changing the laser focusing. This is an important consideration for MALDI-MSI because it allows correlating laser irradiation coordinates to a corresponding position on the sample. On this basis, it means that the acquisition of a complete image only requires that spectra need to be regularly acquired at the surface of the sample by moving the sample under the laser beam in such a way to cover the whole surface. Each collected spectrum, represents the average of several laser shots in order to obtain a statistical representation of the analyzed area. MALDI-MSI in its most current form is a point-to-point analysis.

MALDI-MSI is composed of four main steps (Fig. 1).

1. Tissue sectioning and preparation
2. Matrix deposit
3. Data acquisition
4. Data processing

The first step concerns tissue sectioning and preparation, which is carried out by traditional means. The main concern is the preservation of tissue integrity while avoiding molecular composition changes, for example, due to enzymatic activation. There is a subtle equilibrium of being able to preserve tissue in a precise state of molecular composition and localization from the time of dissection to its apposition on the experimental plate. As we will discuss later, this is a problem if tissue needs to be kept on long period before analysis.

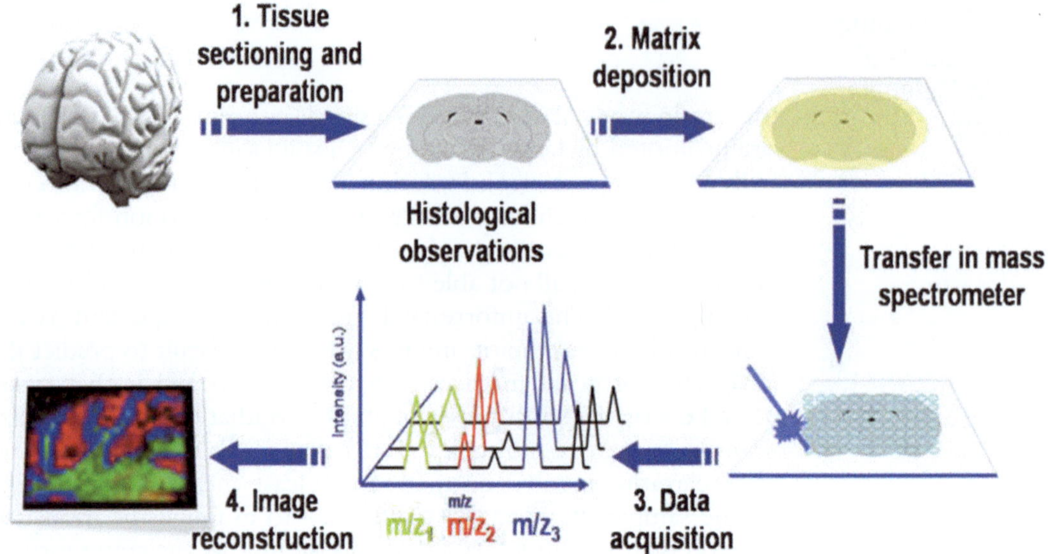

Fig. 1. Schematics workflow for MALDI-MSI experiment.

The second and third steps are based on MALDI-MS technology. The second step, crucial in the analytical sequence, is the deposit of the matrix on the tissue section. The importance of matrix on mass spectra quality is well acknowledged. Choice of matrix and its preparation is a key for successful MALDI. However, in the case of MALDI-MSI, it is very important to take into account, that contrary to classic analyses in solution, the deposit of the matrix must not induce any delocalisation of molecules across the tissue sample. The minimum requirement is that delocalisation must not spread further than the area analyzed within one laser shot. The third step is the acquisition of the data from the tissue. This step is dependent on the mass analyzer used and involves variable parameters which are set in order to achieve the best analytical capacity. The acquisition step requires automation of the analytical process. The final step involves informatics tools that are designed to process the data, including algorithms to reconstruct images.

As with any technology, we must understand the limitations of MALDI-MSI. By making a simple parallel with photography, it is very easy to pinpoint where our efforts should be directed. In photography, the expectation is to reflect an exact copy of a scene viewed by the human eye. The image must have enough definition to see fine details and have enough contrast to observe objects with different sizes, shapes, colours or brightness. Similarly with MALDI-MSI, obtaining a highly resolved image will be defined by the pixel size and the density of pixels of the image. For MALDI-MSI, this means obtaining the smallest irradiated area, while preserving analytical performance and obtaining the highest number of analysis spots, that is, at maximum contiguous pixels, that is, if considering a circular area a distance between two spots that is minored to the diameter of the circle. Being able to obtain an exact copy of whatever we are observing means that we will be able to analyze any class of molecule with the same capacities, independently of the size, amount of this molecule and whatever its place in the system and whatever the kind of sample to be studied. Here are clearly subjacent the challenges for MALDI-MSI: Being able to deal with all samples independently of their preparation and by extension conservation and analyse inside all classes of biomolecules from peptides, proteins to lipids, sugars or oligonucleotides sequences and so even if they are only present at a very low level. Looking at one of the first images performed with MALDI-MSI and publishing and comparing it to a recent image (Fig. 2) directly will help answer some of the questions. And now, a decade after its introduction, MALDI-MSI has greatly improved its capacities and has already broken some of its own challenges by the constant efforts of several research groups really involved in this field.

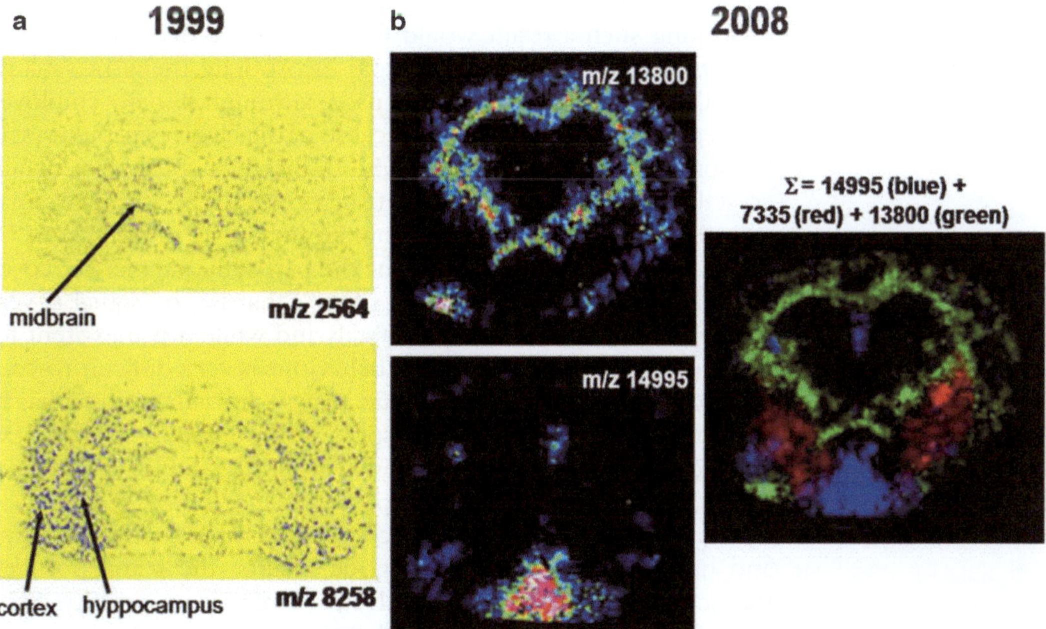

Fig. 2. Evolution of MALDI-MSI from beginning to present day. (**a**) The first MALDI images presented by Caprioli et al. (5, 6) at the *47th ASMS Conference,*1999, Dallas, TX, June 13–17 and (**b**) Molecular images obtained by our group (example of Mice stem cells injected brain tissue sections Wisztorski, Meriaux, et al. Unpublished results).

4. MALDI-MSI a Continuously Evolving Technology

4.1. MALDI Matrix: The Cornerstone of MALDI-MSI

4.1.1. What Is the Best Matrix for MALDI-MSI?

There is insufficient knowledge of MALDI mechanisms to allow theoretical and predictive considerations, and thus we depend solely on empiric experimental data to establish the best matrix. At least three types of matrix are already known for MALDI analysis of biomolecules including sinapinic acid (SA), α-cyano 4-hydroxycinnamic acid (HCCA) and 2,5-dihydroxybenzoic acid (2,5-DHB). While all three can be use for both peptides and proteins analysis, SA is generally preferred for proteins in terms of signal intensity, resolution, signal-to-noise ratio and number of detected compounds, whereas HCCA is generally used for peptides. On the other hand, 2,5-DHB possesses a broader range of analysis and is either used for peptides or proteins. In MALDI-MSI, the situation is more complex due to the interaction between the matrix and the tissue. Based on experimental data where matrix solutions are deposited on a tissue section, it can be observed that HCCA yields good signals for peptides up to m/z 5,000, but only very few and weak signal intensities above this limit. SA yields better signals in the range of m/z 5,000–30,000. However, even if SA is better suited for higher masses on tissue as it is in classical MALDI, mass ranges are more limited on tissue. Thus, when comparing classic solution analysis to tissue analysis, proteins up to a 100,000 Da (e.g., antibodies) cannot be analyzed. This limitation in tissue sections is yet to be overcome. The most likely hypothesis for explaining such a result would be the difficulty to readily extract high mass proteins from tissue and incorporate them into matrix crystals. 2,5-DHB has been less used, although initially employed for peptides/proteins analysis because of the heterogeneous crystallization of this matrix. Indeed, 2,5-DHB, when spotted, generally crystallizes in fine long needles at the rim of the spot leaving small crystals in the inner parts that contain high level of salts. MALDI signal is obtained by irradiating the rim part of the sample. Moreover, 2,5-DHB is known to have "hot spots", that is, the signal is very strong in some parts of some crystals and while it is inexistent for other parts. This type of crystal behaviour is very difficult to consolidate with MALDI direct tissue analysis for which signal has to be reproducible for all studied spots of the sample. Nonetheless, satisfactory results may be obtained when 2,5-DHB is used in a spray or micro-spotting deposit procedure.

Even if classic MALDI matrices are applicable for tissue analysis and have been shown to give good results, the search for new matrices better suited for MALDI-MSI remains a priority. Solid ionic matrices have been found to give good results for tissue analysis. Lemaire et al. (11) showed that novel solid ionic matrices were particularly well suited for peptides analysis up to m/z 10,000.

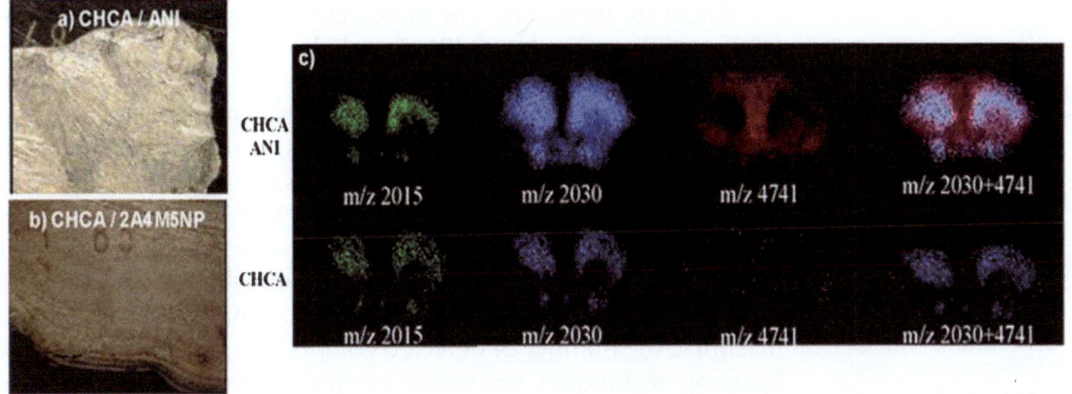

Fig. 3. (**a**) HCCA/Aniline and (**b**) HCCA/2-amino 4-methyl 5-nitropyridine show very homogenous crystallization. (**c**) Comparison of MALDI Molecular Imaging using HCCA/aniline and HCCA.

HCCA/aniline and HCCA/2-amino 4-methyl 5-nitropyridine were best for tissue analysis compared to HCCA, with increased signal intensities, signal-to-noise ratios and number of detected compounds. The principal advantage of these matrices is their homogeneous crystallization (Fig. 3) with very limited delocalisation of peptides/proteins even if they are depositing without any specific system. These studies also showed that ionic matrices are very stable under vacuum conditions and lead to very low material ablation rates allowing the use of the sample for an extended period inside the instrument while performing several acquisitions with no decrease in the performance.

Even if classic MALDI matrices provide good performance, MALDI-MSI analytical performances must be maximized in order to increase the dynamic mass range and the number of analytes observed. Thus the effort to find new matrices continues.

4.1.2. Which Method Is Optimal for Matrix Deposit?

It is clear that adding matrix on top of the tissue sample is the delicate part of sample preparation as several criteria must be fulfilled for successful MALDI-MSI. The first question is whether the solvent used for matrix solubilisation induces a delocalization of analytes in the time needed for crystallization? In most cases, this occurs and needs to be limited. Depositing matrix solution using a micropipette on a tissue section is thus not very satisfactory. In one case, Stoeckli et al. (12) demonstrated that matrix crystals were migrating over 400 μm on the tissue section before solidification. In a typical preparation, matrix crystals can range from 10 to 100 μm. The direct consequence of these observations is that the ideal matrix should have a minimal amount of solvent and produce the smallest size of crystals possible. This point is critical in order to obtain reliable images results as well as image resolution.

There are three different strategies to circumvent this problem (1) spray the matrix solution (2), deposit the matrix solution as discrete spots or (3) blot the analyte on a membrane by transfer (passive or active). Blotting was one of the first strategies tested (13, 14). By adsorbing proteins on membrane, matrix can easily be deposited. Conductive polyethylene membranes were successfully used for direct analysis of a wide range of proteins up to 100,000 Da. Membrane transfers present limitations in term of variability and global transfer yields, especially between different proteins. At the present time, analyses performed directly on the tissues are preferable in order to maximize the number of detected compounds as well as sensitivity. Spraying techniques or matrix application as discrete spots are thus preferred. Spraying of the matrix is now widely used. It is one of the easiest and cheapest ways to carry out MALDI-MSI. Spraying is generally carried out using either a pneumatic sprayer such as TLC sprayers or airbrush systems. Such systems minimize solvent quantities and matrix crystals size but are often lacking in reproducibility as optimal spraying conditions are difficult to control. For peptides/proteins, matrix crystallization must lead to the inclusion of molecules inside the crystals. This is a paradox, since good incorporation of molecules in matrix crystals is better achieved with a reasonable amount of solvent, yet experimental data has confirmed that with little amounts of solvent an important loss of signal is observed. Thus, if spraying conditions are not well controlled, the sample is either too "dry" (not enough solvent) and does not provide good incorporation of analytes in matrix crystals or if the sample is too "wet", this can lead to delocalisation and heterogeneous crystallization. Distance between the sprayer and the sample, flow rate of the spray, time of a spray cycle and number of cycles applied are difficult to control by hand. Therefore, even if in many cases, the microscopic observation of the deposited matrix layer is satisfying, resulting mass spectra are of poor quality in terms of peaks intensity, signal-to-noise or number of detected compounds. Recently, a robot sprayer system was developed by Bruker Daltonics (15) that offers very good reproducibility of the MALDI mass spectra (Fig. 4). This system monitors the growth of matrix layer using the diffraction of matrix crystals combined with a fine control of all parameters allowing for high optimization of spraying conditions for solvents and matrices. In this system, average droplet size is very small (~25 μm). As for discrete deposit of matrix, recent developments have also improved this method which is more complex as it requires deposit of picoliter amounts of matrix according to a raster without mixing in order to cover the whole tissue section (Fig. 4). This strategy requires the use of a robot system that can plot all matrix points. For example, to cover a whole rat brain tissue section of 2 cm by 1 cm requires 5,000 discrete matrix spots. Picoliters of matrix solution are generally not

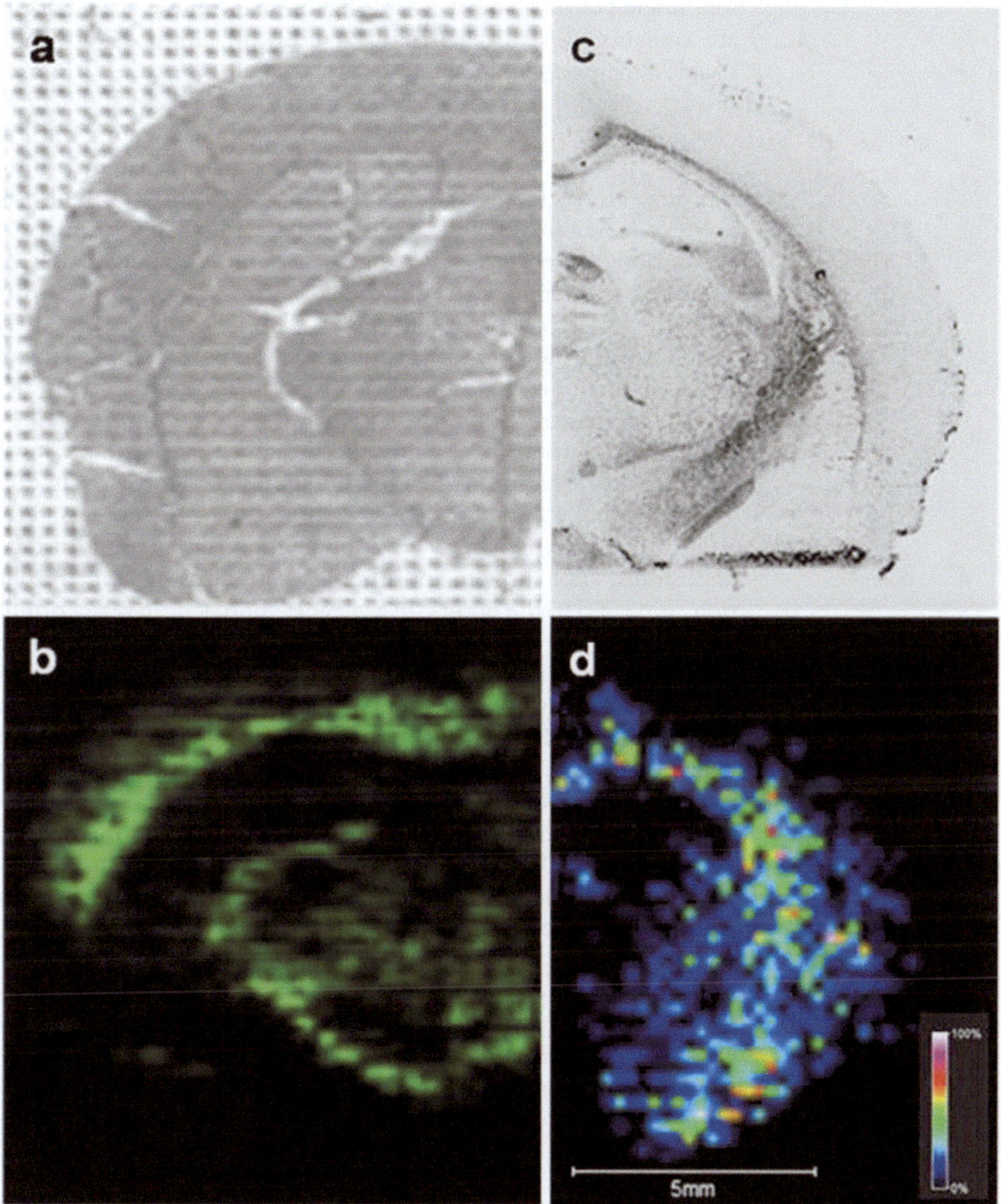

Fig. 4. Image of two sections of rat brain tissue after trypsic digestion and matrix deposit using CHIP-1000 spotter (**a**) optical image, (**b**) MALDI image and ImagePrep sprayer (**c**) optical image, (**d**) MALDI-Image.

sufficient to provide good extraction and incorporation of analytes into matrix crystals and printing systems generally do not allow for a higher concentration of matrix solution (i.e., results in clogging problems of the system). It is therefore necessary to perform several cycles of deposit to obtain good spectra quality but the time for sample preparation becomes long. Several commercial platforms are available to perform picoliter deposit of solution. One of the simplest methods is to use a fuse silicate capillary to generate a very small droplet that is gently deposited onto the samples such as systems used for nano-LC fraction collection. The solution volume and the spot size is then directly dependent on the internal diameter of the capillary. Piezo electric systems used for micro arrays, can also be used for MALDI-MSI. Finally, a system based on acoustic ejection of solutions presents the advantage of avoiding clogging problems. Each of these strategies produce spot sizes ranging from 100 to 500 μm but typically resolutions of 150–200 μm are used for MALDI-MSI. Discrete spots application can be quite long (several hours) and require conditions to be very finely optimized for each solvent and matrix solution in order to obtain optimal MALDI spectra. At a first glance, it may be thought that discrete deposit provides less resolution due to the spot size. However, it does guarantee that compound delocalization cannot be more than the size of the spots. Nonetheless, experimental results using discrete deposit does not show major differences in terms of image resolution compared to the others. In a cheaper approach, an inkjet printer with refilled ink cartridges with matrix solution can be used for matrix application (16). This approach gives good results for monitoring lipids with 2,5-DHB as matrix but was unsuccessful for deposit of HCCA that required to be solubilized in a high percentage of acetonitrile. Alternatively, matrix application by sublimation is also suitable for MALDI imaging (17). This method was previously used for analyzing polymers and presents the advantage of being solvent-free. Matrix application is obtained by heating the solid matrix until sublimation occurs and deposit occurs by recondensation on the tissue sample. This strategy provides extremely fine and homogeneous layers of matrix while avoiding delocalization of samples, leading to highly resolved images. Although, this method is well adapted for lipids detection, it is unusable for peptides and proteins, as the MALDI process requires a real incorporation of peptides and proteins in matrix crystal for the desorption/ionization process to be successful (i.e., peptides/proteins are too polar to be easily transferred into the gas phase).

4.2. Sample Preparation: Tissue Conservation and Imaging Strategies

In molecular imaging technologies, preparation and conservation of samples is crucial. Conservation must ensure that molecules do not move or undergo degradation during conservation. Conservation methods are well established for histology, although compatibility for MALDI-MSI conservation can give rise to analytical difficulties. Samples can be preserved under three forms after dissection, including direct freezing (in isopentane for better preservation of structure integrity), freezing after use of a fixative (formalin, Bouin, Bouin Hollande) and conservation after fixation followed by paraffin embedding. Directly frozen samples create the least analytical problems for mass spectrometry analysis. However, frozen samples must be cut on a cryostat instrument and require storage in freezer systems, preferably at –80°C.

4.3. Acquisition Time

Acquisition time of data for images reconstruction is a parameter sustained by the user with little choice. Acquisition time is principally dependent upon the spatial resolution and the laser repetition rate and does benefit from developments of new laser with higher performances.

Presently lasers with repetition rates up to 1,000 Hz are available on commercial MALDI-TOF systems. This can also be limited by the mass analyzer duty cycle time and the number of duty cycles to be performed for good spectral quality. For TOF analyzer, only one cycle per laser shot is required and limitations due to the speed capacities of the TDC (Time to Digital Converter) system. Therefore, if the mass range to be recorded is too broad, laser shots repetition would become faster than time to digital conversion. In such cases, it is necessary to either decrease the acquisition mass range or the laser repetition rate. On IT, limitations can be the number of duty cycles to be performed to record a mass spectrum.

Typically MALDI-MSI works in a microprobe manner and therefore the acquisition time is dependent on the size of the sample. MALDI-MSI data collection can range from 15 min to greater than 24 h, depending on the size of the sample, the number of points to be acquired (i.e., spatial resolution), the laser repetition rate and the time for moving the sample between each analysis point. For example, for a rat brain section measuring 15 × 10 mm, with a raster of 100 μm in *x* and *y* direction, such that 150 × 100 spots (i.e., 15,000 spots to be analyzed), with 300 laser shots for each spot to insure good statistics, then this acquisition time would be >20 h using a 10 Hz laser repetition rate. With a 1,000 Hz laser rate, the same spectra could be obtained in less than 1 h. If better images are required, that is, by decreasing the footstep between two points, or by increasing the number of laser shots to be averaged, then the time will again increase. If MALDI-MSI is to be an important application in pharmaceutical industry or health research, then acquisition times need to be minimized. The new generation of MALDI-TOF/TOF is now selling with a 1,000 Hz laser rate and the next will be 5,000–10,000 Hz.

4.4. Image Resolution

As previously mentioned, image resolution is dependent on the size of images pixels and the number of pixels in the area studied. The number of pixels is limited by the precision of the (*x*, *y*) table controlling the sample plate. (*x*, *y*) tables present very high precision that insures a good reproducibility within a sample of 1–2 μm. It is also limited by data-processing capacities. Indeed, each pixel added to a sequence corresponds to a full mass spectrum of data that contains thousands of values for intensity versus m/z. The size of the data will thus depend on the mass analyzer used for the MALDI-MSI sequence. Huge amounts of data are generated for an imaging sequence no matter the type of instrument used. For example, an image with a raster of 100×100 µm of a whole rat brain section (roughly 1.5–2 cm by 1–1.5 cm) can take more than six Go. Present computers equipping mass spectrometers are not well adapted to treat such enormous amounts of data and specific systems must be used.

The real challenge for improving MALDI-MSI resolution is to decrease the size of the pixel, that is, decrease the area irradiated by the laser. For typical MALDI analysis, the size of the irradiated area is not of great concern and does not require to be minimized. Thus, most existing systems, whether home-made or commercial instruments, generally are provided with a laser focused to achieve an irradiated area of roughly 100–150 µm in diameter. This produces images with a resolution R of 10,000 pixels/cm^2. In the field of mass spectrometry, this is a good starting but for biologists this resolution is insufficient as this corresponds a least to an irradiation area of 5×5 cells at once. A single cell resolution for MALDI-MSI would require at least 20–25 µm. It is noted that the number of pixels N is proportional to R^2. Thus to obtain the maximum information, an increase by a factor of 2 in resolution will lead to an increase of 4 the number of pixels and data to be processed. The only direct way to decrease size of the irradiated area is to better focus the laser beam. While this is may be a challenge, it may well be achievable for laser physicists. A very high resolution system was tested by Spengler and colleagues for MALDI-MSI (18). Using a very specific set up of lens that were added outside and inside the instrument on the laser beam pathway, a minimum resolution of ~0.5 µm of pixel size was theoretically reachable. Experimentally, the resolution was very close to this value for classic conditions of laser fluence and a homogeneous laser beam; however, it can be larger under specific conditions. This resolution is far below the size of an average cell and would allow for sub-cellular imaging and the relatively fine observation of organelles. While such fine resolution is possible, actually obtaining data for peptides and proteins from tissue sections at such a resolution is another matter. The first consideration is that decreasing the pixel size is also an important decrease in the number of molecule copy available for analysis, meaning that taking into account the amount of proteins available on such a surface will probably be far under our actual detection threshold. Moreover, basic studies have shown that a drastic decrease in the ion yield is observed when decreasing the size of the irradiated area (19). Practically a size of 50 µm is reasonable but ion yields are well below this limit. MALDI is well known for its poor ability to produce ions. Only a small fraction of the molecules ejected from the solid will be present as ions for mass analysis. Even if our knowledge of MALDI processes has considerably improved over the past decade, we have not yet been able to increase ion yield. Thus, MALDI-MSI resolution could only progress from fundamental understanding in this field. Presently, dedicated systems for MALDI-MSI present better focusing system of the laser beam capable of reaching up to 10 μm pixel size and should at least provide 50–75 μm pixel size while maintaining good analytical capacities (20). Recently Chaurand et al. (21) have designed a new system that allows obtaining proteins images at a resolution of 10 μm (Fig. 5). The experimental set up combines a very carefully drawn system of focusing lens and an iris aperture for finely controlling the laser beam size with a source geometry using co-axial illumination of the sample reducing radial distributions of ejected molecules and ions formed and increasing uptake of the ion for the mass analysis. Although, such systems are only in a developmental phase and are not commercially available, other systems attempt

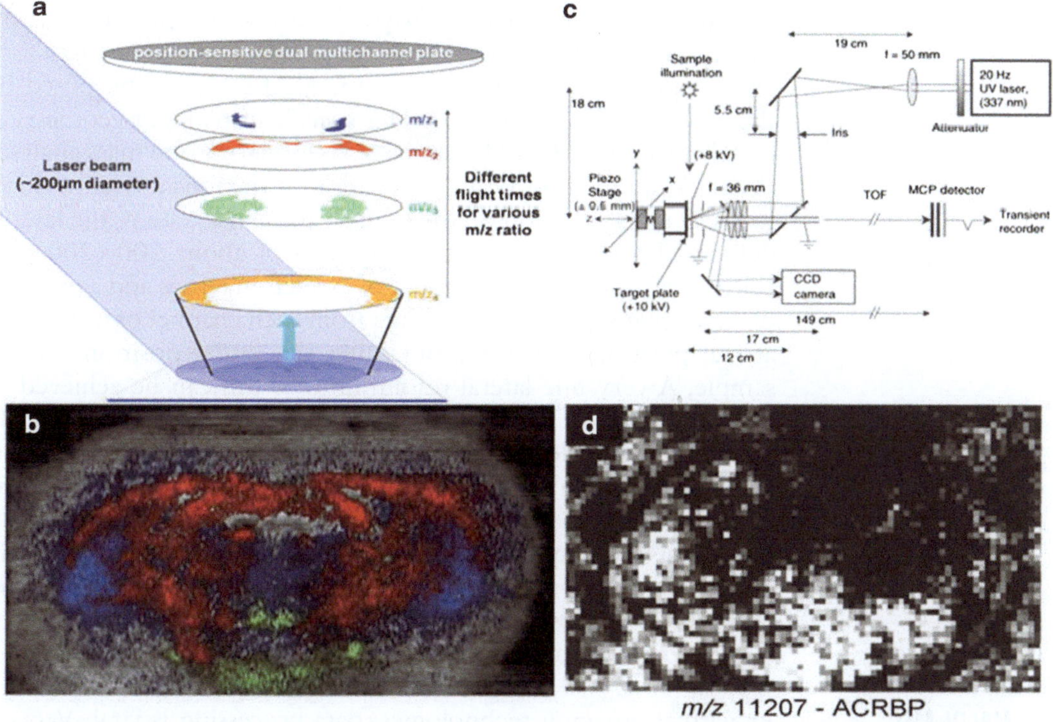

Fig. 5. Comparison of stigmatic mass microscope and a specific MALDI source design for imaging mass spectrometry with a scanning resolution of 10 μm. (**a**) Schematics of the approach using a two-dimensional detector on MALDI Stigmatic imaging. (**b**) High-resolution microscope mode total-ion-count (TIC) image of a rat brain tissue section with overlaid microprobe *m/z* data (*m/z* 1,085 (vasopressin) in *green*, *m/z* 2,030 in *red* and *m/z* 1,431 in *blue*) from (155). (**c**) Schematic of the MALDI TOF mass spectrometer designed for high-resolution imaging mass spectrometry from (21). (**d**) Ion density maps for a secretory epididymal proteins, highly expressed within the lumen of the epididymal tubule ACRBP using this new MALDI source.

increasing images resolution using alternative methods. One way was proposed by Sweedler et al. (22) by performing an overlapping of pixels during the data acquisition. If the sample is irradiated sufficiently so that all matrix material is removed from a spot, and if the sample is moved by only a fraction of irradiated area diameter, for example, half the diameter in a simple case, then for the consecutive spot material and ions can only originate from the half part that was not previously irradiated. This very simple method allows for increasing spatial resolution up to about 25 μm and can be used on all existing instruments. This, however, results in an important increase in the time of acquisition. Other methods that have been designed to be adaptable to all instruments are currently under study. Wisztorski and col. have proposed using mask systems. Again, a very complex system of laser focusing is not necessarily required if covering the tissue with a mask presenting a network of apertures of defined size. The laser beam is simply cut off when getting through the mask system and the irradiated area can be very well controlled. These systems allow a reduction in pixel size up to 30 μm but maintain a high enough ion yield for analyzing peptides and proteins. Masks systems are of great interest as they lead to an increase in the ion yield with some specific aperture shapes. Finally, a very different system of MALDI-MSI in ion microprobe mode, the so-called "stigmatic MALDI", was proposed by the group of Heeren (Fig. 5). This specific instrument geometry is closely related to the old systems of laser microprobe mass analyser (LMMA) and is based on a correlation of the ions arrival position to the original position on the sample. In such a system, the laser is defocused in order to irradiate an area of about 200 × 200 μm. With a very specific arrangement of extraction lenses and a position detector that separately treats data from each channel of its surface, arrival position allows for obtaining the initial position on the sample. A very fine lateral resolution of 4 μm can be achieved in the 200 × 200 μm^2 area in a timeframe of less than 1 ms since only one shot acquisition is needed for this surface. Reconstructing a whole image with high resolution will thus only require contiguously acquiring small areas over the whole sample and then summing them. This technique also has a great potential as shown by recently published reports (23) studying the distribution of peptides in the rat pituitary tissues. This unique system has not yet been setup for high throughput applications.

4.5. MALDI-MSI and Bioinformatics

As with all imaging technologies, data processing is vital. Various software exists for image reconstruction with no dedicated solutions ware (or available) for imaging. Even in the early years of MALDI-MSI, automation of acquisition was proposed on most instruments. Even if they could be used for acquisition, they were not well suited and considerable time was lost for setting the acquisition. Moreover, no tools were available for post-acquisition data processing. The simplest software required to perform acquisition of data with the specific instrumental settings according to a raster defined by the user, to create an average spectrum from all the collected spectra, to extract the intensity of a specific m/z value from each data on user inquiry and to report the intensity in a colour scale as a function of the corresponding (x, y) coordinates on the sample. First usable software following these steps was used by Stoeckli and colleagues. Stoeckli tested several different software allowing for full automation of the data acquisition for different MALDI-TOF instruments. "Biomap", a software dedicated to images reconstruction from whatever MALDI data obtained, offers much different functionalities and is available for free at http://www.maldi-msi.org/.

Optimal data processing can be the result of very different solutions with the final goal of producing images that are nearest of reality. But is reality better described by measuring intensity or peak areas? Must mass spectra data be normalized? Must a minimum threshold of intensity be defined? Must a maximum intensity be defined? How well can a mass calibration be performed? Can image resolution be increased by extrapolation of signal between two data points? What should be the algorithms? Can better tools be designed for biological applications using classification parameters? Overall, two different processing issues must be distinguished, including the necessary processing that best describes the data and the statistical processing for sample classification when performing differential display analysis. Optimization of this work is balanced between obtaining increasing information extraction and inducing false information.

The influence of data processing and classically used data processing are discussed in a recent paper (24). Statistical analysis and clustering is already used to process proteomics data. Clustering methods were used in MALDI-MSI and direct analysis for prospecting human glioma and other brain tumours types compared to normal brain samples (25). The study demonstrated that tumour tissue can be easily discriminated from non-tumour ones with very unique protein profiles that can be extracted by statistical analysis. Moreover, the different grades of tumours were also differentiated combining molecular profiling and clustering. Assessment of protein patterns to disease was reviewed by P. Chaurand et al. (26). Multivariable analysis and clustering can also be used to find regions of interest (ROI). ROI represent the area on the sample were some of the analyzed molecules are differentially expressed in a specific sample compared to another. An image can then be reconstructed from the whole signals, weighed to better observe the differences between the samples and help to identify the spatial correlation of the mass spectra (27). Very recently, researchers have also demonstrated the applicability of principal component analysis (PCA) algorithms for MALDI-MSI (28–30)

5. MALDI-MSI: A General Technology for All Types of Biomolecules?

Endogenous molecules constituting cells represent a large variety of compound families ranging from peptides-proteins, oligonucleotides (DNA, RNA), saccharides, lipids, salts and small organic compounds such as neurotransmitters, ATP, ADP, NO etc. Each of these families is in itself composed of molecules that can present a very large range of physico-chemical properties in term of polarity, hydrophobicity, solubility, molecular mass, and acido-basic properties. A perfect molecular imaging technique should be able to give the distribution of all these compounds at equal levels. None of the tools available are be capable of such a task, but a full understanding of living systems requires a better understanding of the interactions between these very different families. It is therefore worthwhile attempting to analyse them with mass spectrometry.

5.1. Imaging of Peptides and Proteins

Because of MALDI's capacity to obtain ions from compounds of various polarities and to reach very high molecular masses, MALDI-MSI is naturally well designed for monitoring peptides and proteins. Moreover, peptide and protein analysis is currently the main field of applications for such an ion source. The specificity of MALDI compared to other MSI techniques such as SIMS-MSI which presents a very high resolution (1 μm lateral resolution routinely) but which is only well adapted for studying small organic compounds such as lipids and does not allow for analysing peptides or proteins. Thus, it was natural that first efforts and attempts were dedicated to improve MALDI-MSI in the proteomic field. However, even in proteomics, MALDI-MSI has limitations. One of the most striking limitations is the mass range. Experiments have shown that a clear limitation in the mass range is observed for direct analysis of proteins from tissue section with a cut-off near 30,000–35,000 Da which is very different from classic MALDI. The reason for this limitation is not yet clear. Reasonable hypotheses are difficulties of incorporation of higher mass proteins in matrix crystals that are less soluble in the solvent used for matrix solubilisation. More fundamental reasons involve less energy transfer of these proteins from tissue samples preventing the desorption process to occur.

5.2. Imaging of Lipids

Small endogenous compounds can represent a difficulty for MALDI for either practical or fundamental reasons. Compounds of Mw < 1,000 Da can be difficult because of the presence in MALDI of matrix signals which are highly abundant and numerous. On the other hand, some of these compounds are non polar which raises difficulties in the formation of enough ions. Saccharides are very difficult for MALDI in general and have not yet been analyzed by MALDI-MSI.

After proteins, lipids are the family of endogenous molecules that are the most studied. Many efforts were made by the groups of Woods (31–35) and Yost (36, 37) for improving MALDI direct analysis of lipids. Lipids are very complicated to identify from tissue sections as lipids masses can present very small variation even if their composition is very different and they do not even belong to the same family. For example, phosphatidylserine PS 40:6 (Mw = 835.54 Da) and ST 20:0 (Mw = 835.59 Da) are from very different families of lipids but present only a mass difference of 0.05 Da. Moreover, several lipids present the exact same Mw because they do have the same atomic composition and will only vary in their structure by the position of unsaturated bond. Moreover, the molecular weight of lipids are generally lying in a range of a 100–1,000 Da and their signal detection will be hampered by the presence of matrix peaks. For the direct analysis of lipids, imaging protocols used for classic MALDI have been shown to work, that is, using 2,5-DHB as matrix or more specific matrices such a DHAP (2,6-drihydroxyacetophenone) or ATT (6-aza-2-thiothymine). However, it was shown that in the case of DHAP, the matrix was unstable under vacuum and therefore this matrix is not well suited for MALDI-MSI experiments (35, 38). However, it was shown that the addition of heptafluorobutyric acid (HFBA) in the matrix solution increases the stability of DHAP under vacuum allowing MALDI-MSI experiments (39). Recently, several groups have shown that the use of ionic matrices could improve the detection of some class of lipids including gangliosides (40) and phospholipids (41). Whatever the matrix used, analyses are conducted in both positive and negative mode depending on the class of lipids searched.

Great care has to be taken for lipids with matrix deposit because such small organic compounds are easily spread out on the tissue section. For this reason, matrix is generally deposited using a pneumatic spray system or an airbrush for obtaining homogenous crystals repartition while avoiding big quantities of solvents. Other strategies were found to be efficient including solvent-free procedure (42–44) or matrix application by sublimation (17, 43) providing very homogenous matrix coverage on tissue. Concerning microspotting preparation, the use of liquid ionic matrices was found to greatly decrease the time for sample preparation which is generally the critical point with this procedure (41).

Finally, lipid imaging generally requires MS^2 analyses for confirming identification and real assignment. As mentioned, lipids are very small in mass and are well adapted for a variety of mass analyzers. Thus, different instruments have been used for lipids analysis ranging from Q-TOF, TOF-TOF, IT to IM-o-TOF or orbitrap and FT-ICR. All analyzers that present the MS/MS mode are suitable for lipids. Ion Trap (IT) do not give high mass precision but allow for performing MS^n sequence and ease identification by structural elucidation; on the other hand, FT-ICR give highly resolute peaks with a high precision of measure but remain very expensive and difficult instruments. Recently, imaging and direct identification of lipids were carried out on a LTQ Orbitrap leading to the detection of lipids with a very high mass resolving power. Moreover, MS^n experiments were performed with a sub-ppm mass accuracy allowing a better assignment of lipids (45).

5.3. Imaging of Drugs

Imaging of small exogenous compounds such as drugs has recently been developed. Drugs are generally compounds that present a good response in MALDI; although, the same analytical problems as for lipids can be encountered. For drugs, the major difficulty remains interference with matrix ions. It sometimes occur that matrix peaks totally overlap with the drug peak. Therefore, MS/MS is required for determining what part of the signal is to be attributed to the drug. However, the ease of MALDI-MSI as well as its capacity for allowing drugs and metabolites detection, identification and imaging greatly overcomes these analytical difficulties.

5.4. Imaging of Oligonucleotides

Oligonucleotides are also very difficult for MALDI analysis due to phosphate groups with highly complex salts and induce a high instability of the complexes in the gas phase. This leads to the observation of very weak and large peaks with an important decrease with increasing oligonucleotide masses. Under such conditions, imaging of mRNA by MALDI is compromised; even, if some progress was made in this field by studying oligonucleotides under IR-MALDI conditions.

5.5. Towards Specific MALDI-MSI

MALDI-MSI is a powerful strategy that allows for monitoring peptides, proteins and lipids but does not work for oligonucleotides or proteins that exhibit drastic physico-chemical properties. Moreover, the possible correlation of transcripts to their corresponding proteins would be of great advantage for marker validation and pathology prognosis. A novel concept was proposed for imaging a specific mRNA and/or protein (46). This method is an indirect imaging method based on the molecular recognition of a target by a specific probe, using a probe that is modified for MS detection. It is now possible to simultaneously obtain MALDI molecular images specific proteins or genes of interest. The power of such a technique is obtained by multiplexing, using tagged antibodies, tagged lectines for glycoproteins, tagged aptamers for proteins, drugs imaging and tagged nucleotide for transcriptomic studies(46) The method has been demonstrated for proenkephalin mRNA expression in brain, the localization of a membrane enzyme of 180 kDa (46). While this method uses photocleavable tags, another technique (TAMSIM) based on prompt fragmentation in the case of laser desorption has been recently published (47, 48).

TAMSIM is based on an antibody linked with a reporter that is cleaved during laser shots and does not use matrices. This technique has the advantage of not having peptide localization but the disadvantage being less sensitive. It also does not allow producing MALDI images at the same time allowing the detection of the partner surrounding the molecule of interest.

6. Images One Point: Identification of the Major Point

Direct identification of biomolecules is the key point to increase MALDI-MSI potential. The most straightforward strategy consists of identifying molecules directly from tissue section without any procedures involving extraction and separation. Concerning small compounds including lipids or drugs, in situ identification can be easily achieved notably with the use of devices providing high resolving power and MSn capabilities. In the case of proteins, strategies for direct identification have to be developed. The best procedure would be to fragment intact protein in a time scale compatible with the mass spectrometer. For example, by using FT-ICR instrument equipped with a ESI source, electron capture dissociation (ECD) (49, 50) could fragment intact protein presenting high charged state which is not compatible with MALDI sources where ions have a low charged state even for ions generated from proteins.

By taking into account instrumental specificities of MALDI-TOF instrument, in source decay (ISD) (51, 52) is the only approach allowing "Top-Down" experiment. The second strategy requires the development of an in situ enzymatic digestion using micro-spotter allowing "Bottom-Up" experiment.

6.1. Top Down Strategy

Fragmentation along the protein backbone in the MALDI source was observed in first by the team of Brown and Lennon (51, 52). The time scale between ionization of proteins and their extraction from the source is large enough to allow their fragmentation and lead to the formation of z and c fragment ions according to the Roepstorff's nomenclature (53). The N or C-terminus moiety of the protein is then easily achieved and therefore after databank interrogation the corresponding protein could be indentified. The main drawback of ISD is the lack of selection of precursor ion. For this reason, the protein of interest need to be purified to avoid the detection of ISD fragments ions from several proteins in the same mass spectrum. However, due to the fact that ISD fragment ions are detected as intact ions, a pseudo MS3 on these fragments called T3 sequencing (54) can be performed allowing sequencing of the N-terminus or the C-terminus moiety of the protein. Moreover, this strategy could be adapted for MALDI-MSI experiments where mixture of proteins is detected on each pixel. The choice of matrix is a very important parameter for the success of the ISD experiment. Indeed, ISD involves the transfer of a radical proton from the matrix to the proteins and 2,5-DHB was found to be efficient (51). Recently, Demeure and colleagues have shown that 1,5-DAN was more efficient for ISD experiments (55) providing a better fragmentation yield.

Very recently, Debois and colleagues have introduced the in situ identification of proteins directly from porcine eye lens tissue section and rat brain tissue section using ISD strategy (56). The beta-crystalline B2 as protein was identified directly from porcine eye lens from the investigation of ISD fragments ions after data bank interrogation and T3 sequencing. As described in this chapter, a very simple and fast localization and identification of proteins directly from tissue section can be achieved in one acquisition step. However, this strategy suffers from the inability to study formalin-fixed paraffin-embedded (FFPE) tissues on which proteins are reticulated between them. The proteins cannot be directly transferred in gas phase preventing their detection and therefore their fragmentation. On the contrary, the Bottom-Up strategy allows the detection and the identification of proteins whatever the mode of conservation.

6.2. Bottom-Up Strategy

The second approach allowing the direct in situ identification of proteins is based on the classical Bottom-Up strategy. Basically, a solution of enzyme is deposited on a region of interest or on a whole tissue section using a micro-spotter. Peptides are then generated from the digestion of proteins allowing the localization and the identification of the corresponding proteins after MALDI-MSI and MS/MS experiments. This strategy was introduced by Lemaire and colleagues (57) and has allowed the detection and the identification of proteins directly from FFPE tissue sections. The team of Caprioli has then improved the procedure by using a micro-spotter leading to a better and more reproducible application of trypsin on a fresh rat brain tissue section (58). Several teams have then used this procedure notably in the case of clinical application including cancer research (59–61) or model animals of Parkinson disease (62) from FFPE tissue sections. To date, it was clearly demonstrated that to retrieve information from FFPE tissue, the more suitable procedure remains the Bottom-Up strategy. However, many efforts were undertaken to improve the analysis of FFPE tissues from which the in situ enzymatic digestion remains hard to perform owing to the residual hydrophobic feature of FFPE tissue even after paraffin removal. Several strategies involving antigen-retrieval strategies were then developed and allow improving the in situ enzymatic digestion on FFPE tissue (63) and therefore the detection and the identification of proteins.

However, due to the presence of many fragment ions from different series after MS/MS experiment, the corresponding protein is sometimes not identified after databank interrogation. Moreover, on each position, a mixture of digested proteins is detected and therefore no PMF from a specific protein is available. This implies that identification of proteins is based on the fragmentation of one or two peptides without any PMF. This could explain why the corresponding protein is sometimes not clearly identified. To overcome this drawback, an in situ N-Terminal derivatization strategy was recently introduced after the in situ enzymatic digestion in order to orientate fragmentation towards a unique series (64). This way the MS/MS spectra are easier to interpret and therefore the protein assignment is greatly improved. It was shown that the better candidate was the N-(succinimidyloxycarbonylmethyl)-tris(2,4,6-trimethoxyphenyl)phosphonium bromide (TMPP) which allow a fast N-Terminal derivatization of tryptic peptides at room temperature. This reagent leads to the detection of a strong a_i^+ series of fragment ions after MS/MS experiment.

7. From Basic Developments to Neurosciences

The field of clinical proteomics has grown tremendously in the last 10 years and in this field MSI opens the door to histopathology proteomics. The goal of clinical proteomics is to characterize cellular circuitry and to understand the impact of disease and therapy on cellular networks by getting access to information of how the disease is detected, treated, and managed. The major technological advancements that can be done with MALDI-imaging is the direct identification of novel markers and in situ characterization from fresh sections/biopsy embedded in paraffin (e.g., including archived material) (65, 66). Several clinical and pathological studies in neurodegenerative diseases provide evidence that MALDI imaging is a key technology for biomarkers hunting, localization and cross-validation (66–77). The use of archived materials in paraffin blocks from hospital pathology departments represents a "gold mine" of existing information (57). The application of MALDI-imaging to such archived materials could lead to the creation of an international disease marker database, and would allow the elaboration of early diagnostics for various pathologies as well as a follow up in disease progression.

7.1. Invertebrate Nervous System Investigation by MALDI-MSI

The earliest peptide profiling experiments on invertebrate nervous systems using MALDI-MS were carried out on mollusks, first on the gastropod *Lymnaea stagnalis* (1, 78–81) and later on several cephalopods (82). The experimental strategy in these studies was the comparison of peptide mass spectra patterns obtained from different parts of the nervous system, for example, neuronal somata vs. neurohemal organ axon terminals (1, 78–81). This approach resulted in the detection of novel peptides, in addition to peptides previously identified by conventional molecular biological and peptide chemistry methods. In this manner, complex peptide processing and expression patterns could be predicted that were not detected with more conventional methods.

Such a strategy, combining peptide fingerprinting of single neurons by MALDI, molecular cloning, peptide chemistry, and electrospray ionization mass spectrometry, has been generalized to study the intricate processing pattern of a preprohormone expressed in identified neurons. In *L. stagnalis*, some experiments were conducted on neuroendocrine light yellow cells (LYCs) or caudodorsal cells. The LYCs are known to express a precursor, named prepro-LYCP (LYCPs, light yellow cell peptides). Prediction of its processing into three peptides, LYCP I, II, and III, at conventional dibasic processing sites flanking the peptide domains on the precursor, were confirmed by mass spectrometry. However, MALDI analysis of single LYCs revealed trimmed variant peptides derived from LYCP I and II. The variants were much more abundant than the intact peptides, indicating that LYCP I and II serve as intermediates in a peptide-processing sequence (80). Furthermore, MALDI also allows detecting colocalization of novel peptides with the LYCPs (83). Caudodorsal cells of Lymnaea are known to initiate and coordinate ovulation and egg mass production and associated behaviours through the release of a complex set of peptides that are derived from the caudodorsal cell hormone-I (CDCH-I) precursor. Fingerprinting by MALDI of peptides in the commissure demonstrated the presence of all sequenced peptides and, in addition, could identify two other peptides derived from pro-CDCH-1, the beta 1- and beta 3-peptides (80). Recently, Sweedler and colleagues, studying the bee *Apis mellifera* genome, showed that 200 neuropeptides can be predicted, of which 100 were confirmed by mass spectrometry. Moreover, this study opens the door of future molecular studies with the identification of 36 genes, 33 of which were previously unreported (84, 85). Using time-of-flight secondary ion mass spectrometry (ToF-SIMS) and MALDI-MS sample preparation methods, molecular ion maps with a high spatial resolution of cholesterol and the neuropeptide APGWamide were obtained (23, 86). APGWamide was predominantly localized in the cluster of neurons that regulate male copulation behaviour of Lymnaea which is in line with its biological activity (87). Clearly, MALDI peptide profiling gives access to the most complete peptide representation in specific areas, and differential analyses of several distinct areas with MSI yields a representative map of all biomolecules present at one time.

In another recent report, MSI of neuropeptides in crustacean neuronal tissues (pericardial organ (PO) and brain) was used to reveal that two RFamide-family peptides and a truncated orcokinin peptide present distinct localization from those of other members of their respective families. Over 30 previously sequenced neuropeptides were identified based on mass measurement. MSI study at the organ-level study elucidated the spatial relationships between multiple neuropeptide isoforms of the same family as well as the relative distributions of neuropeptide families (88, 89).

In insects, thanks to the genome sequencing, MALDI is now more extensively used. In the honeybee, over 450 neuropeptides have been discovered (85). In addition to the MS methods, MALDI-MSI have been performed and have allowed to localize all these neuropeptides (85) or venoms peptides (90). Similar studies have been performed in *Tribolium casteneum* (91) or *Anopheles gambiae* (92). In *Schistocerca gregaria*, orcokinins, a family of myotropic neuropeptides, have been identified using such a strategy (93).

In leeches, our team has recently begun a series of peptidomic analyses followed by MALDI-MSI experiments of embryonic and adult medicinal leeches (94, 95) (Figs. 6–9), One of our goals was to get information about regeneration processes in adult by suspecting the re-expression of embryonic factors in course of the regeneration process. For this purpose, we first tried to obtain molecular maps from whole mounted, opened embryos at different stages of development, in order to obtain maps of when and where specific proteins and peptides are first expressed and whether such expression is stable or variable in time and space. Peptidomic analyses of whole embryos at different stages of developments show differences between early (6, 8) and later (12a, 12b) embryonic stages (Fig. 6) (96). The peptidomic pattern is completely different with smaller peptides in early stages than in later ones. However, an interesting point is the fact that the peptide pattern is more similar between adult leech cord in course of regeneration and the early stages of embryos. The peptides in common have been identified and are presented in (Table 1) are known to be involved in regeneration in planarian and vertebrates (97–100). Shot gun analyses after trypsin digestion of adult nerve cord after 6 H regeneration compared to whole E12 stages embryos extracts show the presence of a list of proteins implicated in neurogenesis that are re-expressed in adults in course of regeneration processes, for example, LAR-interacting protein (Liprin), chemoattractants factors (IL-16 and C1q related proteins), IgG superfamily molecule (Tractin, LeechCAM), factor affection cytokinesis (Hillarin), actin-binding proteins (filamin, Gliarin, Macrolin), guidance factors (Netrin, syntaxin), antimicrobial-neurotrophic factors (Destabilase, neuromacin), stem cells factor (Lox 2), gap junction proteins (innexins), brain kinase (cytoplasmic SRC) and tyrosine phosphatase receptors (Table 1) (94, 95).

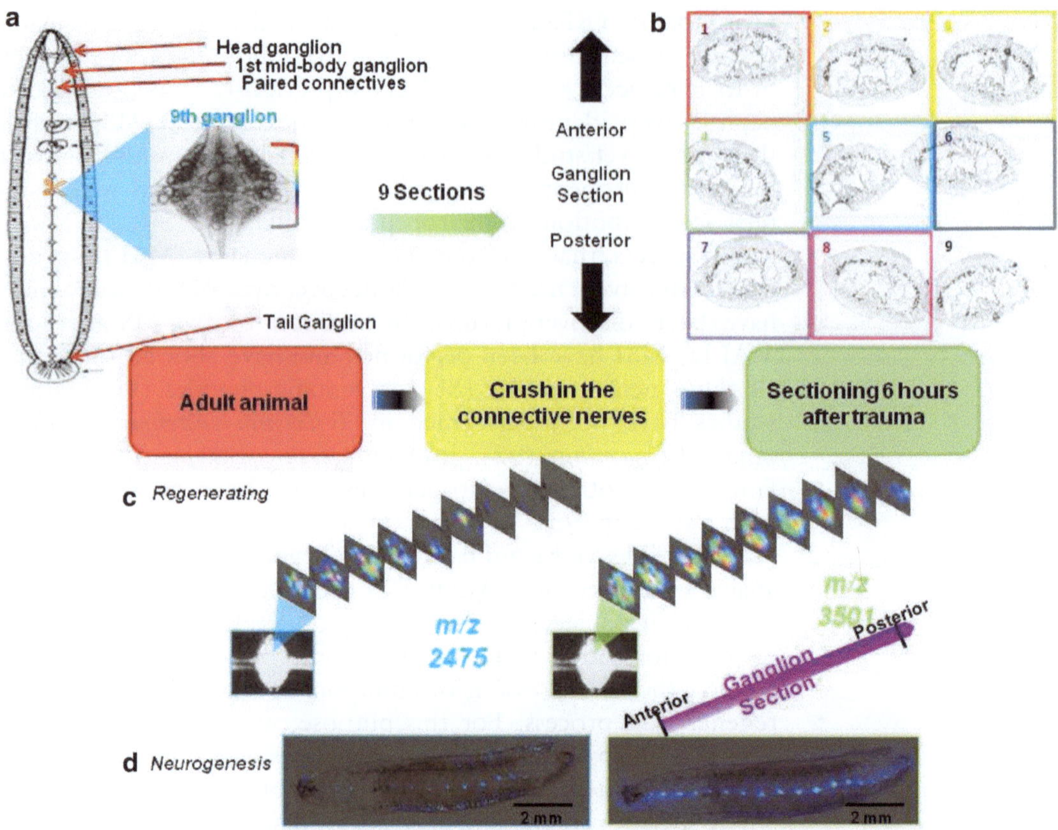

Fig. 6. MALDI-MSI analysis of peptides in sections of regenerating adult CNS. (**a**) Image of the dorsal aspect of a live adult specimen of a medicinal leech (*Hirudo verbana)*, head up (*left part*). Drawing features the ventral nerve cord, from the head ganglion to the tail ganglion, including the 21 midbody ganglia (*right part*). The location of the connective nerve crush, anterior to midbody ganglion 9 (*red scissors*), and the nine cross-sections (panel C2) are indicated. Example of a live midbody ganglion in culture (inset on the *right*). The interganglionic connective nerves and the nerve roots are labeled [19]. (**b**) Two-dimensional representation of all the mass spectra (range *m/z* = 1,000–30,000) corresponding to locations within a ganglion in the nine sections (panel C2) shows variations in protein expression. The spectra are displayed as adjacent parallel lines in bands corresponding to each section (right of the graph). The number of pixels varies among sections, leading to bands of different widths. The spectra are normalized and ionic signal intensity is coded according to the colour scale bar (0%: black to 100%: white) (left of the graph). (Section distribution for individual *m/z* values is diagrammed in Supplemental Fig. 4S). C,D: Expression of ions at *m/z* 2,475 and at *m/z* 3,501 in both regenerating adult CNS segmental ganglia (**c**) and embryonic (**d**). (**c**) Distribution of the *m/z* 2,475 and *m/z* 3,501 ions in sections of the regenerating adult ganglion analyzed in **b**. The inset shows a magnified image of the data with the abundance of the ion colour coded according to the colour bar at right. The peak corresponding to this ion is absent in a control adult, indicating a strong up-regulation of expression following injury. (**d**) Distribution of the *m/z* 2,475 ion in a 12-day-old leech embryo determined by MALDI-MSI of a dorsally-opened, whole mounted specimen. The ion is found at the highest abundance in the segmental ganglia of the ventral nerve cord. Head on the left, tail on the right, dorsal midline on the upper and lower margins of the dissected embryo.

In order to obtain a global map of peptides/proteins that might be involved in leech adult CNS regeneration at the ganglionic level, we performed MALDI MSI of sections of regenerating adult CNS following mechanical damage. Adult experimental animals received a crush in the connective nerves near the anterior margin

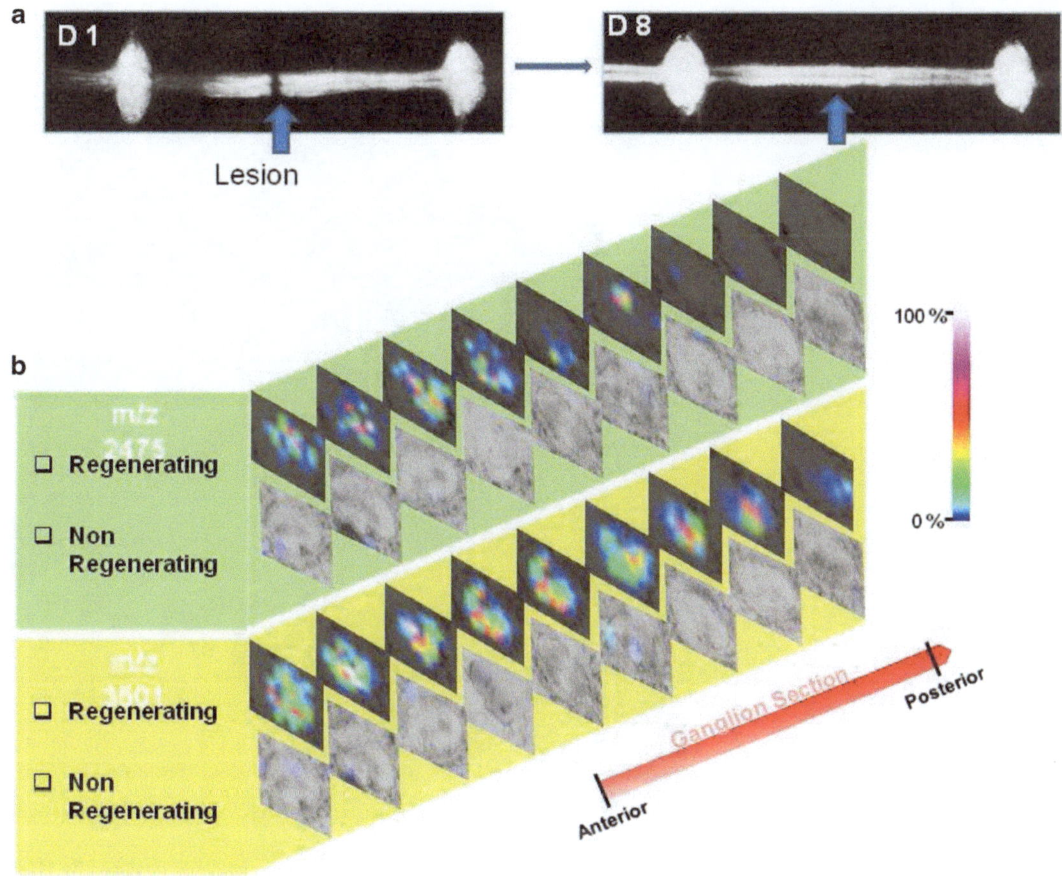

Fig. 7. Expression of ions at *m/z* 2,475 and at *m/z* 3,501 in both regenerating or not adult CNS segmental ganglia. (**a**) Inset picture of leech nerve cord containing deux ganglia connected by a lesionned connect if at Day 1 and D8 regeneration processes. (**b**) Distribution of the *m/z* 2,475 and *m/z* 3,501 ions in sections of the regenerating adult ganglion analyzed in **b**. The *inset* shows a magnified image of the data with the abundance of the ion colour coded according to the colour bar at right. The peak corresponding to this ion is absent in a control adult indicating a strong up-regulation of expression following injury.

of midbody ganglion 9, leaving the rest of the nerves between ganglia intact (Fig. 6). After allowing 6 h for regeneration to be established, frozen 10 μm cross sections of whole animals were cut from anterior of the crush site to posterior of ganglion 9. Nine sections covering ganglion 9 from anterior (Sect. 1) to posterior (Sect. 9) were then imaged with MALDI-TOF in the region of the nervous system (Fig. 6). The spectra were taken in the range of *m/z* from 1,000 to 30,000 Da and normalized. To facilitate the comparison and interpretation of the data, the spectra for all sections were plotted together in a two-dimensional representation, with the spectra displayed as parallel horizontal lines, *m/z* values along the abscissa and intensity at each point represented by colour (blue = low, white = high; scale on the left), and the set of spectra for each section separated from each other by thick lines (Figs. 6 and 7).

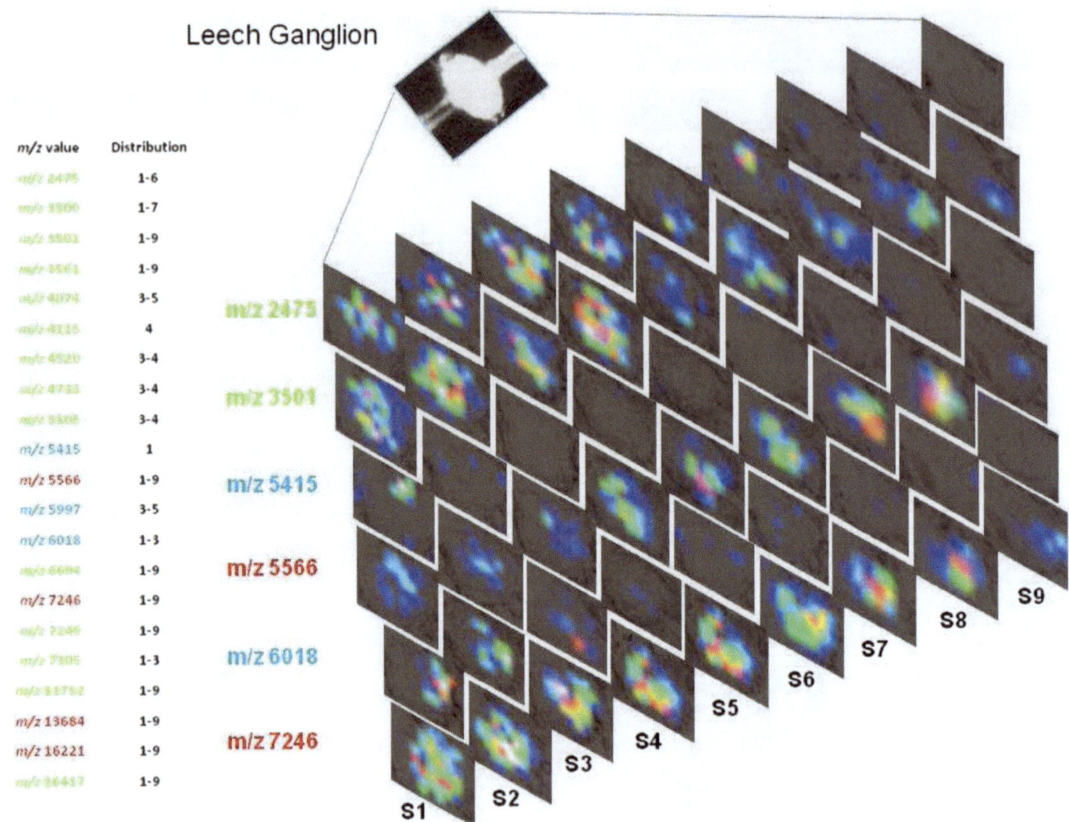

Fig. 8. 3D MSI maps of leech ganglion in course of regeneration reconstructed on the nine sections analyses by MSI in course of regeneration. *Inset table* present identified *m/z* and their localization in leech sections.

The spectra obtained for all nine cross-sections of the ganglion were then subjected to PCA followed by hierarchical clustering. The dendrogram of the clustering results shows that spectra represented by some branches of the dendrogram correspond to more anterior locations while others appear to be more posterior and are more numerous. These statistical analyses highlight two distinct areas, corresponding to the anterior (red, orange) and the posterior (blue) parts of the ganglion, and show significant differences in terms of their nature and level expression (94, 95). This suggests that peptides are produced by neurons that reside closer to the lesion or that factors are differentially transported towards the damaged area. For example, the peptide with the *m/z* of 2,475, previously detected in the embryo nerve cord (Fig. 6), is also present in the adult regenerating ganglion with an anterior expression bias (Fig. 6), whereas it is absent in controls (non regenerating adult CNS segmental ganglia, Fig. 7). This peptide, which has been recently identified as a fragment of a novel intermediate filament protein, *Hm*IF4 (95) and has the N-terminal sequence

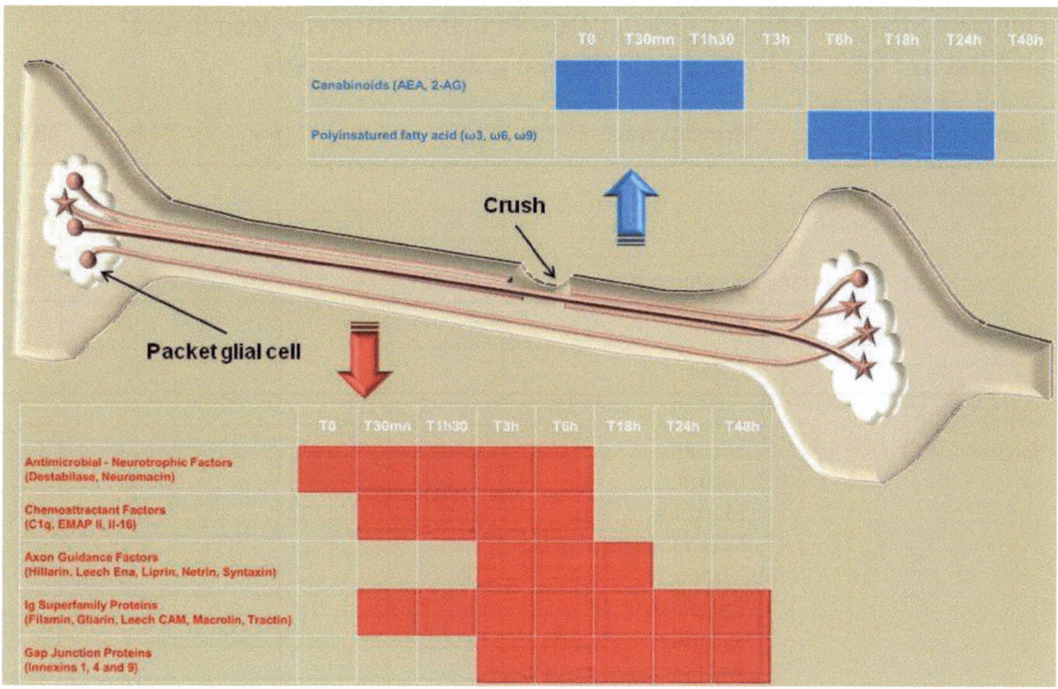

Fig. 9. Scheme of peptides/proteins and lipids identified in lesioned leech cord in course of regeneration processes.

GTRTMERSVRTSSQYASGGPMPN, provides evidence for the idea that embryonic factors are re-expressed or up-regulated during the process of regeneration (94, 95). Based on such strategy, 3D MSI was performed at the level of leech adult ganglion and presented in Fig. 8.

MSI coupled with proteomic and lipidomic strategies performed on adult leech ganglion in regeneration do not clearly establish the peptidomic pattern present in the ganglion in course of the regeneration process (Fig. 9 and Table 1). The data reflect that while many of these peptides/proteins are probably housekeeping and general maintenance molecules that are required to reconstruct the damaged tissues, we expect that some will be specialized signals, neurotrophic and guidance factors, and neurospecific molecules that are key to the re-establishment of a unique network of connections. It is the latter that we are particularly interested in identifying among the many *m/z* values present in the dataset. Our data confirm the presence, and in some cases the up- or down-regulation, of different functional groups of proteins in the regenerating adult leech brain that are also expressed at significant levels in the embryonic CNS (Fig. 6 (101–103)). Among these are proteins implicated in cytoskeletal remodeling, including the Intermediate Filament (IF) proteins Gliarin and Macrolin (104) and the actin-binding protein Filamin (105). Both IFs contain the coiled-coil rod domain typical of the superfamily of IF

Table 1
Protein identification in time course of regeneration processes based on complementary techniques (2D-Gel, DD-HPLC, Bottom-up proteomic and soustractive DNA libraries) which are also present in embryos

Class of proteins	Protein name	0 h	1 h	6 h	24 h	48 h
Immune factors						
	IL-16, EMAP II	■	■	■		
	Antimicrobial	■	■	■	■	
	C1q	■	■	■		
	CRIP				■	
Cytoskeleton						
Intermediate Fil.	Filamin			■	■	■
	Gliarin		■	■	■	■
	Macrolin			■	■	■
	Leech CAM			■	■	■
	Tractin			■	■	■
	Protein 4.1.				■	
	Synapsin			■	■	
	ReN3				■	
Microtubules	Tubulins				■	
Axon guidance						
Ig superfamily	Hillarin			■	■	
	Leech ENA			■	■	
	Liprin			■	■	
Chemotrophic F.	Netrin			■	■	■
	Syntaxin			■	■	■
	Destabilase		■	■	■	■
TRP	LAR2 e.d			■	■	■
HSP and chaperones						
	Cyclophilin			■		
	PDI				■	
	PPI				■	
	HSP90				■	
Metal oxidation						
	Neurohemerythin		■	■	■	■
	COXI		■		■	
Metabolism						
AA/metabolism	AA dehydrogenase			■	■	
	ATPase inhibitor				■	
Energy	ATP synthase			■	■	
Morphogenesis						
Homeobox gene	LOX2			■	■	■
Gap junction	Innexins			■	■	■
Calcium						
Calcium sensor	NCS2		■			
	Neurocalcineurin			■		
Others	Calmodulin-like				■	

proteins flanked by unique N- and C-terminal domains, but Gliarin is found in all glial cells, including macro- and microglial cells (102, 104), whereas Macrolin is expressed in only a single pair of giant connective glial cells (104). In contrast to Gliarin and Macrolin, *Hm*IF4 appears to be highly expressed in neurons (95). Filamin, with two calponin homology domains and 35 filamin/ABP-repeat domains, has been implicated along with Tractin in muscle development and nerve formation (105). Another interesting functional group of proteins identified in this proteomic study as potentially involved in both neural development and regeneration is comprised of several neural members of the Ig superfamily (IgSF), specifically Tractin, Hillarin and the receptor tyrosine phosphatase *Hm*LAR2. Related to the last of these, we have recently reported that *Hm*LAR1, a sibling RPTP of *Hm*LAR2, is upregulated in specific neurons in response to a nerve crush, and that the regeneration of severed axonal projections is significantly impaired when RNAi is used to block this upregulation (106) Tractin and LeechCAM have been implicated in neurite outgrowth in the course of neurogenesis by Johansen, Zipser and collaborators (105, 107–110). Tractin is widely expressed and is differentially glycosylated in sets and subsets of peripheral sensory neurons that form specific fascicles in the central nervous system (CNS). Additional proteins identified in this screen that appear to be involved in neural regeneration include several previously identified in leech brain, including Netrin (111), Hillarin (112) and Lena (leech homolog of Enabled) (113). Hillarin is localized to the axon hillock of leech neurons and affects cell and axonal cytokinesis through its interactions with septins (114, 115). Lena, a cytosolic protein implicated in actin-based cell motility (113) has been shown to associate in the leech with the *Hm*LAR receptors, whose ectodomains are thought to promote an adhesive interaction that enhances neuronal sprouting (116). Something interesting and somewhat unexpected among our observations is the adult expression of the homeobox gene LOX2, which in the embryo is expressed in 25–30 pairs of neurons repeated in the posterior two thirds of the midbody ganglia (117, 118). Possibly this transcription factor regulates a set of specific growth responses that are triggered by neuronal damage, a hypothesis that has not been tested experimentally at this time. Of particular novelty is the dynamic expression of three antimicrobial peptides in leech brain (Fig. 8) as a result of mechanical injury: *Hm*-neuromacin, *Hm*-lumbricin (119), and the novel one, *Hm*AMP3. These antimicrobial peptides produced by neurons and microglia have been recently shown by our group to promote the regeneration of neurites in axotomized leech CNS (119), indicating that they have multiple functional roles in the CNS. Moreover, other immune factors appear to also participate in the neuroregenerative process. We recently demonstrated that several leech CNS immune factors identified here (Table 1), such as the cytokines related to EMAP-II (120) and IL-16 (121) as well as the complement factor C1q (122), exert chemotactic effects towards leech microglia. In mammals, C1q is known to be synthesized and released by activated microglia in order to maintain and balance microglial activation in diseased CNS tissue (123).

Neuroinflammation is another key aspect of the neuroregenerative process. Molecules related to cyclophilin, ERp60 (102) in conjunction with cytokines like those related to IL-16, IL-17 and IL-25, are implicated in the control of inflammation (124–128) and the presence of such molecules in leech CNS (121, 129) after trauma suggests a generality and convergence of such a biological phenomena at early stage of the regeneration process.

All these results demonstrate that MSI allows obtaining neuropeptides/neurohormones molecular maps reflecting the physiological state of the animal and that adult leech brain regeneration recapitulate embryonic neurogenesis.

Neuroinflammation seems to be necessary for initiating microglial activation (130, 131), but it needs to be then controlled in order to block the apoptotic loss of neurons which may occur following excessive brain inflammation. This could be mediated by specific lipids that participate in the regulation of neuroinflammation, including cannabinoids and omega lipids (132, 133). As we observed through in vivo and in vitro experiments, triacyl-*sn*-glycerols (C10, C14, C16), and omega lipids are synthesized as an early response to the lesion, in conjunction with endocannabinoids (AEA and 2-AG). These factors act as neuroimmunomodulators (134) and in the limited regeneration (135) in the mammalian spinal cord. The fatty acids are the targets of cerebral lipoxygenases that release very powerful anti-inflammatory factors, such as the neuroprotectins or the resolvins. In the series ω6, arachidonic acid is produced by all cells from the action of phospholipases and is the precursor of several neuroprotective agents, such as the endocannabinoids, which are synthesized in the brain and have anti-inflammatory properties. The major lipid in the ω9 series is oleic acid (C18:1), which also protects the nervous system by blocking the resulting inflammation after excessive stimulation (exitotoxicity). Sulfatides and gangliosides constitute another class of lipids preserving the CNS in vertebrates. Some studies have shown that their diminution at the cerebral level is directly associated with the manifestation of neurodegenerative conditions such as are observed in Alzheimer's disease (136–138). In our context, stearic acid and phosphatidylinositol, as well the mono-unsaturated omega-9 fatty acid (oleic acid) are produced early after trauma and each show a specific pattern of expression, spatially and temporally. At the level of the trauma, oleic acid and phosphatidylinositol migrated from connective to neurons (packet cell) whereas stearic acid accumulates in connective and less in ganglion. This can be explained by the fact that oleic acid regulates inflammation at the level of neurons (139) whereas stearic acid seems to play a role in tissue repair and plasticity, near the lesionned connective (140, 141).

We propose that these data can be thought of together through the following model. Cannabinoids are more implicated at later stages of the regeneration/repair process. In fact, the neurite outgrowth tests showed that, among them, AEA is more important in scar formation whereas 2-AG appears to be more involved in axon extension. Cannabinoids are part of the regeneration process along with peptides and proteins, and all need to be taken into account together to achieve a deeper understanding of the whole of the biological process. In addition, microglial cells, in conjunction with neurons and blood cells are able to regulate neural inflammation very quickly and to stimulate neurite outgrowth, also with the release of cannabinoids, which at later stages appear to act along gangliosides in apoptosis regulation, neurite outgrowth and the release of axonal guidance factors. In the same time window, 3 h after lesion, embryonic factors' re-expression possibly occurs through homeobox gene activation, axon guidance and neurotrophic factors released. This is also linked to a close interaction between cells and the implication of intermediate filaments and cell–matrix interactions performing a net where neurites are able to sprout and receive some positive and negative factors such as neuroregulins (erb-2 like factors, Cuvillier-Hot, unpublished data), inhibitors of NOGO and inhibitory factors.

In summary, proteomic and lipidomic approaches were employed to profile and identify different lipids and proteins in leech embryos as well as normal and regenerating adult leech nervous system. These are proposed as strong candidates for important roles in the mechanisms of neural regeneration (Fig. 9). The overlap between these profiles observed in regeneration in response to physical damage and in the neuroimmune response to bacterial insult suggests that these complex dynamic processes, involving many different types of cells and mediators, have much in common. Moreover, the overlap between molecular profiles observed in neural development and adult regeneration suggests that a significant recapitulation of neurogenic programs is present in the course of regeneration. The data presented here are only a beginning, but they already identify similarities between the molecular underpinnings of invertebrate and vertebrate responses to trauma, similarities that can be exploited in furthering our understanding of the reasons for the limited capacity to regenerate neurites in the mammalian CNS.

7.2. Applications to Vertebrates' Central Nervous System

Because its anatomy has been extensively characterized, the rat brain was the first biological model used in MALDI MSI studies, and several molecular maps of different neuropeptides have been reported (142, 143) as the one we recently performed on spinal cord (Fig. 10). The spinal cord is organized into discrete, anatomically

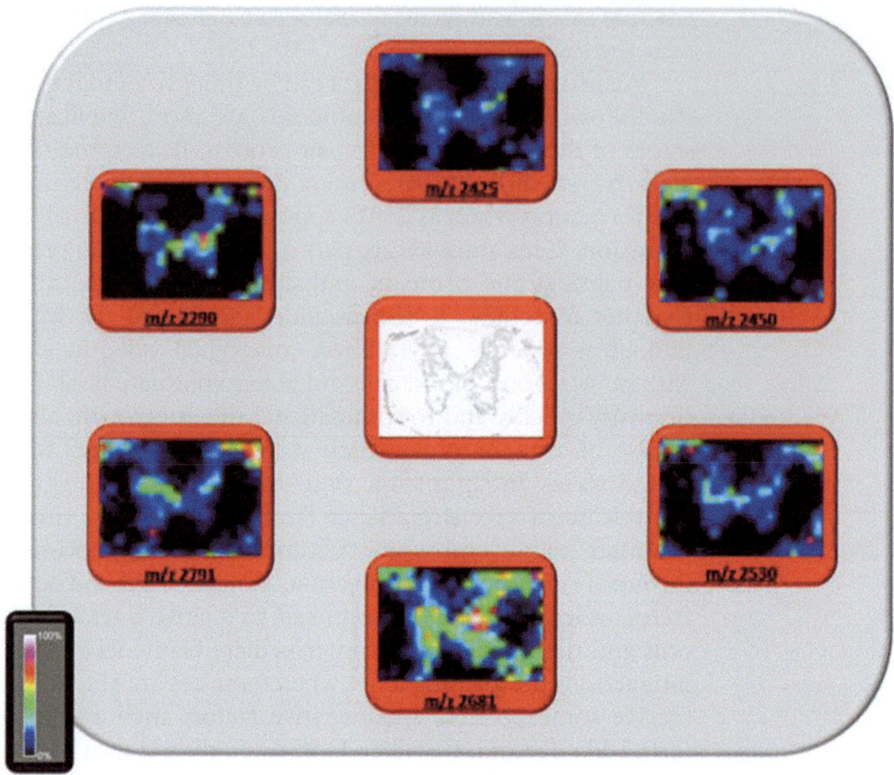

Fig. 10. Cellular and molecular maps of rat spinal cord peptides using MALDI-MSI technology.

defined areas that include motor and sensory networks composed of chemically diverse cells. The MSI data presented here reveal the spatial distribution of multiple neuropeptides obtained within single, 10 μm sections of rat spinal cord (Fig. 10). These MSI analyses reveal new insights into the chemical architecture of the spinal cord and set the stage for future imaging studies of the chemical changes induced by pain, anesthesia and drug tolerance. Similarly, it can be possible to obtain molecular peptides maps in embryonic mouse brain (Fig. 11) or molecular lipid maps of gangliosides which are particularly abundant in the CNS and thought to play important roles in memory formation, neuritogenesis, synaptic transmission and other neural functions (144). In these conditions, the C18-species was widely found distributed throughout the frontal brain whereas the C20-species selectively localized along the entorhinal-hippocampus projections, especially in the molecular layer (ML) of the dentate gyrus (DG). This points out to a specific localization of glangliosides in the brain. Taking the above view into account, it is important to develop MSI for pathological diseases like neurodegenerative disease.

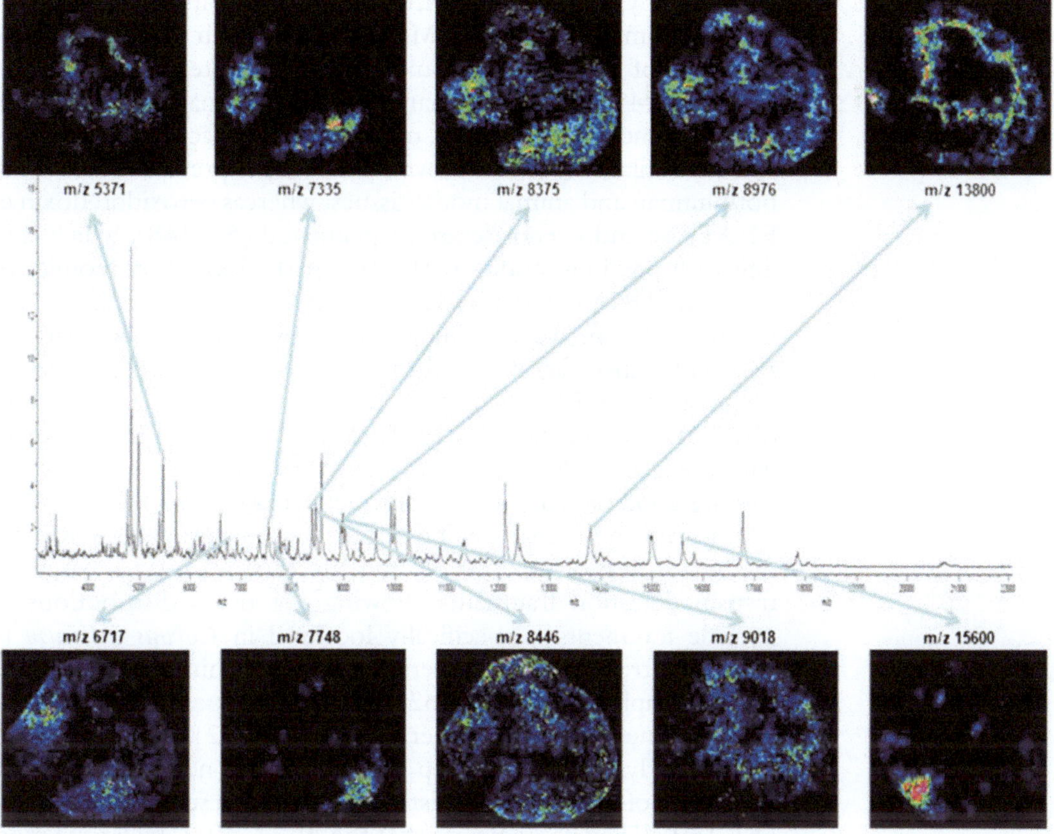

Fig. 11. Molecular images of mice embryos in MALDI-MSI.

7.3. Applications to Brain Diseases

Neurodegeneration diseases induce proteome changes in the brain that can be investigated using neuroimaging techniques. For example, Parkinson's disease (PD) is the second most common neurological disease after Alzheimer's disease. PD is characterized by a selective degeneration of dopaminergic neurons in the *substantia nigra pars compacta* and by cytoplasmic inclusions (Lewy Bodies) where specific proteins are stored like the α-synuclein (145). The first tissue profiling studies by MALDI on 6-OHDA Parkinson model have been performed by Per Andrén group (146). Several differences were found in the dopamine-depleted side of the rat brain when compared to the corresponding intact side, in calmodulin, cytochrome c, and cytochrome c oxidase, for example, implicating denervation of dopamine neurons in the regulation of ubiquitin pathways, at least in a classical animal model of PD (74). This study also emphasizes the utility of molecular profiling with MALDI-MSI because it has the capacity to distinguish between metabolic fragments, conjugated proteins and posttranslational modifications (74).

Recently, we examined MALDI tissue profiling combining the use of automatic spotting of MALDI matrix with in situ enzymatic digestion of FFPE tissue from 6-OHDA unilaterally treated animals (57, 147) followed by nanoLC/MS-MS analysis. These analyses confirmed that ubiquitin, trans-elongation factor 1, hexokinase and neurofilament M are down-regulated, as previously shown in both human and animal model tissues, whereas peroxidoredoxin 6, F1 ATPase and α-enolase are up-regulated (62, 148), which is in line with previous studies performed with classical proteomics or genomic (DNA microarray) approaches.

In addition, we identified three novel putative biomarkers, *trans* elongation factor 1 (eEF1) and the collapsin response mediator proteins, (CRMP-1 and -2) using protein libraries (62). Our observation of increases in CRMP-1 and CRMP-2 is in agreement with previous molecular data (149). We speculate that CRMP factors are good biomarkers for neurodegenerative diseases like PD or AD, a hypothesis that can be tested by comparing the CRMP-2 mRNA expression in controls and the MALDI images based on trypsin digestion fragments showing the tissue distributions of peptide fragments is specifically localized in *Corpus callosum* in 6-OHDA treated animal whereas in control animals the distribution is completely different (62). This specific localization is in line with the ones of neurodegenerative diseases (62).

Similarly, Stoeckli's group has applied this new technology to study amyloid beta peptide distribution in brain sections from mice (12, 150). They demonstrate that the Aβ-(1–40) is by far the most abundant amyloid peptide. Three main regions can be distinguished: two very intense areas are located in the parietal and the occipital cortical lobes and the third one close to the lower part of the Sylvian fissure, that is, in the hippocampus region. The normalized distributions of Aβ-(1–40) and Aβ-(1–42) show that they are the most abundant amyloid peptides. MSI gives access to the levels of known targets but also allows the mapping of the different targets with great accuracy, which is not possible when whole-brain extracts are analyzed (12, 150). These and other results mentioned above establish the great potential of MALDI MSI as a new tool for the study of the consequences of neurodegenerative disease.

MALDI MSI is also a very appropriate tool for assaying the distribution of pharmaceuticals in rat brain tissue slices, which is critical information for new drug development. In fact, some studies of clozapine (36) and repiridone (151) have recently been performed by MALDI MSI. The data confirm that chronic risperidone treatment, which is accompanied by a behavioural phenotype of extrapyramidal origin, produces alterations in the striatal protein profile, possibly subsequent to blockade of dopaminergic systems. These results suggest that possible mechanisms involved in APD-induced EPS include metabolic dysfunction and oxidative stress (151).

8. The Out Coming Fields of Applications

After 10 years of developments (Fig. 12), novel directions of MSI are its linking to positron emission tomography (PET), X-ray CT instrumentation, magnetic resonance imaging (MRI) for both pre-clinical and clinical research. The complementarities between non-invasive techniques and molecular data obtained by MALDI-MSI will result in more precision for a better diagnosis. Moreover, the introductions of new in vivo techniques such as Desorption/ionization electrospray (DESI) offer the possibility to perform surface analyses of tumours. Based on the assumption that protein patterns in tumours compared to normal tissue is different, the in vivo surface analyses will give access to real-time diagnosis and will help surgeons in the future for removing the entire tumour and the cells that are changing their phenotype from benign to carcinoma. We can expect that a more resolute technique such as Jet desorption/ionisation technology will offer rapid in vivo molecular diagnosis, thus assisting surgeons in order to know the exact size of the tumour that can be removed. Finally, diagnosis from coelioscopy or smears based on data obtained from direct tissue analysis by MALDI (152–154) coupled to tissue arrays with tagged apatmers, antibodies, lectins will offer access to disease prognostics. MALDI-MSI will

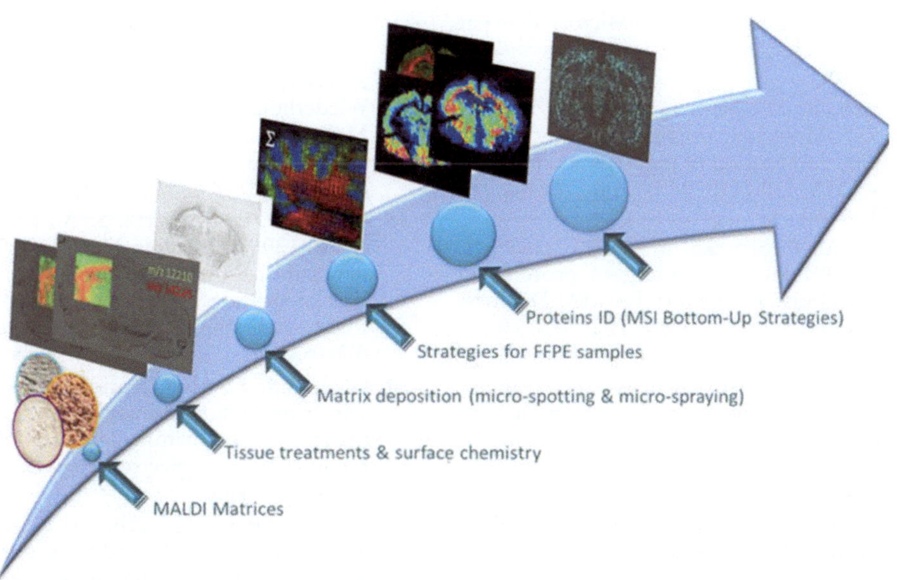

Fig. 12. Ten years of developments including new solid [11] or liquid [41] ionic matrices, matrices deposition [38], tissue treatments (washing procedures [57]) of frozen or FFPE [57], high mass protein demasking [156, 157], bottom-up strategy on FFPE tissue [62, 147, 148] or with derivatization [64], Specific Tag mass MALDI-MSI [46]. All these points show that this technology still need developments and the major point for the future is the in tissue direct peptides/proteins structure determination.

become a routine technology utilized in a clinical setting and an important complement for pathologists in order to perform molecular histo-pathological diagnoses.

Acknowledgements

Research from our laboratories mentioned here has been supported by grants from the Centre National de la Recherche Scientifique Département de la politique industrielle (to MS and IF), Ministère de L'Education Nationale, de L'Enseignement Supérieur et de la Recherche, Agence National de la recherche (to IF), ARCIR Region Nord Pas de Calais (to IF).

References

1. Jimenez, C. R., van Veelen, P. A., Li, K. W., Wildering, W. C., Geraerts, W. P., Tjaden, U. R., and van der Greef, J. (1994) Neuropeptide expression and processing as revealed by direct matrix-assisted laser desorption ionization mass spectrometry of single neurons, *J Neurochem 62*, 404–407.
2. Verbueken, A. H., Bruynseels, F. J., and Van Grieken, R. E. (1985) Laser microprobe mass analysis: a review of applications in the life sciences, *Biomed Mass Spectrom 12*, 438–463.
3. Castaing, R. a. S., G. (1962) Microanalyse par emission ioinque secondaire, *Microscopie 1*, 395–410.
4. Caprioli, R. M., Farmer, T. B., and Gile, J. (1997) Molecular imaging of biological samples: localization of peptides and proteins using MALDI-TOF MS, *Anal Chem 69*, 4751–4760.
5. Chaurand, P., Stoeckli, M., and Caprioli, R. M. (1999) Direct profiling of proteins in biological tissue sections by MALDI mass spectrometry, *Anal Chem 71*, 5263–5270.
6. Stoeckli, M., Farmer, T. B., and Caprioli, R. M. (1999) Automated mass spectrometry imaging with a matrix-assisted laser desorption ionization time-of-flight instrument, *J Am Soc Mass Spectrom 10*, 67–71.
7. Karas, M., and Kruger, R. (2003) Ion formation in MALDI: the cluster ionization mechanism, *Chem Rev 103*, 427–440.
8. Knochenmuss, R., and Zenobi, R. (2003) MALDI ionization: the role of in-plume processes, *Chem Rev 103*, 441–452.
9. Fournier, I., Marinach, C., Tabet, J. C., and Bolbach, G. (2003) Irradiation effects in MALDI, ablation, ion production, and surface modifications. Part II. 2,5-dihydroxybenzoic acid monocrystals, *J Am Soc Mass Spectrom 14*, 893–899.
10. Fournier, I., Tabet, J. C., and Bolbach, G. (2002) Irradiation effects in MALDI and surface modifications Part I : Sinapinic acid monocristals, *Int JMS 219*, 1515–1523.
11. Lemaire, R., Tabet, J. C., Ducoroy, P., Hendra, J. B., Salzet, M., and Fournier, I. (2006) Solid ionic matrixes for direct tissue analysis and MALDI imaging, *Anal Chem 78*, 809–819.
12. Stoeckli, M., Staab, D., Staufenbiel, M., Wiederhold, K. H., and Signor, L. (2002) Molecular imaging of amyloid beta peptides in mouse brain sections using mass spectrometry, *Anal Biochem 311*, 33–39.
13. Chaurand, P., Schwartz, S. A., and Caprioli, R. M. (2002) Imaging mass spectrometry: a new tool to investigate the spatial organization of peptides and proteins in mammalian tissue sections, *Curr Opin Chem Biol 6*, 676–681.
14. Chaurand, P., and Caprioli, R. M. (2002) Direct profiling and imaging of peptides and proteins from mammalian cells and tissue sections by mass spectrometry, *Electrophoresis 23*, 3125–3135.
15. Schuerenberg, M., Luebbert, C., Deininger, S. O., Ketterlinus, R., and Suckau, D. (2007) MALDI tissue imaging: mass spectrometric localization of biomarkers in tissue slices, *Nature Methods.*
16. Baluya, D. L., Garrett, T. J., and Yost, R. A. (2007) Automated MALDI Matrix Deposition

Method with Inkjet Printing for Imaging Mass Spectrometry, *Anal Chem*.
17. Hankin, J. A., Barkley, R. M., and Murphy, R. C. (2007) Sublimation as a Method of Matrix Application for Mass Spectrometric Imaging, *J Am Soc Mass Spectrom*.
18. Spengler, B., and Hubert, M. (2002) Scanning microprobe matrix-assisted laser desorption ionization (SMALDI) mass spectrometry: instrumentation for sub-micrometer resolved LDI and MALDI surface analysis, *J Am Soc Mass Spectrom 13*, 735–748.
19. Dreisewerd, K. S., M.; Karas,M.; Hillenkamp, F. (1995) Influence of the laser intensity and spot size on the desorption ofmolecules and ions in matrix-assisted laser desorption/ionization with a uniform beam profile, *Int J Mass Spectrom 141*, 127–148.
20. Holle, A., Haase, A., Kayser, M., and Hohndorf, J. (2006) Optimizing UV laser focus profiles for improved MALDI performance, *J Mass Spectrom 41*, 705–716.
21. Chaurand, P., Schriver, K. E., and Caprioli, R. M. (2007) Instrument design and characterization for high resolution MALDI-MS imaging of tissue sections, *J Mass Spectrom 42*, 476–489.
22. Jurchen, J. C., Rubakhin, S. S., and Sweedler, J. V. (2005) MALDI-MS imaging of features smaller than the size of the laser beam, *J Am Soc Mass Spectrom 16*, 1654–1659.
23. Altelaar, M. A. F. T., I.M.;,McDonnell, L.A.;Verhaert, P.D.E.M.; De Lange, R.P.J.; Adanc, R.A.H.; Mooid, W.J., and Heeren, R. M. A. P. S. R. (2007) High-resolution MALDI imaging mass spectrometry allows localization of peptide distributions at cellular length scales in pituitary tissue sections, *Int J Mass Spectrom 260*, 203–211.
24. Klerk, L. A. B., A.; Fletcher, I.W.; van Liere, R.; Heeren, R.M.A. . (2007) Extended data analysis strategies for high resolution imaging MS: New methods to deal with extremely large image hyperspectral datasets, *Int J Mass Spectrom 260*, 222–236.
25. Schwartz, S. A., Weil, R. J., Johnson, M. D., Toms, S. A., and Caprioli, R. M. (2004) Protein profiling in brain tumors using mass spectrometry: feasibility of a new technique for the analysis of protein expression, *Clin Cancer Res 10*, 981–987.
26. Chaurand, P., Schwartz, S. A., and Caprioli, R. M. (2004) Assessing protein patterns in disease using imaging mass spectrometry, *J Proteome Res 3*, 245–252.
27. McCombie, G., Staab, D., Stoeckli, M., and Knochenmuss, R. (2005) Spatial and spectral correlations in MALDI mass spectrometry images by clustering and multivariate analysis, *Anal Chem 77*, 6118–6124.
28. Deininger, S. O., Ebert, M. P., Futterer, A., Gerhard, M., and Rocken, C. (2008) MALDI Imaging Combined with Hierarchical Clustering as a New Tool for the Interpretation of Complex Human Cancers, *J Proteome Res*.
29. Van de Plas, R., Ojeda, F., Dewil, M., Van Den Bosch, L., De Moor, B., and Waelkens, E. (2007) Prospective exploration of biochemical tissue composition via imaging mass spectrometry guided by principal component analysis, *Pac Symp Biocomput*, 458–469.
30. Walch, A., Rauser, S., Deininger, S. O., and Hofler, H. (2008) MALDI imaging mass spectrometry for direct tissue analysis: a new frontier for molecular histology, *Histochem Cell Biol 130*, 421–434.
31. Jackson, S. N., Wang, H. Y., and Woods, A. S. (2005) In situ structural characterization of phosphatidylcholines in brain tissue using MALDI-MS/MS, *J Am Soc Mass Spectrom 16*, 2052–2056.
32. Wang, H. Y., Jackson, S. N., McEuen, J., and Woods, A. S. (2005) Localization and analyses of small drug molecules in rat brain tissue sections, *Anal Chem 77*, 6682–6686.
33. Jackson, S. N., Wang, H. Y., and Woods, A. S. (2005) Direct profiling of lipid distribution in brain tissue using MALDI-TOFMS, *Anal Chem 77*, 4523–4527.
34. Jackson, S. N., Wang, H. Y., Woods, A. S., Ugarov, M., Egan, T., and Schultz, J. A. (2005) Direct tissue analysis of phospholipids in rat brain using MALDI-TOFMS and MALDI-ion mobility-TOFMS, *J Am Soc Mass Spectrom 16*, 133–138.
35. Wang, H. Y., Jackson, S. N., and Woods, A. S. (2007) Direct MALDI-MS analysis of cardiolipin from rat organs sections, *J Am Soc Mass Spectrom 18*, 567–577.
36. Hsieh, Y., Chen, J., and Korfmacher, W. A. (2007) Mapping pharmaceuticals in tissues using MALDI imaging mass spectrometry, *J Pharmacol Toxicol Methods 55*, 193–200.
37. Garrett, T. J. P.-C., M.C.;Kovtoun, V.;Bui, H.; Izgarian, N.; Stafford, G.; Yost, R.A. (2007) Imaging of small molecules in tissue sections with a new intermediate-pressure MALDI linear ion trap mass spectrometer, *Int J Mass Spectrom 260*, 166–176.
38. Franck, J., Arafah, K., Barnes, A., Wisztorski, M., Salzet, M., and Fournier, I. (2009) Improving tissue preparation for matrix-assisted laser desorption ionization mass spectrometry imaging. Part 1: using microspotting, *Anal Chem 81*, 8193–8202.

39. Colsch, B., and Woods, A. S. Localization and imaging of sialylated glycosphingolipids in brain tissue sections by MALDI mass spectrometry, *Glycobiology 20*, 661–667.
40. Chan, K., Lanthier, P., Liu, X., Sandhu, J. K., Stanimirovic, D., and Li, J. (2009) MALDI mass spectrometry imaging of gangliosides in mouse brain using ionic liquid matrix, *Anal Chim Acta 639*, 57–61.
41. Meriaux, C., Franck, J., Wisztorski, M., Salzet, M., and Fournier, I. Liquid ionic matrixes for MALDI mass spectrometry imaging of lipids, *J Proteomics 73*, 1204–1218.
42. Puolitaival, S. M., Burnum, K. E., Cornett, D. S., and Caprioli, R. M. (2008) Solvent-Free Matrix Dry-Coating for MALDI Imaging of Phospholipids, *J Am Soc Mass Spectrom.*
43. Bouschen, W., Schulz, O., Eikel, D., and Spengler, B. Matrix vapor deposition/recrystallization and dedicated spray preparation for high-resolution scanning microprobe matrix-assisted laser desorption/ionization imaging mass spectrometry (SMALDI-MS) of tissue and single cells, *Rapid Commun Mass Spectrom 24*, 355–364.
44. Trimpin, S. A perspective on MALDI alternatives-total solvent-free analysis and electron transfer dissociation of highly charged ions by laserspray ionization, *J Mass Spectrom 45*, 471–485.
45. Rompp, A., Guenther, S., Schober, Y., Schulz, O., Takats, Z., Kummer, W., and Spengler, B. Histology by Mass Spectrometry: Label-Free Tissue Characterization Obtained from High-Accuracy Bioanalytical Imaging, *Angew Chem Int Ed Engl 49*, 3834–3838.
46. Lemaire, R., Stauber, J., Wisztorski, M., Van Camp, C., Desmons, A., Deschamps, M., Proess, G., Rudlof, I., Woods, A. S., Day, R., Salzet, M., and Fournier, I. (2007) Tag-mass: specific molecular imaging of transcriptome and proteome by mass spectrometry based on photocleavable tag, *J Proteome Res 6*, 2057–2067.
47. Thiery, G., Shchepinov, M. S., Southern, E. M., Audebourg, A., Audard, V., Terris, B., and Gut, I. G. (2007) Multiplex target protein imaging in tissue sections by mass spectrometry--TAMSIM, *Rapid Commun Mass Spectrom 21*, 823–829.
48. Thiery, G., Anselmi, E., Audebourg, A., Darii, E., Abarbri, M., Terris, B., Tabet, J. C., and Gut, I. G. (2008) Improvements of TArgeted multiplex mass spectrometry IMaging, *Proteomics 8*, 3725–3734.
49. McLafferty, F. W., Kelleher, N. L., Begley, T. P., Fridriksson, E. K., Zubarev, R. A., and Horn, D. M. (1998) Two-dimensional mass spectrometry of biomolecules at the subfemtomole level, *Curr Opin Chem Biol 2*, 571–578.
50. Zubarev, R. A., Horn, D. M., Fridriksson, E. K., Kelleher, N. L., Kruger, N. A., Lewis, M. A., Carpenter, B. K., and McLafferty, F. W. (2000) Electron capture dissociation for structural characterization of multiply charged protein cations, *Anal Chem 72*, 563–573.
51. Reiber, D. C., Brown, R. S., Weinberger, S., Kenny, J., and Bailey, J. (1998) Unknown peptide sequencing using matrix-assisted laser desorption/ionization and in-source decay, *Anal Chem 70*, 1214–1222.
52. Reiber, D. C., Grover, T. A., and Brown, R. S. (1998) Identifying proteins using matrix-assisted laser desorption/ionization in-source fragmentation data combined with database searching, *Anal Chem 70*, 673–683.
53. Roepstorff, P., and Fohlman, J. (1984) Proposal for a common nomenclature for sequence ions in mass spectra of peptides, *Biomed Mass Spectrom 11*, 601.
54. Raska, C. S., Parker, C. E., Huang, C., Han, J., Glish, G. L., Pope, M., and Borchers, C. H. (2002) Pseudo-MS3 in a MALDI orthogonal quadrupole-time of flight mass spectrometer, *J Am Soc Mass Spectrom 13*, 1034–1041.
55. Demeure, K., Quinton, L., Gabelica, V., and De Pauw, E. (2007) Rational selection of the optimum MALDI matrix for top-down proteomics by in-source decay, *Anal Chem 79*, 8678–8685.
56. Debois, D., Bertrand, V., Quinton, L., De Pauw-Gillet, M. C., and De Pauw, E. MALDI-In Source Decay Applied to Mass Spectrometry Imaging: A New Tool for Protein Identification, *Anal Chem.*
57. Lemaire, R., Desmons, A., Tabet, J. C., Day, R., Salzet, M., and Fournier, I. (2007) Direct analysis and MALDI imaging of formalin-fixed, paraffin-embedded tissue sections, *J Proteome Res 6*, 1295–1305.
58. Groseclose, M. R., Andersson, M., Hardesty, W. M., and Caprioli, R. M. (2007) Identification of proteins directly from tissue: in situ tryptic digestions coupled with imaging mass spectrometry, *J Mass Spectrom 42*, 254–262.
59. Groseclose, M. R., Massion, P. P., Chaurand, P., and Caprioli, R. M. (2008) High-throughput proteomic analysis of formalin-fixed paraffin-embedded tissue microarrays using MALDI imaging mass spectrometry, *Proteomics 8*, 3715–3724.
60. Ronci, M., Bonanno, E., Colantoni, A., Pieroni, L., Di Ilio, C., Spagnoli, L. G., Federici, G., and Urbani, A. (2008) Protein

unlocking procedures of formalin-fixed paraffin-embedded tissues: Application to MALDI-TOF imaging MS investigations, *Proteomics 8*, 3702–3714.

61. Djidja, M. C., Francese, S., Loadman, P. M., Sutton, C. W., Scriven, P., Claude, E., Snel, M. F., Franck, J., Salzet, M., and Clench, M. R. (2009) Detergent addition to tryptic digests and ion mobility separation prior to MS/MS improves peptide yield and protein identification for in situ proteomic investigation of frozen and formalin-fixed paraffin-embedded adenocarcinoma tissue sections, *Proteomics 9*, 2750–2763.
62. Stauber, J., Lemaire, R., Franck, J., Bonnel, D., Croix, D., Day, R., Wisztorski, M., Fournier, I., and Salzet, M. (2008) MALDI imaging of formalin-fixed paraffin-embedded tissues: application to model animals of Parkinson disease for biomarker hunting, *J Proteome Res 7*, 969–978.
63. Gustafsson, J. O., Oehler, M. K., McColl, S. R., and Hoffmann, P. Citric acid antigen retrieval (CAAR) for tryptic peptide imaging directly on archived formalin-fixed paraffin-embedded tissue, *J Proteome Res 9*, 4315–4328.
64. Franck, J., El Ayed, M., Wisztorski, M., Salzet, M., and Fournier, I. (2009) On-tissue N-terminal peptide derivatizations for enhancing protein identification in MALDI mass spectrometric imaging strategies, *Anal Chem 81*, 8305–8317.
65. Cornett, D. S., Reyzer, M. L., Chaurand, P., and Caprioli, R. M. (2007) MALDI imaging mass spectrometry: molecular snapshots of biochemical systems, *Nat Methods 4*, 828–833.
66. Lemaire, R., Menguellet, S. A., Stauber, J., Marchaudon, V., Lucot, J. P., Collinet, P., Farine, M. O., Vinatier, D., Day, R., Ducoroy, P., Salzet, M., and Fournier, I. (2007) Specific MALDI Imaging and Profiling for Biomarker Hunting and Validation: Fragment of the 11 S Proteasome Activator Complex, Reg Alpha Fragment, Is a New Potential Ovary Cancer Biomarker, *J Proteome Res 6*, 4127–4134.
67. Chaurand, P., DaGue, B. B., Pearsall, R. S., Threadgill, D. W., and Caprioli, R. M. (2001) Profiling proteins from azoxymethane-induced colon tumors at the molecular level by matrix-assisted laser desorption/ionization mass spectrometry, *Proteomics 1*, 1320–1326.
68. Chaurand, P., Sanders, M. E., Jensen, R. A., and Caprioli, R. M. (2004) Proteomics in diagnostic pathology: profiling and imaging proteins directly in tissue sections, *Am J Pathol 165*, 1057–1068.
69. Fournier, I., Day, R., and Salzet, M. (2003) Direct analysis of neuropeptides by in situ MALDI-TOF mass spectrometry in the rat brain, *Neuro Endocrinol Lett 24*, 9–14.
70. Hintersteiner, M., Enz, A., Frey, P., Jaton, A. L., Kinzy, W., Kneuer, R., Neumann, U., Rudin, M., Staufenbiel, M., Stoeckli, M., Wiederhold, K. H., and Gremlich, H. U. (2005) In vivo detection of amyloid-beta deposits by near-infrared imaging using an oxazine-derivative probe, *Nat Biotechnol 23*, 577–583.
71. Langstrom, B., Andren, P. E., Lindhe, O., Svedberg, M., and Hall, H. (2007) In vitro imaging techniques in neurodegenerative diseases, *Mol Imaging Biol 9*, 161–175.
72. Laurent, C., Levinson, D. F., Schwartz, S. A., Harrington, P. B., Markey, S. P., Caprioli, R. M., and Levitt, P. (2005) Direct profiling of the cerebellum by matrix-assisted laser desorption/ionization time-of-flight mass spectrometry: A methodological study in postnatal and adult mouse, *J Neurosci Res 81*, 613–621.
73. Meistermann, H., Norris, J. L., Aerni, H. R., Cornett, D. S., Friedlein, A., Erskine, A. R., Augustin, A., De Vera Mudry, M. C., Ruepp, S., Suter, L., Langen, H., Caprioli, R. M., and Ducret, A. (2006) Biomarker discovery by imaging mass spectrometry: transthyretin is a biomarker for gentamicin-induced nephrotoxicity in rat, *Mol Cell Proteomics 5*, 1876–1886.
74. Pierson, J., Svenningsson, P., Caprioli, R. M., and Andren, P. E. (2005) Increased levels of ubiquitin in the 6-OHDA-lesioned striatum of rats, *J Proteome Res 4*, 223–226.
75. Reyzer, M. L., and Caprioli, R. M. (2005) MALDI mass spectrometry for direct tissue analysis: a new tool for biomarker discovery, *J Proteome Res 4*, 1138–1142.
76. Schwartz, S. A., Weil, R. J., Thompson, R. C., Shyr, Y., Moore, J. H., Toms, S. A., Johnson, M. D., and Caprioli, R. M. (2005) Proteomic-based prognosis of brain tumor patients using direct-tissue matrix-assisted laser desorption ionization mass spectrometry, *Cancer Res 65*, 7674–7681.
77. Skold, K., Svensson, M., Nilsson, A., Zhang, X., Nydahl, K., Caprioli, R. M., Svenningsson, P., and Andren, P. E. (2006) Decreased striatal levels of PEP-19 following MPTP lesion in the mouse, *J Proteome Res 5*, 262–269.
78. Jimenez, C. R., Li, K. W., Dreisewerd, K., Spijker, S., Kingston, R., Bateman, R. H.,

Burlingame, A. L., Smit, A. B., van Minnen, J., and Geraerts, W. P. (1998) Direct mass spectrometric peptide profiling and sequencing of single neurons reveals differential peptide patterns in a small neuronal network, *Biochemistry 37*, 2070–2076.

79. Li, K. W., Hoek, R. M., Smith, F., Jimenez, C. R., van der Schors, R. C., van Veelen, P. A., Chen, S., van der Greef, J., Parish, D. C., Benjamin, P. R., and et al. (1994) Direct peptide profiling by mass spectrometry of single identified neurons reveals complex neuropeptide-processing pattern, *J Biol Chem 269*, 30288–30292.
80. Li, K. W., Jimenez, C. R., Van Veelen, P. A., and Geraerts, W. P. (1994) Processing and targeting of a molluscan egg-laying peptide prohormone as revealed by mass spectrometric peptide fingerprinting and peptide sequencing, *Endocrinology 134*, 1812–1819.
81. Li, L., Garden, R. W., and Sweedler, J. V. (2000) Single-cell MALDI: a new tool for direct peptide profiling, *Trends Biotechnol 18*, 151–160.
82. Li, L., Romanova, E. V., Rubakhin, S. S., Alexeeva, V., Weiss, K. R., Vilim, F. S., and Sweedler, J. V. (2000) Peptide profiling of cells with multiple gene products: combining immunochemistry and MALDI mass spectrometry with on-plate microextraction, *Anal Chem 72*, 3867–3874.
83. Li, K. W., van Golen, F. A., van Minnen, J., van Veelen, P. A., van der Greef, J., and Geraerts, W. P. (1994) Structural identification, neuronal synthesis, and role in male copulation of myomodulin-A of Lymnaea: a study involving direct peptide profiling of nervous tissue by mass spectrometry, *Brain Res Mol Brain Res 25*, 355–358.
84. Hummon, A. B., Amare, A., and Sweedler, J. V. (2006) Discovering new invertebrate neuropeptides using mass spectrometry, *Mass Spectrom Rev 25*, 77–98.
85. Hummon, A. B., Richmond, T. A., Verleyen, P., Baggerman, G., Huybrechts, J., Ewing, M. A., Vierstraete, E., Rodriguez-Zas, S. L., Schoofs, L., Robinson, G. E., and Sweedler, J. V. (2006) From the genome to the proteome: uncovering peptides in the Apis brain, *Science 314*, 647–649.
86. Altelaar, A. F., van Minnen, J., Jimenez, C. R., Heeren, R. M., and Piersma, S. R. (2005) Direct molecular imaging of Lymnaea stagnalis nervous tissue at subcellular spatial resolution by mass spectrometry, *Anal Chem 77*, 735–741.
87. de Lange, R. P., and van Minnen, J. (1998) Localization of the neuropeptide APGWamide in gastropod molluscs by in situ hybridization and immunocytochemistry, *Gen Comp Endocrinol 109*, 166–174.
88. DeKeyser, S. S., Kutz-Naber, K. K., Schmidt, J. J., Barrett-Wilt, G. A., and Li, L. (2007) Imaging mass spectrometry of neuropeptides in decapod crustacean neuronal tissues, *J Proteome Res 6*, 1782–1791.
89. DeKeyser, S. S., and Li, L. (2007) Mass spectrometric charting of neuropeptides in arthropod neurons, *Anal Bioanal Chem 387*, 29–35.
90. Francese, S., Lambardi, D., Mastrobuoni, G., la Marca, G., Moneti, G., and Turillazzi, S. (2009) Detection of honeybee venom in envenomed tissues by direct MALDI MSI, *J Am Soc Mass Spectrom 20*, 112–123.
91. Amare, A., and Sweedler, J. V. (2007) Neuropeptide precursors in Tribolium castaneum, *Peptides 28*, 1282–1291.
92. Dani, F. R., Francese, S., Mastrobuoni, G., Felicioli, A., Caputo, B., Simard, F., Pieraccini, G., Moneti, G., Coluzzi, M., della Torre, A., and Turillazzi, S. (2008) Exploring proteins in Anopheles gambiae male and female antennae through MALDI mass spectrometry profiling, *PLoS One 3*, e2822.
93. Hofer, S., Dircksen, H., Tollback, P., and Homberg, U. (2005) Novel insect orcokinins: characterization and neuronal distribution in the brains of selected dicondylian insects, *J Comp Neurol 490*, 57–71.
94. Meriaux, C., Arafah, K., Tasiemski, A., Wisztorski, M., Bruand, J., Boidin-Wichlacz, C., Desmons, A., Debois, D., Laprevote, O., Brunelle, A., Gaasterland, T., Macagno, E., Fournier, I., and Salzet, M. Multiple changes in peptide and lipid expression associated with regeneration in the nervous system of the medicinal leech, *PLoS One 6*, e18359.
95. Bruand, J., Sistla, S., Meriaux, C., Dorrestein, P. C., Gaasterland, T., Ghassemian, M., Wisztorski, M., Fournier, I., Salzet, M., Macagno, E., and Bafna, V. (2011) Automated Querying and Identification of Novel Peptides using MALDI Mass Spectrometric Imaging, *J Proteome Res*, In Press.
96. Wisztorski, M., Croix, D., Macagno, E., Fournier, I., and Salzet, M. (2008) Molecular MALDI imaging: An emerging technology for neuroscience studies, *Dev Neurobiol 68*, 845–858.
97. Montolio, M., Messeguer, J., Masip, I., Guijarro, P., Gavin, R., Antonio Del Rio, J., Messeguer, A., and Soriano, E. (2009) A

semaphorin 3A inhibitor blocks axonal chemorepulsion and enhances axon regeneration, *Chem Biol 16*, 691–701.

98. Cebria, F., Nakazawa, M., Mineta, K., Ikeo, K., Gojobori, T., and Agata, K. (2002) Dissecting planarian central nervous system regeneration by the expression of neural-specific genes, *Dev Growth Differ 44*, 135–146.
99. Astic, L., Pellier-Monnin, V., Saucier, D., Charrier, C., and Mehlen, P. (2002) Expression of netrin-1 and netrin-1 receptor, DCC, in the rat olfactory nerve pathway during development and axonal regeneration, *Neuroscience 109*, 643–656.
100. Shifman, M. I., and Selzer, M. E. (2000) Expression of the netrin receptor UNC-5 in lamprey brain: modulation by spinal cord transection, *Neurorehabil Neural Repair 14*, 49–58.
101. Blackshaw, S. E., Babington, E. J., Emes, R. D., Malek, J., and Wang, W. Z. (2004) Identifying genes for neuron survival and axon outgrowth in Hirudo medicinalis, *J Anat 204*, 13–24.
102. Vergote, D., Macagno, E. R., Salzet, M., and Sautiere, P. E. (2006) Proteome modifications of the medicinal leech nervous system under bacterial challenge, *Proteomics 6*, 4817–4825.
103. Vergote, D., Sautiere, P. E., Vandenbulcke, F., Vieau, D., Mitta, G., Macagno, E. R., and Salzet, M. (2004) Up-regulation of neurohemerythrin expression in the central nervous system of the medicinal leech, Hirudo medicinalis, following septic injury, *J Biol Chem 279*, 43828–43837.
104. Xu, Y., Bolton, B., Zipser, B., Jellies, J., Johansen, K. M., and Johansen, J. (1999) Gliarin and macrolin, two novel intermediate filament proteins specifically expressed in sets and subsets of glial cells in leech central nervous system, *J Neurobiol 40*, 244–253.
105. Venkitaramani, D. V., Wang, D., Ji, Y., Xu, Y. Z., Ponguta, L., Bock, K., Zipser, B., Jellies, J., Johansen, K. M., and Johansen, J. (2004) Leech filamin and Tractin: markers for muscle development and nerve formation, *J Neurobiol 60*, 369–380.
106. Sethi, J., Zhao, B., Cuvillier-Hot, V., Boidin-Wichlacz, C., Salzet, M., Macagno, E. R., and Baker, M. W. (2010) The receptor protein tyrosine phosphatase HmLAR1 is up-regulated in the CNS of the adult medicinal leech following injury and is required for neuronal sprouting and regeneration, *Mol Cell Neurosci 45*, 430–438.
107. Huang, Y., Jellies, J., Johansen, K. M., and Johansen, J. (1997) Differential glycosylation of tractin and LeechCAM, two novel Ig superfamily members, regulates neurite extension and fascicle formation, *J Cell Biol 138*, 143–157.
108. Huang, Y., Jellies, J., Johansen, K. M., and Johansen, J. (1998) Development and pathway formation of peripheral neurons during leech embryogenesis, *J Comp Neurol 397*, 394–402.
109. Jie, C., Xu, Y., Wang, D., Lukin, D., Zipser, B., Jellies, J., Johansen, K. M., and Johansen, J. (2000) Posttranslational processing and differential glycosylation of Tractin, an Ig-superfamily member involved in regulation of axonal outgrowth, *Biochim Biophys Acta 1479*, 1–14.
110. Xu, Y. Z., Ji, Y., Zipser, B., Jellies, J., Johansen, K. M., and Johansen, J. (2003) Proteolytic cleavage of the ectodomain of the L1 CAM family member Tractin, *J Biol Chem 278*, 4322–4330.
111. Gan, W. B., Wong, V. Y., Phillips, A., Ma, C., Gershon, T. R., and Macagno, E. R. (1999) Cellular expression of a leech netrin suggests roles in the formation of longitudinal nerve tracts and in regional innervation of peripheral targets, *J Neurobiol 40*, 103–115.
112. Ji, Y., Schroeder, D., Byrne, D., Zipser, B., Jellies, J., Johansen, K. M., and Johansen, J. (2001) Molecular identification and sequence analysis of Hillarin, a novel protein localized at the axon hillock, *Biochim Biophys Acta 1519*, 246–249.
113. Biswas, S. C., Dutt, A., Baker, M. W., and Macagno, E. R. (2002) Association of LAR-like receptor protein tyrosine phosphatases with an enabled homolog in Hirudo medicinalis, *Mol Cell Neurosci 21*, 657–670.
114. Finger, F. P., Kopish, K. R., and White, J. G. (2003) A role for septins in cellular and axonal migration in C. elegans, *Dev Biol 261*, 220–234.
115. Ji, Y., Rath, U., Girton, J., Johansen, K. M., and Johansen, J. (2005) D-Hillarin, a novel W180-domain protein, affects cytokinesis through interaction with the septin family member Pnut, *J Neurobiol 64*, 157–169.
116. Baker, M. W., and Macagno, E. R. (2010) Expression levels of a LAR-like receptor protein tyrosine phosphatase correlate with neuronal branching and arbor density in the medicinal leech, *Dev Biol 344*, 346–357.
117. Aisemberg, G. O., Wong, V. Y., and Macagno, E. R. (1995) Genesis of segmental identity in the leech nervous system, *EXS 72*, 77–87.
118. Aisemberg, G. O., Wysocka-Diller, J., Wong, V. Y., and Macagno, E. R. (1993) Antennapedia-class homebox genes define diverse neuronal sets in the embryonic CNS of the leech, *J Neurobiol 24*, 1423–1432.

119. Schikorski, D., Cuvillier-Hot, V., Leippe, M., Boidin-Wichlacz, C., Slomianny, C., Macagno, E., Salzet, M., and Tasiemski, A. (2008) Microbial challenge promotes the regenerative process of the injured central nervous system of the medicinal leech by inducing the synthesis of antimicrobial peptides in neurons and microglia, *J Immunol 181*, 1083–1095.

120. Schikorski, D., Cuvillier-Hot, V., Boidin-Wichlacz, C., Slomianny, C., Salzet, M., and Tasiemski, A. (2009) Deciphering the immune function and regulation by a TLR of the cytokine EMAPII in the lesioned central nervous system using a leech model, *J Immunol 183*, 7119–7128.

121. Croq, F., Vizioli, J., Tuzova, M., Tahtouh, M., Sautiere, P. E., Van Camp, C., Salzet, M., Cruikshank, W. W., Pestel, J., and Lefebvre, C. (2010) A homologous form of human interleukin 16 is implicated in microglia recruitment following nervous system injury in leech Hirudo medicinalis, *Glia 58*, 1649–1662.

122. Tahtouh, M., Croq, F., Vizioli, J., Sautiere, P. E., Van Camp, C., Salzet, M., Daha, M. R., Pestel, J., and Lefebvre, C. (2009) Evidence for a novel chemotactic C1q domain-containing factor in the leech nerve cord, *Mol Immunol 46*, 523–531.

123. Farber, K., Cheung, G., Mitchell, D., Wallis, R., Weihe, E., Schwaeble, W., and Kettenmann, H. (2009) C1q, the recognition subcomponent of the classical pathway of complement, drives microglial activation, *J Neurosci Res 87*, 644–652.

124. Guo, L. H., Mittelbronn, M., Brabeck, C., Mueller, C. A., and Schluesener, H. J. (2004) Expression of interleukin-16 by microglial cells in inflammatory, autoimmune, and degenerative lesions of the rat brain, *J Neuroimmunol 146*, 39–45.

125. Kleinschek, M. A., Owyang, A. M., Joyce-Shaikh, B., Langrish, C. L., Chen, Y., Gorman, D. M., Blumenschein, W. M., McClanahan, T., Brombacher, F., Hurst, S. D., Kastelein, R. A., and Cua, D. J. (2007) IL-25 regulates Th17 function in autoimmune inflammation, *J Exp Med 204*, 161–170.

126. Liebrich, M., Guo, L. H., Schluesener, H. J., Schwab, J. M., Dietz, K., Will, B. E., and Meyermann, R. (2007) Expression of interleukin-16 by tumor-associated macrophages/activated microglia in high-grade astrocytic brain tumors, *Arch Immunol Ther Exp (Warsz) 55*, 41–47.

127. Melzer, N., Meuth, S. G., Torres-Salazar, D., Bittner, S., Zozulya, A. L., Weidenfeller, C., Kotsiari, A., Stangel, M., Fahlke, C., and Wiendl, H. (2008) A beta-lactam antibiotic dampens excitotoxic inflammatory CNS damage in a mouse model of multiple sclerosis, *PLoS One 3*, e3149.

128. Mueller, C. A., Schluesener, H. J., Conrad, S., Pietsch, T., and Schwab, J. M. (2006) Spinal cord injury-induced expression of the immune-regulatory chemokine interleukin-16 caused by activated microglia/macrophages and CD8+ cells, *J Neurosurg Spine 4*, 233–240.

129. Macagno, E. R., Gaasterland, T., Edsall, L., Bafna, V., Soares, M. B., Scheetz, T., Casavant, T., Da Silva, C., Wincker, P., Tasiemski, A., and Salzet, M. (2010) Construction of a medicinal leech transcriptome database and its application to the identification of leech homologs of neural and innate immune genes, *BMC Genomics 11*, 407.

130. Batchelor, P. E., and Howells, D. W. (2003) CNS regeneration: clinical possibility or basic science fantasy?, *J Clin Neurosci 10*, 523–534.

131. Popovich, P. G. (2000) Immunological regulation of neuronal degeneration and regeneration in the injured spinal cord, *Prog Brain Res 128*, 43–58.

132. Salzet, M., and Macagno, E. (2009) Recent Advances on Development, Regeneration and Immune Responses of the Leech Nervous System. In: Shain DH, editor. Annelids as Models Systems in the Biological Sciences, *Wiley Blackwell*, 156–185.

133. Tasiemski, A., and Salzet, M. (2010) Leech Immunity : From brain to peripheric responses. In: Söderhäll K, editor. Invertebrate Immunity. Advances in Experimental Medicine and Biology. Uppsala: Department of Comparative Physiology, *Landes Bioscience 708*.

134. Farooqui, A. A., Ong, W. Y., Horrocks, L. A., Chen, P., and Farooqui, T. (2007) Comparison of biochemical effects of statins and fish oil in brain: the battle of the titans, *Brain Res Rev 56*, 443–471.

135. Lin, Y. H., and Salem, N., Jr. (2007) Whole body distribution of deuterated linoleic and alpha-linolenic acids and their metabolites in the rat, *J Lipid Res 48*, 2709–2724.

136. Calon, F., and Cole, G. (2007) Neuroprotective action of omega-3 polyunsaturated fatty acids against neurodegenerative diseases: evidence from animal studies, *Prostaglandins Leukot Essent Fatty Acids 77*, 287–293.

137. Calon, F., Lim, G. P., Yang, F., Morihara, T., Teter, B., Ubeda, O., Rostaing, P., Triller, A., Salem, N., Jr., Ashe, K. H., Frautschy, S. A., and Cole, G. M. (2004) Docosahexaenoic acid protects from dendritic pathology in an Alzheimer's disease mouse model, *Neuron 43*, 633–645.

138. Mazza, M., Pomponi, M., Janiri, L., Bria, P., and Mazza, S. (2007) Omega-3 fatty acids and antioxidants in neurological and psychiatric diseases: an overview, *Prog Neuropsychopharmacol Biol Psychiatry 31*, 12–26.

139. Pereira, L. M., Hatanaka, E., Martins, E. F., Oliveira, F., Liberti, E. A., Farsky, S. H., Curi, R., and Pithon-Curi, T. C. (2008) Effect of oleic and linoleic acids on the inflammatory phase of wound healing in rats, *Cell Biochem Funct 26*, 197–204.

140. Drouet, J., Dupont, J., Nguyen, T. L., Fournie, J., and Goyffon, M. (1972) Fatty acids and phospholipids in repair tissue of a fracture site in irradiated mice effect of cysteamine, *C R Seances Soc Biol Fil 166*, 1585–1591.

141. Gorio, A. (1986) Ganglioside enhancement of neuronal differentiation, plasticity, and repair, *CRC Crit Rev Clin Neurobiol 2*, 241–296.

142. Salzet, M. (2005) Neuropeptide-derived antimicrobial peptides from invertebrates for biomedical applications, *Curr Med Chem 12*, 3055–3061.

143. Salzet, M. (2006) Invertebrate Neuropeptides and Hormones: Basic Knowledge and Recent Advances *Res. Signpost 1*, 17–37.

144. Sugiura, Y., Shimma, S., Konishi, Y., Yamada, M. K., and Setou, M. (2008) Imaging mass spectrometry technology and application on ganglioside study; visualization of age-dependent accumulation of C20 ganglioside molecular species in the mouse hippocampus, *PLoS One 3*, e3232.

145. Beal, M. F., and Hantraye, P. (2001) Novel therapies in the search for a cure for Huntington's disease, *Proc Natl Acad Sci U S A 98*, 3–4.

146. Pierson, J., Norris, J. L., Aerni, H. R., Svenningsson, P., Caprioli, R. M., and Andren, P. E. (2004) Molecular profiling of experimental Parkinson's disease: direct analysis of peptides and proteins on brain tissue sections by MALDI mass spectrometry, *J Proteome Res 3*, 289–295.

147. Wisztorski, M., Franck, J., Salzet, M., and Fournier, I. MALDI direct analysis and imaging of frozen versus FFPE tissues: what strategy for which sample?, *Methods Mol Biol 656*, 303–322.

148. Stauber, J., Macaleese, L., Franck, J., Claude, E., Snel, M., Kukrer Kaletas, B., Wiel, I. M., Wisztorski, M., Fournier, I., and Heeren, R. M. (2009) On-Tissue Protein Identification and Imaging by MALDI-Ion Mobility Mass Spectrometry, *J Am Soc Mass Spectrom.*

149. Barzilai, A., Zilkha-Falb, R., Daily, D., Stern, N., Offen, D., Ziv, I., Melamed, E., and Shirvan, A. (2000) The molecular mechanism of dopamine-induced apoptosis: identification and characterization of genes that mediate dopamine toxicity, *J Neural Transm Suppl*, 59–76.

150. Stoeckli, M., Knochenmuss, R., McCombie, G., Mueller, D., Rohner, T., Staab, D., and Wiederhold, K. H. (2006) MALDI MS imaging of amyloid, *Methods Enzymol 412*, 94–106.

151. O'Brien, E., Dedova, I., Duffy, L., Cordwell, S., Karl, T., and Matsumoto, I. (2006) Effects of chronic risperidone treatment on the striatal protein profiles in rats, *Brain Res 1113*, 24–32.

152. Chen, H., Talaty, N. N., Takats, Z., and Cooks, R. G. (2005) Desorption electrospray ionization mass spectrometry for high-throughput analysis of pharmaceutical samples in the ambient environment, *Anal Chem 77*, 6915–6927.

153. Kauppila, T. J., Wiseman, J. M., Ketola, R. A., Kotiaho, T., Cooks, R. G., and Kostiainen, R. (2006) Desorption electrospray ionization mass spectrometry for the analysis of pharmaceuticals and metabolites, *Rapid Commun Mass Spectrom 20*, 387–392.

154. Talaty, N., Takats, Z., and Cooks, R. G. (2005) Rapid in situ detection of alkaloids in plant tissue under ambient conditions using desorption electrospray ionization, *Analyst 130*, 1624–1633.

155. Altelaar, A. F., Luxembourg, S. L., McDonnell, L. A., Piersma, S. R., and Heeren, R. M. (2007) Imaging mass spectrometry at cellular length scales, *Nat Protoc 2*, 1185–1196.

156. Franck, J., Longuespee, R., Wisztorski, M., Van Remoortere, A., Van Zeijl, R., Deelder, A., Salzet, M., McDonnell, L., and Fournier, I. MALDI mass spectrometry imaging of proteins exceeding 30,000 daltons, *Med Sci Monit 16*, BR293-299.

157. van Remoortere, A., van Zeijl, R. J., van den Oever, N., Franck, J., Longuespee, R., Wisztorski, M., Salzet, M., Deelder, A. M., Fournier, I., and McDonnell, L. A. MALDI imaging and profiling MS of higher mass proteins from tissue, *J Am Soc Mass Spectrom 21*, 1922–1929.

Chapter 12

Profiling of HIV Proteins in Cerebrospinal Fluid

Melinda Wojtkiewicz and Pawel Ciborowski

Abstract

HIV-1 proteins are rarely identified during mass spectrometry-based proteomic profiling studies of body fluids from HIV-1-infected people even when elaborated fractionation schema and highly sensitive instruments are used. Genotyping of HIV-1 isolated from body fluids does not provide exact information about characteristics of circulating proteins and is a limiting factor in expanding an important segment of our knowledge about the course of infection. Therefore, we propose that in vitro amplification of freshly isolated virus followed by sucrose cushion purification will yield sufficient amounts of viral proteins for mass spectrometric characterization. This chapter provides protocols for virus propagation using CD4+ T cell line or human macrophages, virus purification, and preparation of samples for two-dimensional electrophoresis and mass spectrometry analyses.

Key words: Cerebrospinal fluid, Proteomics, Biomarkers, HIV proteins

1. Introduction

HIV proteins have been at the center of research interest for many years as potential antigens for developing a protective vaccine, diagnostic purposes, as well as explaining mechanisms of HIV pathogenesis with special emphasis on neurological complications (1, 2). It has been shown for more than a decade that viral proteins such as envelope glycoproteins gp120, gp160, and gp41; the non-structural protein Nef; the trans-activating gene regulatory protein Tat; and HIV-1 accessory protein vpr are neurotoxic (3). Further on, it has been shown that viral proteins can compromise the integrity of blood–brain barrier (BBB) (4). The exact mechanism of neurotoxicity is not fully understood, and many direct and indirect molecular mechanisms have been proposed (5). Despite many studies, a correlation between molecular structure of these viral proteins and the risk of developing cognitive impairment has not

Yannis Karamanos (ed.), *Expression Profiling in Neuroscience*, Neuromethods, vol. 64,
DOI 10.1007/978-1-61779-448-3_12, © Springer Science+Business Media, LLC 2012

been proven. Genetic studies of HIV *env* gene led to the classification of the viral types and subtypes (6); however, correlation between subtypes and neurological complications has not been proven (7). Moreover, the envelope protein has a high degree of sequence variability. In recent years, glycosylation of gp120 protein which may have a profound effect on protein folding demonstrated that conformational changes in the structure of gp120 lead to increased susceptibility to virus-neutralizing antibodies (8). Whether changes in glycosylation lead to changes in immunogenicity and/or antigenicity or is related to other pathogenic effects is an open question (8–10). After introduction of combination antiretroviral therapy (cART, formerly HAART), a productive infection measured by RT assay is suppressed to the borderline levels of detection. Low levels of viral copies such as less than 20/mL lead to production of some level of proteins and maintain chronic immune activation (11). Viral proteins have been shown to be produced and located inside infected cells, on their surface, and in CSF and plasma (12). However, levels of these proteins circulating in these two body fluids are very low, and more extensive structural investigations are not possible without propagation and isolation of the virus. Respective protocols are provided in this chapter.

2. Materials

General Laboratory Materials

10-mL regular tip serological pipette, sterile and disposable (BD Biosciences, San Jose, CA)

15-mL BD Falcon™ conical tube (BD Biosciences)

25-mL regular tip serological pipet, sterile and disposable (BD Biosciences)

50-mL BD Falcon™ conical tube (BD Biosciences)

0.5-mL microcentrifuge tubes (Thermo Fisher Scientific)

Dry ice

Ethanol

Razor blade

Equipment

Isoelectric focusing (IEF) system [e.g., IPGphor II apparatus (GE Healthcare, Piscataway, NJ)]

Gel electrophoresis system [e.g., Ettan DaltSix Electrophoresis System™ (GE Healthcare)]

System to scan fluorescence [e.g., Typhoon 9410 Variable Mode Imager (GE Healthcare)]

Software for gel image analysis [e.g., DeCyder 2D 6.5™ software (GE Healthcare)]

Speed-Vac

4800 MALDI TOF–TOF (AB Sciex)

Tempo LC with MALDI spotter (AB Sciex)

Mascot or GPS Explorer

LTQ Orbitrap (Thermo Scientific) with NanoLC system (Eksigent)

Proteome Discoverer 1.0 or BioWorks 3.3.1 (Thermo Scientific)

2.1. Virus Propagation

CD4+ T cell line, e.g., H9

Seed stocks of T cell line-adapted viruses

25-cm^2 tissue culture flasks (Fisher Scientific)

RPMI medium (Sigma-Aldrich, St. Louis, MO)

Fetal bovine serum (Sigma-Aldrich)

RPMI medium/10% fetal bovine serum

Cryovials (Fisher Scientific)

Ciprofloxacin (Sigma-Aldrich)

Gentamicin (Sigma-Aldrich)

Macrophage colony-stimulating factor (MCSF) (Sigma-Aldrich)

Heat-inactivated human serum (ΔHS, Sigma-Aldrich)

Dulbecco's modified Eagle's medium (DMEM) with phenol red (Invitrogen)

DMEM without phenol red (Invitrogen)

L-glutamine (Invitrogen)

Media A with phenol red

(a) To 1 L of DMEM with phenol red, add the following:
   - 400 μL of ciprofloxacin (400 μL aliquots stored at −20°C)
   - 2 mL of gentamicin (final conc. 0.2%; stock kept at 4°C)
   - 100 mL of ΔHS, thawed and centrifuged at 3,000 rpm for 8 min (final conc. 10%; stored at −20°C)

(b) If L-glutamine is not already in the DMEM, add 10 mL (final conc. 1%; 10 mL aliquots stored at −20°C)

(c) 1 mL MCSF (for media A only; 1 mL aliquots stored at −20°C)

(d) Store media at 4°C.

Media B with phenol red

(a) To 1 L of DMEM with phenol red, add the following:
   - 400 μL of ciprofloxacin (400 μL aliquots stored at −20°C)
   - 2 mL of gentamicin (final conc. 0.2%; stock kept at 4°C)
   - 100 mL of ΔHS, thawed and centrifuged at 3,000 rpm for 8 min (final conc. 10%; stored at −20°C)

(b) If L-glutamine is not already in the DMEM, add 10 mL (final conc. 1%; 10 mL aliquots stored at –20°C)

(c) Store media at 4°C.

Media B without phenol red

(a) To 500 mL of DMEM without phenol red, add the following:
- 200 μL of ciprofloxacin (400 μL aliquots stored at –20°C)
- 1 mL of gentamicin (final conc. 0.2%; stock kept at 4°C)
- 5 mL of glutamine (final conc. 1%; 10 mL aliquots stored at –20°C)
- 50 mL of ΔHS, thawed and centrifuged at 3,000 rpm for 8 min (final conc. 10%; stored at –20°C)

(b) Store media at 4°C.

2.2. Virus Purification

Sucrose (Sigma-Aldrich, S7903)

Phosphate-buffered saline (PBS) powder (Sigma-Aldrich)

Sterile 1× PBS

Milli-Q Ultrapure water

60% Sucrose in PBS

50% Sucrose in PBS

40% Sucrose in PBS

30% Sucrose in PBS

20% Sucrose in PBS

10% Sucrose in PBS

2.3. Two-Dimensional Electrophoresis

Lysis buffer (30 mM Tris–HCl, pH 8, 7 M urea, 2 M thiourea, and 4% CHAPS (w/v)]

Sample buffer (7 M urea, 2 M thiourea, 4% CHAPS (w/v), 2% pharmalyte, and 130 mM dithiotreitol (DTT)]

Tris-(2-carboxyethyl)phosphine hydrochloride (TCEP)

CyDye DIGE fluor labeling kit (GE Healthcare)

Immobilized pH gel (IPG) strips. we used Immobiline™ DryStrip gel (24 cm) (GE Healthcare)

Equilibration solution for CyDye-labeled samples (50 mM Tris–HCl, pH 8.8, 6 M urea, 30% glycerol, 2% sodium dodecylsulfate (SDS), and 0.01% bromophenol blue (w/v)]

DTT (Sigma-Aldrich)

Iodoacetamide

12% Polyacrylamide gels

0.5% Agarose (w/v)

2.4. Preparation of Viral Proteins for Mass Spectrometry Analysis

Acetonitrile (ACN) (Fisher Scientific, Optima)

Ammonium bicarbonate (NH_4HCO_3) (Sigma-Aldrich)

Trifluoroacetic acid (TFA)(99%, Sigma-Aldrich)
Trypsin (Promega, Inc.)
C18 ZipTip (Sigma-Aldrich)
2,5-Dihydroxybenzoic acid (DHB) (Sigma-Aldrich)
alpha-Cyano-4-hydroxycinnamic acid (CHCA) (Sigma-Aldrich)
Formic acid (99%, Sigma-Aldrich)
Water, HPLC grade (Fisher Scientific)
Dry ice/ethanol bath
Wash buffer: 100 mM ammonium bicarbonate (NH_4HCO_3)
Wash solution: 50% ACN
Wash solution 2: 50% ACN, 50 mM NH_4HCO_3
Wash solution 3: 50% ACN, 10 mM NH_4HCO_3
Digestion buffer: 10 mM NH_4HCO_3
Extraction buffer: 60% ACN and 0.1% TFA
Wetting solution: 100% ACN
Resuspension buffer: 0.5% TFA
Equilibration/wash solution: 0.1% TFA
Elution solution: 50% ACN and 0.1% TFA
60% ACN with 0.1% TFA
PicoFrit C18 column-emitter (New Objectives)

3. Methods

3.1. Identification of HIV Proteins in Body Fluids

Proteomic profiling of CSF or serum/plasma from HIV-infected patients with relatively high number of viral copies measured by reverse transcriptase (RT) assay does not yield enough material for sequence coverage, not to mention in-depth structural studies of these proteins. Table 1 shows a summary of HIV proteins and respective peptides identified by mass spectrometry during profiling of CSF samples from patients with cognitive impairment. In this early study, samples were obtained from patients with HIV-associated dementia (HAD) in which viral load tests showed the presence of productive infection and viral replication. Since characterization of viral proteins was not an objective of this study, only non-posttranslationally modified peptides were identified using Sequest algorithm to search NCBI database (12). Further studies using plasma/serum did not yield better sequence coverage (13). It is important to note that neurotoxic HIV Tat protein has not been found in any of the tested samples using proteomic profiling approach. Therefore, for characterization of viral proteins, one has to isolate them from body fluids using affinity chromatography, or isolate, propagate, and purify the virus.

Table 1
HIV-1 proteins identified in 12 CSF samples from HIV-1-infected patients

Protein	Peptide sequence	NCBI	z	XCorr	DeltaCn	Ions
Envelope glycoprotein	R.IGPGQAFYATGAIIGDIRQAHCNISSDK.W	58220977	3	3.01	0.32	27/108
Envelope glycoprotein	K.GDMKNCSFNITTNIKGK.M	1845947	3	3.14	0.33	22/64
Envelope glycoprotein	R.SENITNNVKNIIAQLTEPVK.I	62906476	3	3.04	0.41	26/76
Envelope glycoprotein	R.SENITNNAKIIIAHLNESVEINCTR.-	72539346	3	2.97	0.3	14/96
Envelope glycoprotein	K.SVRIGPGQTFYATGDVTGDIR.K	46948945	3	3.1	0.23	25/80
Envelope glycoprotein	gp120 K.EALQEVVEQLR.Q	85662893	2	2.86	0.27	14/20
Envelope glycoprotein	gp120 K.EALQEVVEQLR.Q	85662893	2	2.71	0.3	14/20
Envelope glycoprotein	gp120 K.EALQEVVEQLR.Q	85662893	2	3.09	0.24	14/20
Envelope glycoprotein	gp120 K.EALQEVVEQLR.Q	85662893	2	3.18	0.32	15/20
Envelope glycoprotein	gp120 K.EALQEVVEQLR.Q	85662893	3	3.13	-	19/40
gag–pol fusion polyprotein	R.EFPSEQTRXNSPTR.A	37934098	3	3.01	0.23	25/52
gag–pol fusion polyprotein	3 R.EFPSEQTRXNSPTR.A	37934098	3	3.1	0.22	27/52
gag protein	K.TSITMQRSNFKGPK.R	80975076	2	2.73	0.42	14/26
vpu protein	MLNLQARIDYR.L	13569324	2	2.54	0.35	14/20
pol polyprotein	K.QFTEAVQK.I	13095233	1	2.08	-	10/14
vif protein	K.PPLPSVMKLTEDRWNK.P	2853437	3	3.15	0.31	26/60
Reverse transcriptase	K.LVDXRELNR.R	29650702	2	2.59	0.21	14/16

Adapted from Rozek et al. (12)

3.2. Virus Propagation

Viruses can be propagated using in vitro cultures and many standard protocols have been established (14, 15). High numbers of HIV particles can be generated by using either 293 T cells, human CD4+ T-cell line, or human monocyte-derived macrophages (MDM) (16, 17).

3.2.1. Virus Propagation by Human CD4+ T-Cell Line (Note 1)

1. In a 25-cm^2 tissue culture flask, infect 5×10^6 cells from the CD4$^+$ T-cell line with 0.5–1 mL of viral stock at a multiplicity of infection of 0.01. Incubate for 3 h.
2. Centrifuge the culture for 5 min at 1,000 rpm at room temperature. Remove supernatant and resuspend cells in 5 mL of RPMI 1640 medium containing 10% FBS.
3. Incubate the culture, monitoring for cytopathic effect daily (Fig. 1). Maintain cell concentration at 1×10^6 cells/mL by counting cells and adding medium to appropriate volumes every 48–72 h.
4. Harvest viral supernatant 3–4 days after infection by centrifuging cells for 5 min at 1,500 rpm at room temperature.
5. Transfer the supernatant in 0.5 mL aliquots into cryovials. Snap-freeze in dry ice/ethanol bath and then transfer to a –80°C freezer for long-term storage.

3.2.2. Monocyte Differentiation to MDM and HIV-1 Infection

1. Monocyte differentiation.
   (a) Dilute monocytes to 1×10^6 cells/mL in media A with phenol red. Plate cells on 6-well plates (3×10^6 cells/well). In addition, plate 100,000 cells/well in a 96-well plate if you need small aliquots for future analyses, e.g., RT assay after HIV in vitro infection. Incubate at 37°C with 5% CO_2.
   (b) Exchange half of the media with media A with phenol red on days 2 and 5.
2. MDM infection (day 7)
   (a) Prepare a 1:10 dilution of HIV-1_{ADA} in media B with phenol red. Exchange full media, removing all media from each well.
   (b) Infect half of the 6-well plates with HIV-1 by placing 1 mL of the diluted HIV-1 into each well. Infect four of the eight wells on the 96-well plates by adding 20 μL of the diluted HIV-1 to each well. The remaining 6-well plates are designated as uninfected. Incubate with 1 mL media B with phenol red. Incubate the four uninfected wells on each of the 96-well plates with 20 μL media B with phenol red.
   (c) After incubating for 4 h at 37°C with 5% CO_2, add an additional 3 mL media B with phenol red to each well of the 6-well plates (infected and uninfected). To each of the 8

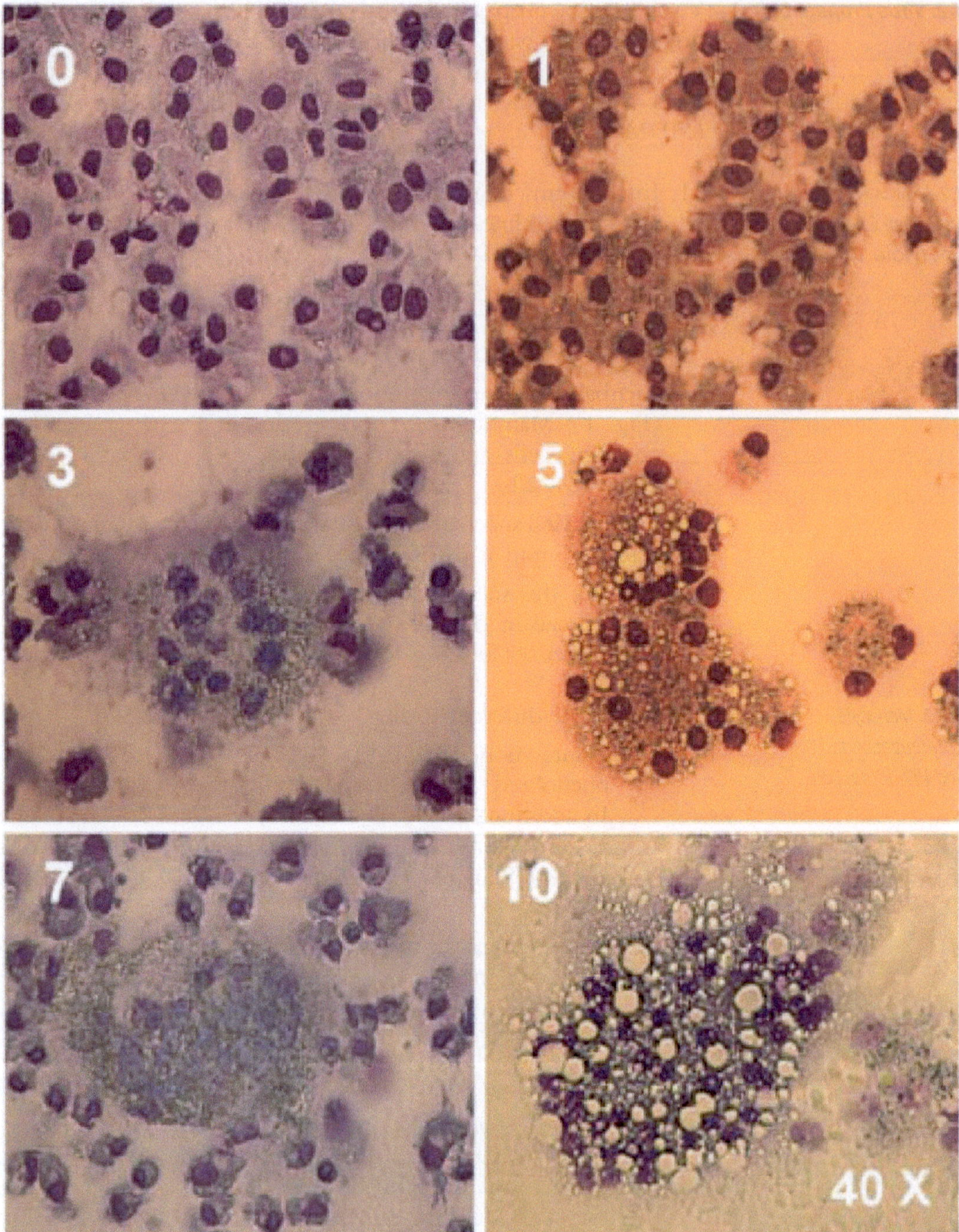

Fig. 1. Progression of in vitro infection of human monocyte-derived macrophages with HIV_{ADA}. Shortly after infection, macrophages start fusing creating multinucleated giant cells (MGC). (Adapted from Ciborowski et al. "Investigating the Human Immunodeficiency Virus Type One- Infected Monocyte-Derived Macrophage Secretome." Virology. 2007, 363:198–209).

wells on the 96-well plates, add 60 μL media B with phenol red.

(d) Incubate all plates at 37°C with 5% CO_2 for an additional 20 h. Total infection time is 24 h.

3. Day 1 postinfection. Perform a full media exchange on each plate. Remove all 4 mL and 80 μL from the 6-well and 96-well plates, respectively. Add 3 mL media B with phenol red to each well on the 6-well plates, and 100 μL to each of the eight wells on the 96-well plates.
4. Day 3 postinfection. Perform half media exchanges to all wells using media B with phenol red. Increase serum to 20%.
5. Day 5 postinfection. Collect medium with propagated virus.

3.3. Virus Purification

Virus purification can be done using a sucrose gradient, or commercially available iodixanol gradients and Sepharose columns (Sigma-Aldrich). There are many versions of gradients available to try, varying in percentages of sucrose and solvent composition (15, 18). Before running through the gradient, viral supernatants should be spun for 10 min to remove cellular debris and then filtered through a 0.2-μm filter.

3.3.1. Sucrose Gradient Centrifugation

1. Prepare sucrose step gradients in 40-mL ultracentrifuge tube by layering the following sucrose solutions into the tube in the following order:
   (a) 1 mL 60% sucrose in PBS
   (b) 1 mL 50% sucrose in PBS
   (c) 1 mL 40% sucrose in PBS
   (d) 1 mL 30% sucrose in PBS
   (e) 1 mL 20% sucrose in PBS
   (f) 1 mL 10% sucrose in PBS.
2. Load sample (>2.5 mg/mL of viral stock) onto the top of the gradient.
3. Centrifuge for 1 h at 25,000 rpm. The virus settles between the 30% and 60% sucrose layers and can be best seen against a dark background.
4. Remove the top layers of sucrose and then collect the virus.
5. Prepare a sucrose cushion in 40-mL ultracentrifuge tube by layering first 2 mL of 70% sucrose solution and then 2 mL of 20% solution into the tube.
6. Centrifuge for 1 h at 25,000 rpm.
7. Remove the top layer of sucrose and then collect the viral pellet at the bottom of the tube.
8. Store viral pellet at –80°C.

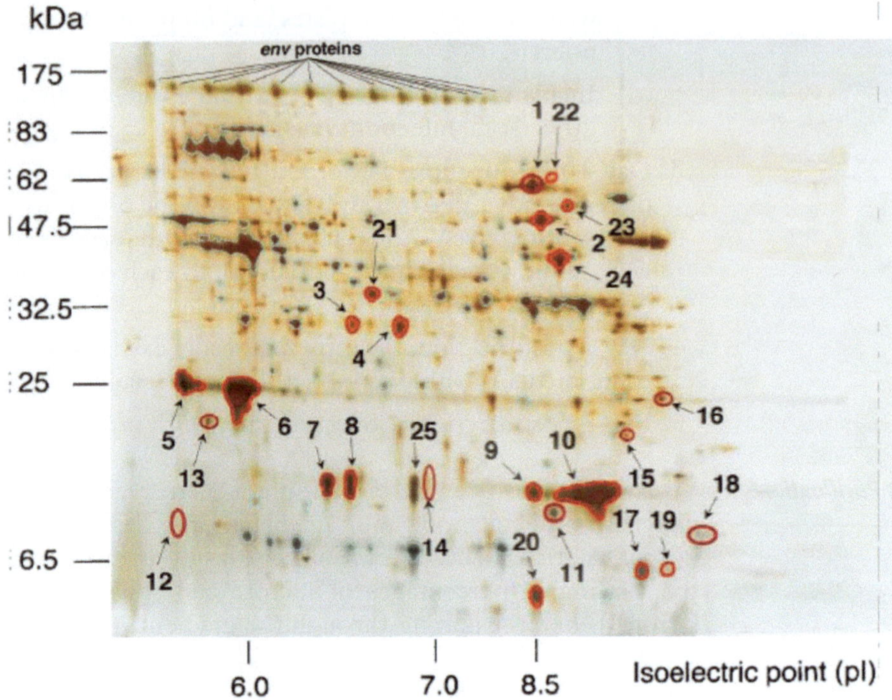

Fig. 2. 2DE separation of HIV_{LAV-1} proteins obtained from isolated and purified virus propagated on T-cell line. (Adapted from Misumi, S. et al. *"Three Isoforms of Cyclophilin Associated with Human Immunodeficiency Virus Type 1 Were Found by Proteomics by Using Two-Dimensional Gel Electrophoresis and Matrix-Assisted Laser Desorption Ionization–Time of Flight Mass Spectrometry."* J. Virol. 2002, 76:10000).

3.4. Two-Dimensional Electrophoresis

Two-dimensional electrophoresis (2DE) consists of two steps. First step (dimension) is a separation based on isoelectric point (IEF). Second step (dimension) is based on molecular weight (Fig. 2). Both dimension separations can be performed using equipment from various manufacturers. In this protocol, we use IPGphor 2 from Amersham/GE Healthcare, Inc.

1. In the IEF tray adapted to the IEF system used, apply the samples along the rows of the tray.
2. With tweezers carefully remove the protective plastic band covering the Immobiline DryStrips gels (24 cm long) with linear immobilized pH gradient 3–10 and place the strip, with the gel facing down, over the sample.
3. Cover the tray with aluminum foil and leave it overnight to rehydrate the gel strip.
4. Place the tray in the IEF apparatus following the manufacturer's recommendation and start the IEF steps (Note 2).
5. After IEF, stop the IEF system and take the tray out of the system.

6. With tweezers take the Immobiline DryStrips, if necessary carefully remove the excess mineral oil with a paper and place the Immobilin DryStrip in an adapted tube and add 10 mL of equilibration solution (50 mM Tris–HCl, pH 8.8, 6 M urea, 30% glycerol, 2% SDS, and 0.01% bromophenol blue) completed with 100 mM DTT. Rock the tube for 15 min at room temperature.
7. Decant the equilibration solution and add 10 mL of equilibration solution completed with 100 mM iodoacetamide. Rock the Immobilin DryStrips for 15 min at room temperature (do not apply the alkylation steps for minimal dye-labeled samples).
8. Load the Immobilin DryStrip on the top of 12% polyacrylamide gels.
9. Overlay the strip with 0.5% agarose.
10. Every electrophoresis system large enough for the migration of 24-cm gels can be used for the second step (we used Ettan Daltsix Electrophoresis System™). Start the second-dimension migration step (see Note 3).

3.5. DeCyder™ Analysis

1. For visualization of protein spots, signals from each cyanine have to be collected with the excitation and emission wavelength corresponding to the cyanine using a scanner for fluorescence. We used a Typhoon 9410 Variable Mode Imager (GE Healthcare) (see Note 4).
2. Gel analysis may be performed with dedicated software. We used the software DeCyder 2D 6.5™ software (GE Healthcare).
3. Gel analysis highlights spots showing a high variation between conditions and a low variability. Selected spots should be cut and analyzed by mass spectrometry.

3.6. Preparation of Viral Proteins for Mass Spectrometry Analysis

Proteins fractionated (separated) in polyacrylamide gel are most commonly fragmented by proteolytic enzymes to short (10–25 a.a.) peptides in preparation for protein identification and other mass spectrometry-based characterization. Trypsin is widely used because of its specificity to cleave peptide bonds at the carboxyl side of lysine and arginine, except when either is followed by proline.

3.6.1. In-Gel Tryptic Digestion

1. Put the isolated gel spot into a centrifuge tube.
2. Add 200 μL of 50% ACN and wash by shaking for 5 min.
3. Remove wash.
4. Add 200 μL 50% ACN and 50 mM NH_4HCO_3 to all samples.
5. Wash for 30 min at room temperature on a tilt table.
6. Remove wash.
7. Add 200 μL 50% ACN and 10 mM NH_4HCO_3 to the gel pieces.

8. Wash for 30 min at room temperature on a tilt table.
9. Remove wash.
10. Speed-Vac gel pieces to complete dryness.
11. Add 10 μL (0.1 μg/μL) modified trypsin to all samples (Note 5).
12. Let stand for 5–10 min to allow enzyme/buffer solution to absorb into the gel.
13. Add an additional 50 μL 10 mM NH_4HCO_3.
14. Incubate at 37°C for overnight (at least 16 h).
15. Extract the peptides by adding 200 μL 60% ACN and 0.1% TFA, and shaking at room temperature for 60 min (use more volume if necessary)
16. Remove buffer containing peptides and add to a new centrifuge tube.
17. Repeat steps 15 and 16 and add to the previous extract.
18. Dry the combined washes by Speed-Vac.

3.6.2. Sample Clean-Up

1. From stock solutions, aliquot the following in PCR tubes:
   (a) 25 μL wetting solution per ZipTip
   (b) 100 μL equilibration/wash solution per ZipTip
   (c) 100 μL elution solution per ZipTip
2. Dry samples, and resuspend in a minimal volume of 10–20 μL of 0.5% TFA.
3. Pipette 10 μL of wetting solution into a ZipTip. Dispense to waste. Repeat.
4. Aspirate equilibration solution into the same ZipTip. Dispense to waste. Repeat.
5. Bind peptides/proteins to ZipTip by fully aspirating and dispensing the sample on the ZipTip 3–10 times.
6. Aspirate wash solution into the ZipTip and dispense to waste. Repeat at least once.
7. Carefully aspirate and dispense 10 μL of elution solution through the pipette tip at least six times without pulling air back into the ZipTip. Dispense the elution into conical autosampler vials.
8. Speed-Vac dry (samples may be stored at −20°C until loading on LC/MS/MS).
9. Immediately before running samples on LC-MS/MS, add 5 μL mobile phase A (0.1% formic acid in HPLC water) and mix well by pipetting.

3.7. Characterization of Viral Proteins by Mass Spectrometry

Two types of ionization are commonly used in mass spectrometry of proteins and peptides: electrospray ionization (ESI) and matrix-assisted laser desorption ionization (MALDI). These complement each other by ionizing non-overlapping subsets of peptides of interest; therefore, we typically use both ionization techniques, and respective instruments are the LTQ Orbitrap (Thermo Scientific) and 4800 MALDI TOF–TOF (AB Sciex).

3.7.1. Electrospray Ionization Mass Spectrometry Using Ion Trap

A variety of HPLC systems are used to achieve a nano-flow rate of 250 nL/min sprayed into the orifice of mass spectrometers. The flow rate was created through a 1/75 split ratio in the flow lines when Surveyor HPLC is used. Eksigent and other newer LC systems do not require split of flow. A 10-port valve equipped with two, alternating, peptide traps is placed at the junction of LC and mass spectrometer so that when one sample is being analyzed, the next sample is loaded onto another trap and washed with mobile phase A, protecting the column from contaminant build-up. Many laboratories use microcapillary C18 emitter columns made in-house, while we used a PicoFrit C18 column-emitter from New Objectives, Inc. Mass spectrometer is tuned using direct infusion of 10 pmol/μL of angiotensin in mobile phase A. A tune file is a set of optimal parameters for detection of peptides. Although we used angiotensin for this purpose, other peptides can be used as well.

1. Use a syringe pump to flow 250 nL/min of angiotensin into a PicoTip emitter the same size as the PicoFrit column. Start manual data acquisition (Tune Plus) with two microscans and a fill time of 10. Typically, the transfer tube temperature is set to 200°C. Adjust spray voltage to somewhere between 1.6 and 2 kV. Use the automatic tune feature and tune on peak 649 *m/z*. Save tune file to be used in instrument method.
2. Create an instrument method. The typical method parameters for a proteomics experiment are summarized in Table 2. Operate acquisition in data-dependent mode (MS/MS). One precursor scan in the Orbitrap is followed by fragmentation of the five most abundant peaks in the LTQ (Fig. 3). For LCQ, we used three most abundant precursor ions.
3. Load peptides onto the peptide trap with 98:2 HPLC water with 1% formic acid: ACN with 1% formic acid and elute using a 90-min linear gradient of 0–80% acetonitrile with 1% formic acid. Typically, one blank should be run to find background peaks for a mass rejection list that is added to the instrument method. A blank should also be run before a set of samples to detect and avoid carryover peaks. A control injection of 25 fmol (LCQDecaPlus) or 10 fmol (LTQ Orbitrap) of digested bovine albumin or other protein digest is used to monitor performance of the nano-LC-MS/MS system.

3.7.2. Database Search and Interpretation

The last several years have led to a rapid development of existing and new software packages for database searches of spectra used from mass spectrometry analyses. Currently, output files provide detailed information about MS/MS fragmentation, mass error, sequence

Table 2
Typical LTQ Orbitrap settings for analysis of five of the most intense precursor ions

Orbitrap		LTQ				
First scan—resolution set to 60,000 scan from 300 to 2,000 m/z		MS/MS of 1st most intense peak in scan 1	MS/MS of 2nd most intense peak in scan 1	MS/MS of 3rd most intense peak in scan 1	MS/MS of 4th most intense peak in scan 1	MS/MS of 5th most intense peak in scan 1
MS parameters		MS/MS parameters				
Monoisotopic precursor selection enabled		Activation type			CID	
No charge state rejection		Min. signal required			50,000	
Dynamic exclusion enabled		Isolation width			2	
Repeat count	2	Normalized coll. energy			35	
Repeat duration	60	Default charge state			2	
Exclusion list size	500	Activation Q			0.250	
Exclusion duration	60	Activation time			30	

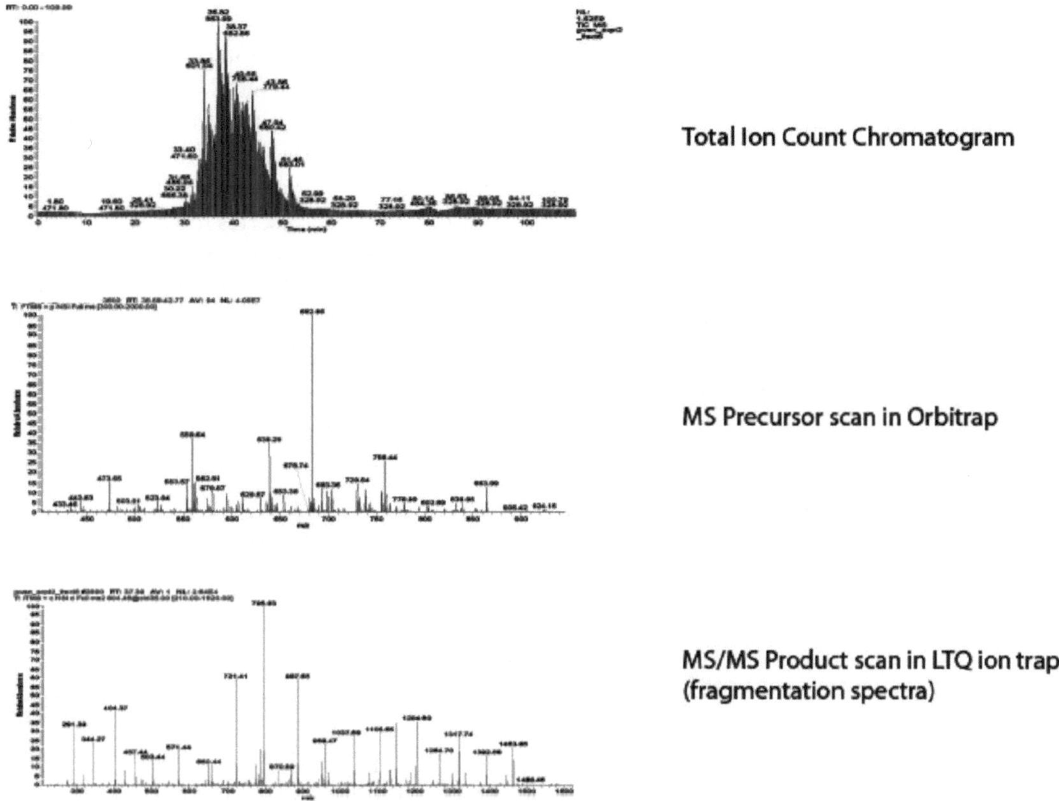

Fig. 3. Example of typical nano-LC-LTQ Orbitrap MS/MS analysis. In this case, peptides are eluted between 20th and 60th min of acetonitrile gradient. Majority of peptides are eluted between 30th and 50th min

coverage, probability of false-positive identifications, etc. More importantly, software provide an ability to set search parameters and filters' thresholds by users based on criteria of their choice, allowing to compare search results performed using various parameters. Taking this under consideration, it is critically important that search parameters are always posted along with search results so that data can be comparable. In this section, we provide guidance for database searches using BioWorks 3.3.1. or Proteome Discoverer 1.0, both packages based on SEQUEST algorithm and offered by Thermo Fisher Scientific along with their ion trap mass spectrometers.

1. Downloading database
   There are many compiled databases that can be used to search data. An automatic download of the NCBI database can be found at ftp://ftp.ncbi.nih.gov/blast/db/FASTA/nr.gz. Once this file is downloaded and unzipped, subdatabases can be created for the requirements for specific needs using key words. We exclude keratins to speed up the search. To accelerate searches, we index our subdatabases. Monoisotopic mass type should be selected for LTQ Orbitrap data and average mass type for older ion traps such as LCQ.

2. Setting search parameters
   Setting search parameters is at the discretion of individual investigators. Search parameters are an important key to reduce false-positive hits (too loose criteria) and at the same time prevent losing valuable data (too tight criteria). For ion trap, we typically use a peptide tolerance of 2.0 amu and fragment ion tolerance of 1 amu, using only b and y ions series in the calculations. Search obtained spectra using the BioWorks 3.3.1 (or Proteome Discoverer 1.0) against a NCBI database indexed for human proteins. No modifications should be set except for a fixed Oxidation (M) and dynamic Carboxylmethyl (C) (see Note 6).
3. Setting filters for output files
   Filtering output data is important to help eliminate excessive data. We use the following filters:

(a)	Peptide	Delta CN	0.100
(b)	Peptide	Sp—Preliminary score	500.0
(c)	Peptide	Xcorr vs. charge state	1.50, 2.00, 2.50, 3.00
(d)	Peptide	Protein probability	0.001

3.7.3. MALDI TOF–TOF

MALDI-TOF mass spectrometry provides an alternative and complementary technique to ESI mode of protein/peptide identification and characterization. It can be accomplished using two approaches. One is to use peptide fingerprinting method based on a set of peak representing peptides resulting from tryptic digest. This approach is suitable for samples containing a mixture of only one or two proteins. It is used very often for protein identification in spots excised from 2DE. In this method, samples are usually deposited manually on the target plate. In this case, fragmentation of peptides is not necessary; however, mass accuracy is critical and number of peaks obtained experimentally has to match 30% or more of theoretical peaks. The other approach is to use MALDI-TOF/TOF mode to fragment peptides and derive their amino acid sequences. Similarly to ESI-nano-LC-MS/MS method, this approach consists of full scan followed by TOF/TOF (MS/MS) scans of most abundant precursors. In this approach, low and high complexity samples can be analyzed. Low complexity samples can be deposited manually and high complexity samples have to be fractionated using nano-LC system and automatically spotted on MALDI target. DHB used in MALDI is especially useful in the analysis of glycopeptides (Fig. 4).

We use an AB 4800 (AB Sciex) MALDI TOF–TOF for analysis. Ionization is supported by drying the sample with alpha-CHCA or some other matrix to form crystals. When these crystals are hit by a laser, the matrix transfers its charge to the peptide. Spots can be made by manually spotting HPLC fractions onto a plate or by utilizing a plate spotter such as the Tempo LC MALDI spotter

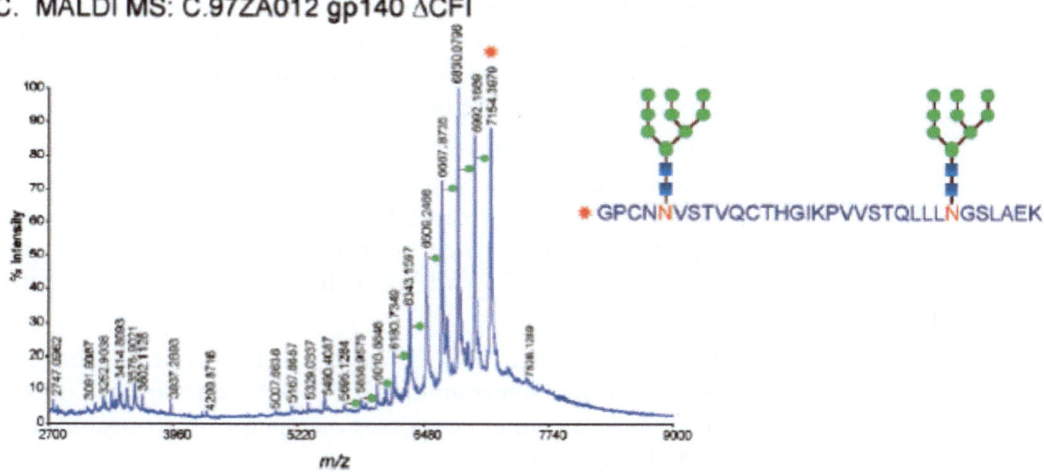

Fig. 4. Example of mapping posttranslational N-glycosylation of gp140 viral protein using MALDI-TOF/TOF approach. C. CON and C97ZA012 proteins used in this study were recombinant expressed in vaccinia virus. (Adapted from Go et al. *"Glycosylation Site Specific Analysis of Clade C HIV-1 Envelope Proteins."* J. Proteome Res. 2009, 8:4231).

(AB Sciex). Like ESI mass spectrometry, tuning for an optimized signal is necessary for good fragmentation results.

1. Mix and spot several calibration spots containing calibration mix 1 and 2 (AB Sciex) in alpha-CHCA. Pre-spotted calibration plates provided by the manufacturer can also be used. Spot manually by pipetting 1.0 μL of sample onto the plate, allow it to dry, and then spot 0.5 μL of matrix on top of it twice.
2. Use a reflector-positive MS acquisition factory method to create a starter method, but save within another folder modifying for a range of 400–3,000 *m/z*. Create both an MS internal and an MS default calibration processing file, including a peak list for the calibration mix. Perform an internal calibration to update the MS operating mode's default calibration. Perform a plate calibration as well.
3. Use a MS/MS 1 kV factory method as a starter method but save within another folder. Create both an MS/MS internal and a default calibration processing method file, including a peak list for the fragmentation of angiotensin (1,296 m/z) in the internal method. Perform an MS/MS internal calibration to update the MS/MS operating mode's default calibration.
4. If there is poor signal in MS/MS mode, or your precursor peak is much larger than your fragmentation peaks, use the automatic tune job to tune the y^2 and x^2 deflectors and the Timed Ion Selector (TIS) offset with the CID off.
5. Use an interpretation factory method to create a starter method but save within another folder, modifying it to exclude common background peaks found in your calibration or blank matrix spots. This will use one precursor scan to pick the five most

abundant peaks and analyze by MS/MS. Dynamically excluded ions are selected twice for one minute. Make sure to include your MS/MS acquisition method and MS/MS default processing method in the interpretation method.

6. Inject samples prepared for analysis onto a C18 peptide trap followed by a microcapillary C18 column. If necessary, an offline C18 column can be used for manual spotting with a pipettor, layering matrix over sample, but the fractions may need to be dried down to volume in a Speed-Vac before spotting.
7. Load peptides with HPLC water with 1% formic acid and elute using a 90-min linear gradient of 0–80% acetonitrile. Collected fragments can either be spotted manually or with an automated MALDI spotter.
8. Prepare a fresh matrix solution: 1:1 mixture of 10 μg/mL each of alpha-CHCA and DHB in 60% ACN with 0.1% TFA. Samples spotted are mixed or layered with equal parts of matrix solution, and spotted on the plate in increments of 0.5 μL twice or automatically mixed on the spotter. Manually spot additional calibration, blank, and BSA 100 fmol spots for calibration, method verification, and a reject peak list. Each sample set should be preceded by a calibration in both MS and MS/MS modes as well as a plate calibration for accuracy. Run the BSA in both interactive and automatic modes to verify your interpretation method and sensitivity of your acquisition methods.
9. Create an automatic job run to use your reflector-positive acquisition method to find precursor masses, and then the interpretation method for peak selection. Fragmentation in MS/MS will run automatically.

4. Notes

1. Appropriate safety precautions should be taken while handling HIV-infected material. Health Protection Agency (2009). Procedure for the care and propagation of cell cultures for virus isolation. National Standard Method, VSOP 39, Issue 2. http://www.hpa-standardmethods.org.uk/pdf_sops.asp.
2. IEF step was carried out at a constant temperature of 20°C with a total 45 kVh (500 V for 0.5 kVh, gradient to 1,000 V for 0.8 kVh, gradient to 8,000 V for 13.5 kVh, and 8,000 V for 30.2 kVh). For the preparative run after saturation labeling, 65 kVh was applied (500 V for 0.5 kVh, gradient to 1,000 V for 0.8 kVh, gradient to 8,000 V for 13.5 kVh, and 8,000 V for 50.2 kVh).

3. Second dimension was carried out with an Ettan Daltsix Electrophoresis System™ (GE Healthcare Piscataway, NJ) at 20°C. Power was held constant to 5 W/gel for the initial 30 min and 17 W/gel until bromophenol blue reached the gel bottom.
4. Signals were collected for Cy2-, Cy3-, and Cy5-labeled samples at an excitation wavelength of 488, 520, and 620 nm, respectively, using Typhoon 9410 Variable Mode Imager (GE Healthcare Piscataway, NJ). Gels were scanned at 100 μm resolution and analyzed using DeCyder 2D 6.5™ software (GE Healthcare Piscataway, NJ). For visualization of protein spots used in identification after minimal labeling, gels were stained with Sypro Ruby and scanned at 400 nm on Typhoon 9410. Spots selected for protein identification after DeCyder analysis were picked from gels by automatic Ettan™ Spot Picker (GE Healthcare, Piscataway) with a 2-mm diameter tip.
5. Gels should be white after drying and transparent when trypsin is absorbed. Use more volume if gel piece is not saturated.
6. If glycopeptides or phosphopeptides are of interest, databases must be set to look for these modifications.

Acknowledgments

We would like to thank Dr. Gwenael Pottiez for help in preparation of this manuscript. This work was partially supported by the National Institutes of Health 1 P20DA026146-01 and 2 P01 NS043985-05.

References

1. Strain, M. C., S. Letendre, S. K. Pillai, T. Russell, C. C. Ignacio, H. F. Gunthard, B. Good, D. M. Smith, S. M. Wolinsky, M. Furtado, J. Marquie-Beck, J. Durelle, I. Grant, D. D. Richman, T. Marcotte, J. A. McCutchan, R. J. Ellis, and J. K. Wong. 2005. Genetic composition of human immunodeficiency virus type 1 in cerebrospinal fluid and blood without treatment and during failing antiretroviral therapy. J Virol 79, 1772–1788.
2. Jurado, A., P. Rahimi-Moghaddam, S. Bar-Jurado, J. S. Richardson, M. Jurado, and A. Shuaib. 1999. Genetic markers on HIV-1 gp120 C2-V3 region associated with the expression or absence of cognitive motor complex in HIV/AIDS. J NeuroAIDS 2, 15–28.
3. Nath, A., and J. Geiger. 1998. Neurobiological aspects of human immunodeficiency virus infection: neurotoxic mechanisms. Prog Neurobiol 54, 19–33.
4. Ricardo-Dukelow, M., I. Kadiu, W. Rozek, J. Schlautman, Y. Persidsky, P. Ciborowski, G. D. Kanmogne, and H. E. Gendelman. 2007. HIV-1 infected monocyte-derived macrophages affect the human brain microvascular endothelial cell proteome: new insights into blood-brain barrier dysfunction for HIV-1-associated dementia. J Neuroimmunol 185, 37–46.
5. Jana, A., and K. Pahan. 2004. Human immunodeficiency virus type 1 gp120 induces apoptosis in human primary neurons through redox-regulated activation of neutral sphingomyelinase. J Neurosci 24, 9531–9540.
6. Simon, F., P. Mauclere, P. Roques, I. Loussert Ajaka, M. C. Muller-Trutwin, S. Saragosti, M.

C. Georges-Courbot, F. Barre-Sinoussi, and F. Brun-Vezinet. 1998. Identification of a new human immunodeficiency virus type 1 distinct from group M and group O. Nat Med 4, 1032–1037.

7. Ciborowski, P. 2009. Biomarkers of HIV-1-associated neurocognitive disorders: challenges of proteomic approaches. Biomark Med 3, 771–785.
8. Go, E. P., J. Irungu, Y. Zhang, D. S. Dalpathado, H. X. Liao, L. L. Sutherland, S. M. Alam, B. F. Haynes, and H. Desaire. 2008. Glycosylation site-specific analysis of HIV envelope proteins (JR-FL and CON-S) reveals major differences in glycosylation site occupancy, glycoform profiles, and antigenic epitopes' accessibility. J Proteome Res 7, 1660–1674.
9. Go, E. P., Q. Chang, H. X. Liao, L. L. Sutherland, S. M. Alam, B. F. Haynes, and H. Desaire. 2009. Glycosylation site-specific analysis of clade C HIV-1 envelope proteins. J Proteome Res 8, 4231–4242.
10. Irungu, J., E. P. Go, Y. Zhang, D. S. Dalpathado, H. X. Liao, B. F. Haynes, and H. Desaire. 2008. Comparison of HPLC/ESI-FTICR MS versus MALDI-TOF/TOF MS for glycopeptide analysis of a highly glycosylated HIV envelope glycoprotein. J Am Soc Mass Spectrom 19, 1209–1220.
11. Ostrowski, S. R., T. L. Katzenstein, B. K. Pedersen, J. Gerstoft, and H. Ullum. 2008. Residual viraemia in HIV-1-infected patients with plasma viral load<or=20 copies/ml is associated with increased blood levels of soluble immune activation markers. Scand J Immunol 68, 652–660.
12. Rozek, W., M. Ricardo-Dukelow, S. Holloway, H. E. Gendelman, V. Wojna, L. Melendez, and P. Ciborowski. 2007. Cerebrospinal fluid proteomic profiling of HIV-1-infected patients with cognitive impairment. J. Proteome. Res. **6**, 4189–99.
13. Wiederin, J., W. Rozek, F. Duan, and P. Ciborowski. 2009. Biomarkers of HIV-1 associated dementia: proteomic investigation of sera. Proteome Sci 7, 8.
14. Gendelman, H. E., J. M. Orenstein, M. A. Martin, C. Ferrua, R. Mitra, T. Phipps, L. A. Wahl, H. C. Lane, A. S. Fauci, D. S. Burke, and et al. 1988. Efficient isolation and propagation of human immunodeficiency virus on recombinant colony-stimulating factor 1-treated monocytes. J Exp Med 167, 1428–1441.
15. van 't Wout, A. B., H. Schuitemaker, and N. A. Kootstra. 2008. Isolation and propagation of HIV-1 on peripheral blood mononuclear cells. Nat Protoc 3, 363–370.
16. Adachi, A., H. E. Gendelman, S. Koenig, T. Folks, R. Willey, A. Rabson, and M. A. Martin. 1986. Production of acquired immunodeficiency syndrome-associated retrovirus in human and nonhuman cells transfected with an infectious molecular clone. J Virol 59, 284–291.
17. Gendelman, H. E., L. M. Baca, H. Husayni, J. A. Turpin, D. Skillman, D. C. Kalter, J. M. Orenstein, D. L. Hoover, and M. S. Meltzer. 1990. Macrophage-HIV interaction: viral isolation and target cell tropism. Aids 4, 221–228.
18. Kadiu, I., M. Ricardo-Dukelow, P. Ciborowski, and H. E. Gendelman. 2007. Cytoskeletal protein transformation in HIV-1-infected macrophage giant cells. J Immunol 178, 6404–6415.

Chapter 13

Proteomic Profiling of Cerebrospinal Fluid

Gwenael Pottiez and Pawel Ciborowski

Abstract

Cerebrospinal fluid (CSF) is a body fluid which has direct contact with the central nervous system, and as such, changes in its composition might be informative about various aspects of the brain. It has been postulated for quite a long time that proteomic analysis of CSF will reveal protein markers related to neurological disorders, their prognosis and early detection, efficacy of treatment, etc. Several proteomic profiling platforms provide tools to determine changes occurring in protein profiles of CSF reflecting physiological and pathological changes. Two major strategies are used. The first strategy is based on determining quantitative changes at the level of intact proteins followed by protein identification by tandem mass spectrometry of in-gel-digested protein spots. Usually, two-dimensional gel electrophoresis with DIGE technology is used. The second strategy is based on tryptic digestion of entire sample, labeling resulting peptides with mass tags and determining quantitative changes in protein content based on relative ratios of peptides. Typically iTRAQ® technology is used. Regardless of the strategy used, samples of CSF need to be simplified by removing most abundant proteins constituting more than 90% of a total pool of proteins. Detailed protocols are presented in this chapter.

Key words: Proteomics, Neuroproteomics, Biomarkers, CSF, Plasma, 2-DE DIGE, iTRAQ, Sample fractionation

1. Introduction

Cerebrospinal fluid (CSF, *Liquor cerebrospinalis*) surrounds the brain and spinal cord and serves as a communication avenue between the blood and cells of the nervous tissues (1). CSF is also in contact with blood through the blood–brain barrier, thus resembling an ultrafiltrate of plasma in its protein constituents. Functions of the CSF include buoyancy, acid–base buffering, and delivery of electrolytes, signaling molecules, transport molecules, and micronutrients to the brain parenchyma (2). CSF contains sugars, lipids, electrolytes, and proteins. The protein concentration in CSF ranges

Yannis Karamanos (ed.), *Expression Profiling in Neuroscience*, Neuromethods, vol. 64,
DOI 10.1007/978-1-61779-448-3_13, © Springer Science+Business Media, LLC 2012

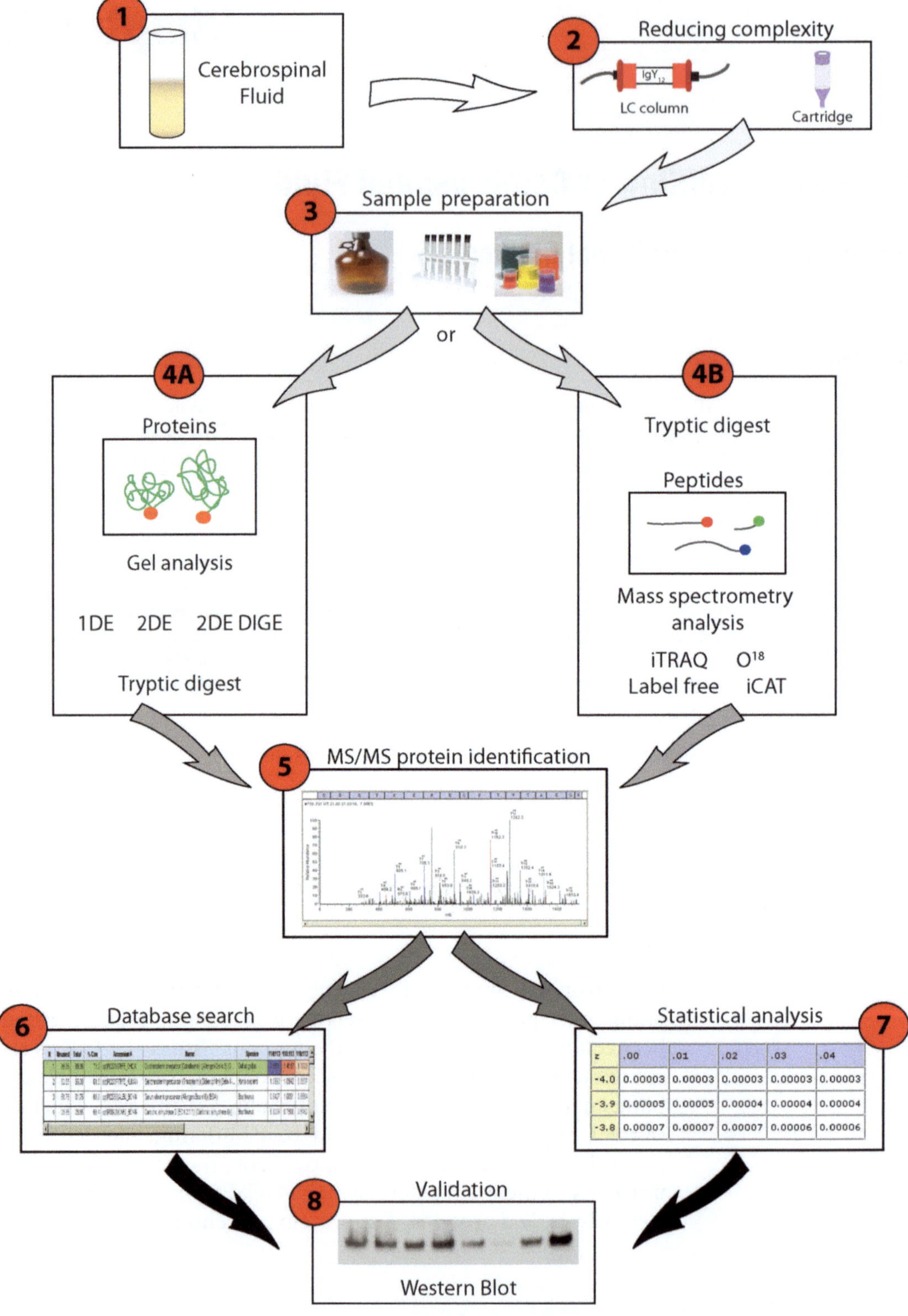

Fig. 1. General workflow of proteomic investigation of CSF. 1. Sample processing. 2. Reducing complexity. 3. Sample clean-up and preparation for labeling. 4A. Protein-based profiling. 4B. Peptide-based profiling. 5. Protein identification using mass spectrometry. 6. Database searches. 7. Statistical analysis. 8. Validation.

from 0.2 to 0.8 mg/mL (0.3–1% of serum protein concentration), with more than 70% of the proteins in CSF being isoforms of albumin, transferrin, and immunoglobulins (3, 4). Production rate under normal physiological conditions is approximately 0.4 mL/min which exceeds 500 mL per day, while total volume in an adult human is 130–150 mL (5).

Changes in composition of CSF are useful tools to gather information about various pathological processes that might be affecting functions of the central nervous system (CNS) (6). The CSF proteome could provide unique biomarkers for early-stage diagnosis or the staging of a neuronal disease, offer potential insight into the biochemical characterization of affected neuronal population, and clarify the molecular basis of CNS pathologies (7–9). In the case of HIV-infected patients, HIV-1 penetrates brain early after infection. As a consequence, approximately half of infected people will develop HIV-associated neurocognitive disorders (HAND) ranging from very mild forms diagnosed by specific psychiatric tests and evaluations to the most severe form of HIV-associated dementia (HAD). Proteomic analysis of CSF samples, from HIV-infected patients at different levels of neurological disorders, offers the opportunity to discover biomarkers of HAND (10). Figure 1 presents general workflow of CSF proteomic analysis used in our studies. This protocol can and should be modified based on specific questions asked in study design.

There are four major limiting factors responsible for the fewer than expected proteomic studies of CSF. First, a spinal tap is an invasive procedure and many patients do not sign consent. Second, a limited amount of CSF can be drawn from one lumbar puncture. Third, CSF contains less proteins (10–100 times) per 1 mL than plasma/serum. Fourth, biomarkers may be present at very low abundance and because of the high turnover rate and dynamic nature of the CSF, there may be limited accumulation of disease-associated proteins at any one time point. Despite these obstacles, more proteomic profiling studies are being and will be performed with more focus on carefully selected cohort of patients, rather than based on pooling multiple samples (11).

2. Materials

2.1. General Laboratory Material

We indicate manufacturers of equipment, supplies, and reagents that we have used; however, other products and sources can also be used.

1. Vortex (Fischer Scientific, Pittsburg, PA).
2. Rocker (Fisher Scientific).
3. Benchtop centrifuge (Fisher Scientific).

4. Centrifuge for 50-mL tubes able to spin at 4000× *g* (Thermo Scientific, Wilmington, DE).
5. Centrifuge for Eppendorf tubes able to spin at 1,500, 4,000, and 13,000× *g* (Eppendorf, Hamburg, Germany).
6. Speed vacuum drying system (Thermo Scientific).
7. Milli-Q (mQ) water purification system (18 MΩ) (Elga LabWater, Woodridge, IL).
8. Power supply for electrophoresis adapted to the two-dimensional gel electrophoresis (2-DE) system (Bio-Rad, Hercules, CA).
9. pH indicator paper, for pH 8–10 and pH 0–3 (Fisher Scientific).

2.2. Required Laboratory Equipment

1. Protein quantification system [we use NanoDrop (Thermo Scientific)] or protein quantification kit.
2. HPLC system [we used LC-20A prominence with EZStart HPCL (Shimazu, Columbia, MD)].
3. Isoelectrofocusing (IEF) system [e.g., IPGphor II apparatus (GE Healthcare, Piscataway, NJ)].
4. Gel electrophoresis system [e.g., Ettan DaltSix Electrophoresis System™ (GE Healthcare)].
5. System to scan fluorescence [we use Typhoon 9410 Variable Mode Imager (GE Healthcare)].
6. Software for gel image analysis [e.g., DeCyder 2D 6.5™ software (GE Healthcare)].
7. Gel spot cutting system [e.g., Ettan Spot Picker (GE Healthcare)] (see Note 1).
8. 3100 OFFGEL Fractionator system (Agilent, Santa Clara, CA).
9. HPLC-tandem mass spectrometry (MS/MS) system [we use either Proteome X system consisting of LCQdecaPlus ion trap and Surveyor HPLC system (ThermoElectron, San Jose, CA) or LTQ-Orbitrap XL mass spectrometer (Thermo-Fisher Scientific) coupled with a nano-LC system (Eksigent, Inc.) and Bioworks 3.2 software (Thermo Electron, San Jose, CA)].
10. LC-plate spotting system [for iTRAQ® labeled samples, we recommend TempoLC™ MALDI plate spotter system (Applied Biosystem (ABI), Carlsbad, CA)].
11. MALDI-TOF/TOF mass spectrometer [e.g., 4800 MALDI TOF/TOF (ABI) with ProteinPilot™ software v2.0.1 (ABI)].

2.3. Sample Processing

1. Cocktail of protease inhibitors for general use (Sigma-Aldrich, St Louis, MO).
2. Triton X-100 (Fisher Scientific, Pittsburg, PA).
3. 0.2-μm filter (Millipore, Billerica, MA).

2.4. Immunodepletion of Samples

1. Multiple Affinity system Human 6 (Agilent Technologies Inc., Santa Clara, CA) with ready-to-use phosphate and elution buffers.
2. Vivaspin 6 (5-kDa cutoff membrane) (Vivascience, Hannover, Germany).

2.5. Two-Dimensional Gel Electrophoresis (2-DE) Differential in Gel Electrophoresis

1. Lysis buffer [30 mM Tris-HCl, pH 8, 7 M Urea, 2 M Thiourea, 4% CHAPS (w/v)].
2. Sample buffer [7 M urea, 2 M thiourea, 4% CHAPS (w/v), 2% pharmalyte, and 130 mM dithiotreitol (DTT)].
3. Tris-(2-carboxyethyl)phosphine hydrochloride (TCEP).
4. CyDye differential in gel electrophoresis (DIGE) Fluor Labeling kit (GE Healthcare).
5. Immobilized pH gel (IPG) strips, we use Immobiline™ DryStrip gel (24 cm) (GE Healthcare).
6. Equilibration solution for CyDye-labeled samples [50 mM Tris-HCl, pH 8.8, 6 M urea, 30% glycerol, 2% sodium dodecylsulfate (SDS), and 0.01% bromophenol blue (w/v)].
7. DTT and iodoacetamide.
   7.1. 12% Polyacrylamide gels.
   7.2. 0.5% Agarose (w/v) in water.

2.6. In-Gel Protein Digest for Mass Spectrometry Analysis

1. 50% Acetonitrile (ACN) (v/v).
2. 50% ACN (v/v), 50 mM NH_4HCO_3.
3. 50% ACN (v/v), 10 mM NH_4HCO_3.
4. Trypsin (Promega, Madison, WI) with trypsin resolubilization solution provided with trypsin vials.
5. 10 mM NH_4HCO_3.
6. 60% ACN (v/v), 0.1% trifluoroacetic acid (TFA) (v/v).
7. 0.5% TFA (v/v).
8. ACN.
9. 50% ACN, 0.1% TFA (v/v).
10. 0.1% Formic acid (v/v).
11. RP-C18 column (New Objective, Woburn, MA).
12. Buffer A: Water:ACN:TFA (98:2:0.1; v/v/v).
13. Buffer B: Water:ACN:TFA (2:98:0.1; v/v/v).

2.7. iTRAQ Labeling

1. Cold (−20°C) 200 proof ethanol.
2. Cold (−20°C) 70% ethanol (v/v).
3. iTRAQ® Reagent application kit—Plasma (AB Sciex, Foster City, CA) (iTRAQ® Reagents contain the following reagents: sample buffer, reducing reagent, labeling reagents (114, 115, 116, and 117 tags), and ethanol).

4. Trypsin (TCPK) (AB Sciex).
5. Iodoacetamide.

2.8. Mix Cation Exchange Sample Cleaning, to Remove Salts and Excess of Labeling Compounds

1. Water and methanol of HPLC grade, formic acid.
2. Water mix cation exchange (MCX) cartridge (Water Corp, Milford, MA).
3. 28% NH_4OH solution (Sigma-Aldrich).

2.9. Peptide Fractionation Using Strong Cation Exchange System

1. Water HPLC grade.
2. Formic acid.
3. Polysulfoethyl A column [100× 2.1 mm, 5 μm, 3 Å (PolyLC, Columbia, MD)].
4. Phase A: 10 mM KH_2PO_4, 25% ACN (v/v), pH 2.7.
5. Phase B: 10 mM KH_2PO_4, 25% ACN (v/v), 500 mM KCl, pH 2.7.

2.10. Peptide Fractionation Using OFFGEL System

1. 0.1% Formic acid (v/v).
2. Agilent 3100 OFFGEL fractionator kit (pH 3–10 12 wells) (Agilent, Santa Clara, CA) (OFFGEL kit contains the following items: Ampholyte solution, 50% glycerol solution, 12-well frame + lid, IPG strips, mineral oil, and electrode pads).
3. PepClean™ C-18 spin columns.
4. 20% ACN (v/v), 2% TFA (v/v).
5. 50% Methanol (v/v).
6. 5% ACN (v/v), 0.5% TFA (v/v).
7. 70% ACN (v/v).

2.11. Tandem Mass Spectrometry (MS/MS) Analysis of iTRAQ®-Labeled Samples

1. 0.1% TFA (v/v) (for LC-MALDI MS/MS).
2. 0.1% Formic acid (v/v) (for LC-ESI-MS/MS).
3. ProteoCol™ C18 trap cartridges (Michrom Biosources, Auburn, CA).
4. LC-MALDI MS/MS analysis.
   - 4.1 Buffer A: Water:ACN:TFA (98:2:0.1).
   - 4.2 Buffer B: Water:ACN:TFA (2:98.0.1).
   - 4.3 Matrix solution: 5 mg/mL α-cyano-4-hydroxycinammic acid (CHCA), 75% ACN (v/v), 0.1% TFA.
5. LC-ESI-MS/MS analysis.
   - 5.1 Buffer A: Water:ACN:formic acid (98:2:0.1).
   - 5.2 Buffer B: Watter:ACN:formic acid (2:98:0.1).

3. Methods

3.1. Processing of CSF Samples

1. Sample of CSF must be frozen at –80°C as soon as possible in the presence of a cocktail of protease inhibitors (Sigma-Aldrich). CSF samples, from HIV-infected and noninfected people, are neutralized with 10% Triton X-100 (v/v) (freshly prepared), to a final concentration of 0.1%.
2. Upon arrival to laboratory (e.g., from clinic), thaw the sample on wet ice.
3. Examine CSF sample visually for the presence of blood. If a sample is contaminated by blood, the sample should be eliminated from the study (see Note 2).
4. Aliquot CSF sample and store at –80°C. Avoid multiple freeze/thaw cycles.
5. Prior to use, the sample requires filtering through 0.2-μm spin filters at 4,000× *g* for 2 min to remove particulate matter.

3.2. Reducing Complexity

The most common method to reduce complexity of CSF samples is immunodepletion of most abundant proteins. A variety of devices in the form of liquid chromatography columns, spin columns, and cartridges are commercially available and are summarized in Table 1.

Immunodepletion using Multiple Affinity Removal System Human 6 (Table 2 shows the yield of the depletion of CSF samples).

1. Wash column with 3 volumes of loading buffer to remove sodium azide at a flow rate of 1 mL/min (any HPLC system supporting flow rate between 0.25 and 1.0 mL/min can be used).
2. Run one cycle of mock immunodepletion using phosphate-buffered saline (PBS) as a sample.
3. Dilute CSF samples with 3 volumes of buffer A (ready-to-use phosphate buffer) and then inject on the column at a flow rate of 0.25 mL/min.
4. Use 1 mL of diluted sample per run. Collect flow-through fractions usually between 3.5 and 10.5 min.
5. Elute bound proteins with buffer B (ready-to-use elution buffer) at the flow rate of 1 mL/min usually for 15–18.5 min.
6. Regenerate the column with buffer A for the next 13.5 min at 1 mL/min flow rate.
7. Concentrate depleted CSF using spin filters Vivaspin 6 with 5-kDa cutoff membrane (Vivascience) by centrifugation at 4,000× *g* at 4°C. Sample volume was reduced to 0.1 mL.
8. Clean samples by adding 10 volumes of cold 200 proof ethanol and vortex (see Note 3).

Table 1
Affinity-based systems to remove most abundant proteins

System	Format	Supplier	Specificity
Proteome Purify™ 12 Human Serum Protein Immunodepletion Resins	Spin filter and resin	EMD/ Calbiochem	Albumin, α-1-Acid Glycoprotein, α-1-Antitrypsin, α-2-Macroglobulin, Apolipoprotein A-I, Apolipoprotein A-II, Fibrinogen, Haptoglobin IgA, IgG, IgM, Transferrin
Seppro® IgY14	Family of columns and spin cartridges	Sigma-Aldrich, Inc.	Albumin, IgG, Fibrinogen Transferrin, IgA, IgM Haptoglobin, α-2-Macro-glubulin, α-1-Acid Glycoprotein, α-1-Antitrypsin, Apo A-I HDL, Apo A-II HDL, Complement C3, LDL (ApoB)
Multiple Affinity Removal LC Column—Human 14	LC column and spin cartridges	Agilent, Inc.	Albumin, IgG, Antitrypsin, IgA, Transferrin, Haptoglobin, Fibrinogen, α-2-Macroglobulin, α-1-Acid Glycoprotein, IgM, Apolipoprotein A-I, Apolipoprotein A-II, Complement C3, Transthyretin
Multiple Affinity Removal LC Column—Human 7	LC column and spin cartridges	Agilent, Inc.	Albumin, IgG, IgA, Transferrin, Haptoglobin, Antitrypsin, Fibrinogen
Multiple Affinity Removal LC Column—Human 6	LC column and spin cartridges	Agilent, Inc.	Albumin, IgG, IgA, Transferrin, Haptoglobin, Antitrypsin
ProteoPrep® 20 Plasma Immunodepletion LC Column	10 mL column	Sigma-Aldrich, Inc.	Albumin, IgG, Transferrin, Fibrinogen, IgA, α_2-Marcroglobulin, IgM, α-1-Antitrypsin, Complement C3, Haptoglobin, Apolipoprotein A-I, Apolipoprotein A-II, Apolipoprotein B, α_1- Acid Glycoprotein, Ceruloplasmin, Complement C4, Complement C1q, IgD, Prealbumin, Plasminogen
ProteoSpin™ Abundant Serum Depletion Kit	Spin cartridges	Norgen Biotech Corp.	Albumin, α-1-Antitrypsin, Transferrin, Haptoglobin
ProteoPrep® Immunoaffinity Albumin and IgG Depletion Kit	Spin cartridges	Sigma-Aldrich, Inc.	Albumin, IgG
Proteome Purify™ 2 Serum Protein Immunodepletion Resins	Single-use resins	R&D Systems	Albumin, IgG

(continued)

Table 1 (continued)

System	Format	Supplier	Specificity
ProteoPrep® Blue Albumin and IgG Depletion Kit	Spin cartridges	Sigma-Aldrich, Inc.	Albumin, IgG
Pierce Albumin/IgG Removal Kit	Spin cartridges	Thermo Scientific Pierce Protein Research Products	Albumin, IgG
ProteoExtract® Albumin/IgG Removal Kit	Resin	EMD/ Calbiochem	Albumin, IgG
Pierce Antibody-Based Albumin/IgG Removal Kit	Spin cartridges	Thermo Scientific Pierce Protein Research Products	Albumin, IgG
Vivapure Anti-HSA Affinity Resin	Spin cartridges or Resin	Sartorius Stedim Biotech	Albumin
POROS® Affinity Depletion Anti-HAS	LC column	Applied Biosciences, Inc.	Albumin
AlbuSorb™ Albumin Depletion Kit	Powder	Biotech Support Group LLC	Albumin

9. Incubate the samples at –20°C until precipitate is formed (≈3 h).
10. Centrifuge at 13,000× *g* for 15 min at 4°C.
11. Decant the supernatant and wash the pellet with cold 70% ethanol.
12. Centrifuge at 13,000× *g* for 5 min at 4°C.
13. Decant the supernatant and speed vacuum dry the pellet.

3.3. Profiling Proteins or Peptides?

Sample preparation is the most important step in a proteomic analysis. Depending on the subsequent steps, sample preparation protocol should be adapted accordingly. Two main profiling approaches are available for the analysis of CSF: (1) protein-based profiling, which consists of full-length protein analysis, and (2) peptide profiling, which starts by the cleavage of the protein into peptides and then analysis of the peptide samples. These two profiling methods require specific sample preparation and treatment described in the following sections.

Table 2
Variation of protein concentration yielded by immunodepletion of CSF samples, from nondemented (ND) patients and patients with HIV-associated dementia (HAD)

	Protein concentration		
Group	Before depletion (μg)	After depletion (μg)	Yield (%)
ND	60	11.0	18.3
ND	55	11.8	21.8
ND	50	6.4	12.8
ND	50	10	20.0
HAD	70	12.7	18.2
HAD	70	21.7	35.1
HAD	60	5.0	8.3
HAD	50	8.5	17.0

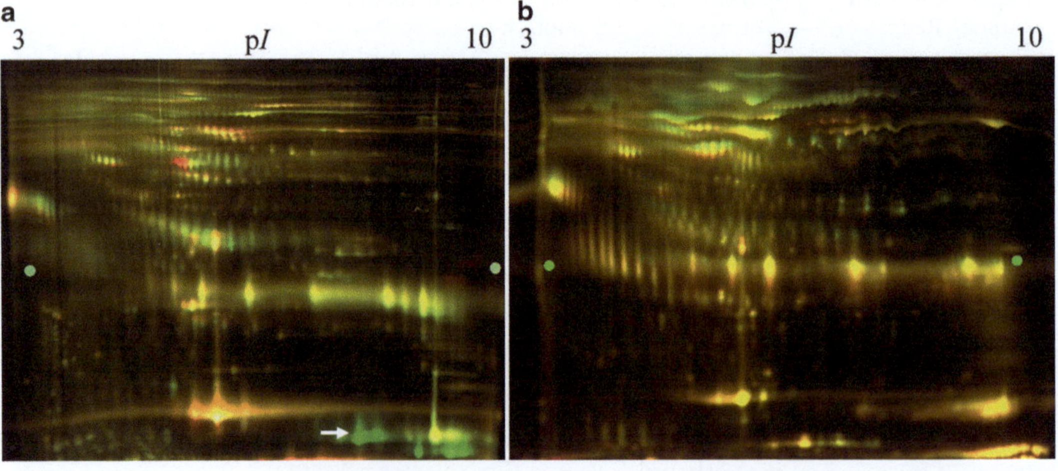

Fig. 2. Examples of gel scans from 2-DE DIGE gels, the *yellow* color shows an equivalent amount of protein from both samples, while *red* and *green* indicate overexpression or under-expression, respectively, of proteins under specific condition. (**a**) 2-DE DIGE (minimal labeling) of human CSF samples. The *white arrow* indicates spots of hemoglobin showing that a sample contaminated by hemoglobin may induce a bias to the gel comparison. (**b**) 2-DE DIGE (saturation labeling) of human plasma samples. (Images adapted from "*Cerebrospinal fluid proteomic profiling of HIV-1-infected patients with cognitive impairment.*" Rozek et al. J Proteome Res. 2007, 6:4189).

3.4. Intact Proteins-Based Profiling

The most commonly used method for profiling of intact proteins is 2-DE. This method is more accurate and sensitive with DIGE technology based on labeling with fluorescent dyes: Cy3, Cy5, or Cy2 (see Fig. 2). Comparisons of 2-DE gels without labeling but based on intensity of spots stained with fluorescent dyes such as

SYPRO® Ruby (Sigma-Aldrich) or Lava Purple (Fluorotechnics, San Francisco, CA) may easily lead to false-positive results due to high variability and is not recommended for CSF samples. There are two 2-DE DIGE labeling techniques available. The first method is minimal labeling of one sample with Cy3 (green), another sample with Cy5 (red), and an equal mixture of both samples with Cy2 (yellow) as an internal standard (see Fig. 2a). One gel contains two samples and one internal standard. The second method is based on saturated labeling of all samples with Cy5 and an equal mixture of all samples with Cy3 (see Fig. 2b). One gel contains one sample and a portion of the internal standard. It has been postulated that the technique based on two dyes has less variability than the first one based on three dyes, although it requires twice as many gels.

3.4.1. Saturation Labeling for Analytical Gels

1. Speed vacuum dry 5 μg of six CSF samples.
2. Resuspend each sample in 4 μL of lysis buffer (30 mM Tris-HCl, pH 8, 7 M urea, 2 M thiourea, 4% CHAPS).
3. Reduce proteins with 1 μL of 2 nM TCEP at 37°C for 1 h.
4. Label proteins with 5 μL of 4 nM of Cy5 saturation dye for 30 min at 37°C. Stop the reaction by adding one volume (10 μL) of sample buffer (7 M urea, 2 M thiourea, 4% CHAPS, 2% pharmalyte, and 130 mM DTT).
5. Prepare an internal standard using 5 μg of each sample mixed together and then speed vacuum dry.
6. Resuspend protein pellet with 4 μL of lysis buffer per 5 μg of proteins.
7. Reduce proteins by adding 1 μL of 2 nM TCEP per 5 μg protein for 1 h at 37°C.
8. Label proteins with 5 μL of 4 nM of Cy3 saturation dye per 5 μg protein for 30 min at 37°C. Add one volume (based on volume of the sample) of sample buffer (7 M urea, 2 M thiourea, 4% CHAPS, 2% pharmalyte, 130 mM DTT) to stop the reaction.
9. Each 5 μg of proteins of CSF, labeled with Cy5, has to be mixed with 5 μg of internal standard, labeled with Cy3.
10. Add rehydration buffer to each sample (7 M urea, 2 M thiourea, 2% CHAPS, 60 mM DTT, and 1% pharmalyte, pH 3–10) to a final volume of 450 μL.

3.4.2. Saturation Labeling for Preparative Gels

Software-based analysis of gels identifies spots with intensities that differ significantly between tested conditions. Corresponding spots excised from preparative gel are used in the subsequent step for protein identification. The latter should contain enough amount of proteins for mass spectrometry-based protein identification (we typically use 250–300 μg of proteins per one preparative gel).

1. Pool CSF sample for a total amount of protein up to 300 μg. Remaining amount of CSF may vary from sample to sample.
2. Speed vacuum dry the samples.
3. Resuspend the pellet in 250 μL of lysis buffer (30 mM Tris-HCl, pH 8, 7 M urea, 2 M thiourea, and 4% CHAPS).
4. Add 1 μL of TCEP 2 nM per 5 μg protein and incubate for 1 h at 37°C.
5. Label proteins by adding Cy3 saturation dye for preparative labeling (4 nM per 5 μg proteins).
6. Incubate for 30 min at 37°C.
7. Stop the reaction by adding sample buffer (7 M urea, 2 M thiourea, and 4% CHAPS) to a final volume of 445.5 μL. Add 4.5 μL of pharmalytes pH 3–10 (1% final concentration) and 4.5 mg of DTT (final concentration of 130 mM).

3.4.3. 2-DE DIGE

2-DE is a two-step procedure. (1) Separation of the proteins according to their global charges, referred to as IEF. (2) Separation of proteins by molecular weight using a detergent such as SDS to unfold and provide a constant charge to the proteins.

1. IEF may be performed with different IEF systems from several manufacturers (we used an IPGphor II apparatus).
2. In the IEF tray adapted to the IEF system used, apply the samples along the rows of the tray.
3. With tweezers carefully remove the protective plastic band covering the Immobiline™ DryStrips gel (24 cm long) with linear immobilized pH gradient 3–10 and place the strip, with the gel facing down, over the sample.
4. Cover the tray with aluminum foil and leave it overnight to rehydrate the gel strip.
5. Place the tray in the IEF apparatus following the manufacturer's recommendation and start the IEF steps (see Note 4).
6. After IEF, stop the IEF system and take the tray out of the system.
7. With tweezers take the Immobiline™ DryStrips and if necessary, carefully remove the excess of mineral oil with a paper and place the Immobilin DryStrip in an adapted tube and add 10 mL of equilibration solution (50 mM Tris-HCl, pH 8.8, 6 M urea, 30% glycerol, 2% SDS, and 0.01% bromophenol blue) completed with 100 mM DTT. Rock the tube for 15 min at room temperature.
8. Decant equilibration solution and add 10 mL of equilibration solution completed with 100 mM iodoacetamide. Rock the Immobiline™ DryStrips for 15 min at room temperature (do not apply the alkylation steps for minimal dye-labeled samples).

9. Load the Immobiline™ DryStrip on the top of a 12% polyacrylamide gel.
10. Overlay the strip with 0.5% agarose.
11. Every electrophoresis system large enough for the migration of 24-cm gels can be used for the second step (we used Ettan Daltsix Electrophoresis System™). Start the second-dimension migration step (see Note 5).

3.4.4. DeCyder™ Analysis

1. For visualization of protein spots, signals from each cyanine have to be collected with the corresponding excitation and emission wavelength using a scanner for fluorescence. We used a Typhoon 9410 Variable Mode Imager (GE Healthcare) (see Note 6).
2. Gel analysis may be performed with dedicated software. We used the DeCyder 2D 6.5™ software (GE Healthcare).
3. Gel analysis highlights spots showing a high variation between conditions and a low variability. Selected spots should be cut and analyzed by mass spectrometry.

3.5. Peptide-Based Profiling

3.5.1. 4-plex iTRAQ Labeling

Protocols developed for profiling of plasma/serum samples are applicable to profiling of CSF samples (see Note 7). The standard iTRAQ labeling protocol provided by AB Sciex for plasma was used with minor modifications (see Note 8). Use of Oasis MCX cartridge (Waters) was necessary to assure high sample quality and reproducible analysis.

1. Aliquot 50 μg of CSF proteins and add 1 mL of 200 proof ethanol.
2. Vortex and incubate at –20°C until formation of a precipitate (≈3 h).
3. Centrifuge at 13,000× *g* for 15 min at 4°C.
4. Decant the supernatant. Add 1 mL of 70% ethanol and vortex shortly to rinse the pellet.
5. Centrifuge at 13,000× *g* for 5 min at 4°C.
6. Decant the supernatant. Speed vacuum dry the pellet to remove residual wash solution.
7. The samples may be stored at –80°C at this step.
8. Add 25 μL of sample buffer—Plasma (AB Sciex), complete with 1 μL of denaturant, and vortex.
9. Add 2 μL of reducing reagent and vortex for 20–30 s. Centrifuge for 1 min using a benchtop centrifuge. This brings the sample to the bottom of the tubes. Incubate at 60°C for 1 h.
10. During the incubation, prepare 84 mM of iodoacetamide (Sigma Aldrich). This cysteine-blocking solution has to be freshly prepared.

11. Add 1 μL of cysteine-blocking solution. Vortex the samples for 20–30 s and centrifuge for 1 min using a microcentrifuge. Incubate in the dark at room temperature for 30 min.
12. To be able to add 10 μL of trypsin (1 μg/μL) to each sample, reconstitute two vials of trypsin (AB Sciex) by adding 25 μL of mQ water (final concentration of 1 μg/μL). Vortex each trypsin vial for 20–30 s and spin for 1 min using a benchtop centrifuge.
13. Complete each sample by adding 10 μL of reconstituted trypsin, vortex, centrifuge, and then incubate at 37°C for 16 h.
14. Bring each iTRAQ® reagent to room temperature for at least 30 min.
15. Dissolve labeling reagent in 70 μL of ethanol (use ethanol provided in the kit) and transfer the contents of one iTRAQ® reagent to the corresponding sample tube. For optimal efficiency of labeling, pH is verified and must be between 8.0 and 10.0. If not, add sample buffer by increments of 0.5 μL up to 5 μL after each addition vortex and check the pH.
16. Incubate the samples at room temperature for 1 h.
17. Add 100 μL of Milli-Q water to quench the reaction and incubate for 30 min.
18. Pool each 4-plex sample from one experiment into one Eppendorf tube.
19. Freeze at −80°C for ~15 min. and Speed vacuum dry the sample (see Note 9).

3.5.2. Post-iTRAQ Labeling Sample Clean-Up

1. Add 1 mL of 0.1% formic acid and verify the pH. If it is more than 3.0, add formic acid up to 0.2% formic acid (v/v) total.
2. Equilibrate each MCX cartridge by passing 1 mL of 50% methanol solution at a flow rate of 1 mL per minute (≈1 drop per second, see Note 10).
3. Apply sample at a flow rate of 1 mL/min.
4. Wash cartridge with 1 mL of 5% methanol, 0.1% formic acid.
5. Wash cartridge with 1 mL of 100% methanol.
6. Elute peptides with freshly prepared MCX elution buffer (50 μL of 28% NH_4OH, 950 μL of methanol).
7. Speed vacuum dry the samples and store at −80°C until first-dimension fractionation.

3.5.3. First-Dimension Fractionation

iTRAQ® samples as described above must be fractionated in two dimensions to reduce the complexity of the sample (not counting immunodepletion as first-dimension fractionation). Typically, liquid chromatography method interfaced with mass spectrometry is a reverse phase (RP) separation and is used as second-dimension fractionation. Then the first fractionation method should be different. Two methods are used, one method is based on LC fractionation using strong cation exchange (SCX) chromatography, and the other method is based on IEF fractionation.

3.5.4. SCX Fractionation

1. Any HPLC system supporting flow rate between 0.1 and 1.0 mL/min can be used and with a 500-μL injection loop, dual pump, and an ultraviolet-visual (UV–Vis) detector set at 215 nm.
2. Use a Polysulfoethyl A column [100×2.1 mm, 5 μm, 3 Å (PolyLC, Columbia, MD)].
3. Prepare mobile phases:
   - Buffer A: 10 mM KH_2PO_4 + 25% ACN, pH 2.7.
   - Buffer B: 10 mM KH_2PO_4 + 25% ACN + 500 mM KCl, pH 2.7.
4. Equilibrate column with mobile phase A for 30 min at 0.25 mL/min.
5. Resuspend each sample in 290 μL of 0.1% formic acid and adjust the pH at less than or equal to 3.0.
6. Set the flow rate at 0.100 mL/min.
7. Fractionate the samples with a continuous salt gradient from 10 to 500 mM of KCl for 60 min.
8. Collect 12 fractions in 3-min intervals between 300 and 500 mM of KCl (around 13–46 min).
9. Desalt the fractions by using RP-HPLC step gradient (see Note 11).
10. Speed vacuum dry samples and store at −80°C until further processing.

3.5.5. OFFGEL Fractionation

OFFGEL fractionation is a method based on the separation of peptides according to their global charge or isoelectric point (p*I*). The fractionation is performed on immobilized ampholyte gel strips. Collection of the samples after fractionation is made possible by positioning a well frame on top of the strip. After migration through the gel, peptides will stabilize at their respective p*I* and diffuse in the well. Afterward, collection of the solution contained in each well provides fractionated samples.

1. Resuspend the sample in 0.1% formic acid for a final concentration of 100 μg of peptide per 360 μL of solution (see Note 12), vortex, and centrifuge.
2. Prepare the OFFGEL stock solution by mixing 600 μL of ampholyte solution and 6 mL of 50% glycerol solution, and complete to 50 mL with mQ water.

3. Mix 480 μL of stock solution with 120 μL of water, to produce the rehydration solution.
4. Mix 360 μL of sample with 1,440 μL of OFFGEL stock solution, vortex, and centrifuge.
5. Set the OFFGEL system according to the protocol provided by the manufacturer (see Note 13).
6. Pipette 40 μL of rehydration solution in each well of the frame. Gently tap the tray on the bench to pull the solution down and incubate for 15 min at room temperature.
7. Pipette 145 μL of sample diluted in OFFGEL solution in each well (the last well may have less volume) and cover the well frame with the lid.
8. Put 2 electrode pads on each side of the strip (if necessary electrode pads have to be changed every 24 h).
9. Set the electrodes according to the manufacturer's recommendations (see Note 14).
10. Cover the row of the tray with mineral oil (400 μL on the anode side and 1,000 μL on the cathode side).
11. Start the preset program for OFFGEL fractionation of peptides with 12-well strips (called OG12-PE00), for a total of 20,000 kVh.
12. After migration, collect each fraction in properly labeled tubes (see Note 15). It is possible to optimize the peptide collection to extract remaining peptides entrapped in the gel and peptides stuck on the plastic well (see Note 16).

3.5.6. OFFGEL Fractions Clean-Up

Due to the presence of glycerol in the samples from OFFGEL, HPLC method to clean up samples is not recommended. Therefore, clean up the samples using PepClean™ C-18 spin columns. However, this clean-up method may also be used for SCX fractions.

1. Measure the volume of each sample and complete them with 1 part of sample buffer (20% ACN, 2% TFA) for 3 parts of sample. This brings the sample to the final composition of 5% ACN and 0.5% TFA.
2. Vortex and spin samples.
3. Tap columns on the bench to bring resin to the bottom of the tube.
4. Put 200 μL of activation solution (50% methanol) in the column and make sure rinsing the wall of the column with the solution.
5. Centrifuge at 1,500× *g* for 1 min at room temperature.
6. Add 200 μL of activation solution on top of the resin and centrifuge.

7. Equilibrate the resin by passing equilibration solution through twice (5% ACN 0.5% TFA).
8. Pipette the sample on top of the resin, place the column in a tube to collect the flow through (it is possible to use the tube containing the sample).
9. Centrifuge at 1,500× *g* for 1 min at room temperature.
10. Pass the sample a second time through the resin and preserve the flow through.
11. Wash the resin twice with 200 μL of wash solution (5% ACN, 0.5% TFA).
12. Elute retained peptides by passing twice through the resin 20 μL of elution solution (70% ACN).
13. Speed vacuum dry all fractions. The dried samples can be stored at −80°C until further experiments.

3.6. Protein Identification by Tandem Mass Spectrometry After 2-DE DIGE

Protein spots with the highest intensity variation between conditions and with a *p*-value less than 0.05 are selected for further analysis by mass spectrometry, in order to identify the proteins related to those spots.

3.6.1. First Step Is Cutting the Spots of Interest

Different methods are used: razor blade, dedicated spot-cutter, or robots. We used an Ettan spot picker (GE Healthcare).

3.6.2. Proteins Entrapped in the Gel Pieces Are Digested, Mainly Using Trypsin

Steps for in-gel digest are presented below.

1. Destain gel pieces with three washing steps (30 min each): (1) 50% ACN, (2) 50% ACN, 50 mM NH_4HCO_3, and (3) 50% ACN, 10 mM NH_4HCO_3 (volume depends on the size of the gel pieces).
2. Speed vacuum dry gel pieces.
3. Add few microliters of trypsin solution (trypsin (Promega) resuspended at 0.1 μg/μL) and make sure that the gel pieces are completely soaked (no white part left in the middle of the pieces).
4. After 10 min of incubation, add 10 μL of 10 mM NH_4HCO_3.
5. Incubate overnight at 37°C.
6. Extract generated peptides by adding 200 μL of extraction solution (60% ACN, 0.1% TFA) and agitate for 60 min.
7. Collect peptide solution in a new tube and speed vacuum dry them.

3.6.3. After In-Gel Digest Clean-Up Peptide Samples Using Zip-Tip® (See Note 17)

1. Resuspend dried peptides in 10 μL of 0.5% TFA, vortex, and spin.
2. Activate Zip-Tip® resin by pipetting 10 μL of wetting solution (ACN).

3. Equilibrate resin with 10 μL of equilibration/wash solution (0.1% TFA).
4. Aspirate and dispense the samples several times (10–20 times to ensure optimal binding).
5. Wash with 10 μL of equilibration/wash solution.
6. Aspirate and dispense elution solution (50% ACN, 0.1% TFA) several times (≈10 times).
7. Speed vacuum dry the samples.

3.6.4. Mass Spectrometry Analysis for Protein Identification with Ion Trap

[We use LCQDecaPlus (ThermoElectron).]

1. Prior to mass spectrometry analysis, resuspend samples in 6 μL of 0.1% formic acid. Inject 5 μL of resuspended peptides in the LC-MS/MS system.
2. Peptides are fractionated on a microcapillary RP-C18 column (New Objective).
3. Buffers for fractionation gradient:
   - Buffer A: water:ACN:TFA (98:2:0.1).
   - Buffer B: water:ACN:TFA (2:98:0.1).

   Fractionation gradient (percentage corresponding to buffer B)
   - 0–5 min 5–15%
   - 5–52 min 15–35%
   - 52–54 min 35–80%
   - 54–64 min 80%
   - 64–65 min 80–5%
   - 65–72 min 5%.
4. Electrospray ionization (ESI) should be set in a nanospray configuration for an optimal MS and MS/MS analysis (For more detailed information, cf. Profiling of HIV protein in CSF).

3.7. MS/MS Analysis for Identification and Quantification with iTRAQ®-Labeled Peptides

iTRAQ® method involves chemical labeling of amino functions of peptides (N-terminal portions and lysines) with a molecule able to be dissociated with high collision energy. Thus, during MS/MS analysis of labeled peptides, a part of the labeling molecule, called tag, will be found in the MS/MS spectrum. Figure 3a shows an example of MS/MS analysis of a labeled peptide using MALDI-TOF/TOF mass spectrometry. Four masses (114.07, 115.07, 116.07, and 117.07) correspond to the four tags. A zoom of this mass range (see Fig. 3b) indicates that these four peaks are in a clear area of the spectrum, and they are placed between two immonium ions without any interference, *m/z* 112.06 corresponding to arginine and *m/z* 120.05 corresponding to phenylalanine. The area under curve of these peaks will be used to relatively quantify the protein expression.

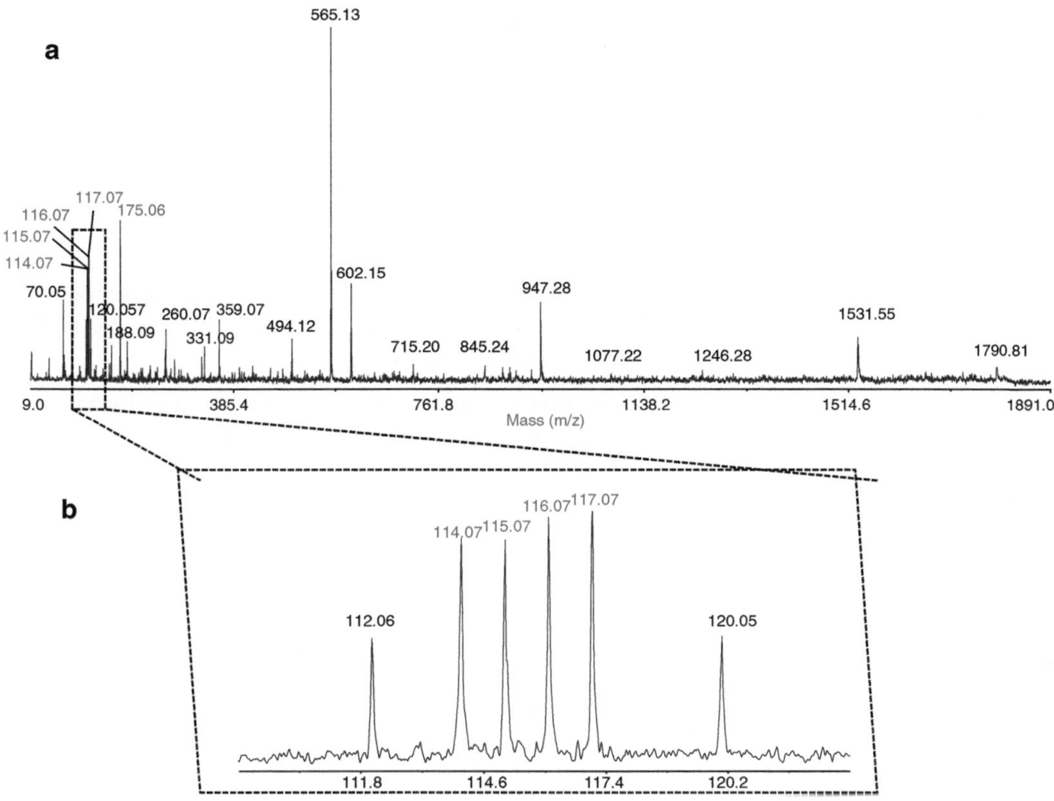

Fig. 3. Representative MS/MS spectrum of an iTRAQ-labeled peptide (1,790.81 *m/z*) (**a**). This MS/MS spectrum shows the fragment ion peaks and in the square the peaks corresponding to the iTRAQ tags (114.07, 115.07, 116.07, and 117.07). A zoom of this spectrum area (**b**) shows that the peaks corresponding to the tags are resolutive. Furthermore, no other ions may interfere with the quantitation due to an alteration of the signal. Only two ions are reported in this region of the spectrum, 112 corresponding to the immonium ion of arginine and 120 the immonium ion of phenylalanine.

MS and MS/MS analysis of iTRAQ®-labeled samples may be performed with MALDI and ESI ion sources. Furthermore, prior to MS analysis, peptide samples should undergo the second dimension of fractionation with RP chromatography. Then, for MALDI analysis, we use off-line LC-MS system. Chromatography is performed with a system separated from the mass spectrometer, and peptides after fractionation are collected by fractions on the MALDI plate where they are mixed with the matrix. On the contrary, ESI source is directly connected to the liquid chromatography column. This on-line system allows mass measurement immediately after RP separation.

3.7.1. LC-MALDI MS/MS Analysis

MALDI-TOF/TOF mass spectrometry is actually the most adapted method for iTRAQ®-labeled sample analysis. Indeed, peptide fragmentation performed by TOF/TOF analyzer provides enough energy to fragment the peptide and the tag molecule. Finally, the information brought by MALDI-TOF/TOF MS/MS analysis allows, with the same fragmentation spectrum, protein identification and peptide quantification using tag peaks (see Fig. 3).

1. Resuspend each fraction from SCX or OFFGEL in 11 μL of 0.1% TFA.
2. Using a Tempo™ LC MALDI spotting system inject 10 μL of resuspended fractions in the system and perform the peptide separation according to the following steps:
   - Load the samples in a ProteoCol™ C18 trap cartridge (Michrom Biosources) and wash the sample for 10 min at a flow rate of 9 μL/min
   - Peptide separation is performed using two buffers:
     - Buffer A: water:ACN:TFA (98:2:0.1).
     - Buffer B: water:ACN:TFA (2:98:0.1).
   - Separation gradient (percentage of buffer B): (see Note 18)
     - 0–5 min 5–15%
     - 5–52 min 15–35%
     - 52–54 min 35–80%
     - 54–64 min 80%
     - 64–65 min 80–5%
     - 65–72 min 5%
3. Eluted peptides are mixed with a matrix solution (5 mg/mL CHCA in 75% ACN and 0.1% TFA) at a flow rate of 1 μL/min (use a Harvard Apparatus syringe pump). Collect fractions every 30 s and apply 2.8 kV to the plate.
4. Acquisition of data for each fraction may be realized on a 4800 MALDI-TOF/TOF mass spectrometer (ABI).
   - MS spectra are acquired from 800 to 4,000 *m/z*, for a total of 1,000 laser shots, with laser intensity fixed for all analyses.
   - MS/MS analyses were realized with 2-kV collision energy with CID gas, and metastable ions were suppressed.

3.7.2. LC-ESI-MS/MS Analysis

In general, ESI ion sources are coupled with a linear ion trap mass spectrometry analyzer, applying collisional energy to perform the fragmentation of the peptides. Linear ion trap is for identification, but not quantification, due to the "one-third" rule. The latter limits the lowest fragments produced to 1/3 the precursor mass, which is above the tag molecule (≈100 Da). On the contrary, methods of fragmentation such as electron transfer dissociation (ETD), pulsed Q dissociation (PQD), and higher energy C-trap dissociation (HCD) are designed to detect low mass information (100–400 *m/z*). Nevertheless, these types of dissociation are not always as informative to allow protein identification.

With the LTQ-Orbitrap (see Note 19), the following strategy is used. For each MS scan, four of the most abundant precursor peaks are selected for fragmentation first with a low-energy fragmentation method (CID) and then with a high-energy collision (HCD). Finally, to complete the protein identification and quantification, both spectra from the same parent ion are combined (see Note 20).

1. Resuspend each fraction from SCX or OFFGEL with 6 μL of 0.1% formic acid.
2. Using a Eksigent nanoLC system, inject 5 μL of resuspended fractions in the system and perform the peptide separation according to the following steps:
   - Load the samples in a ProteoCol™ C18 trap cartridge (Michrom Biosources, Auburn, CA) and wash the sample for at least 30 min at a flow rate of 10 μL/min.
   - Peptide separation is performed using two buffers:
     - Buffer A: water:ACN:formic acid (98:2:0.1).
     - Buffer B: water:ACN:formic acid (2:98:0.1).
   - Separation gradient (percentage of buffer B):
     - 0–65 min 2–65%
     - 66–70 min 65–70%
     - 70–71 min 70–98%
     - 71–110 min 98%
3. The LTQ-Orbitrap XL mass spectrometer (Thermo-Fisher Scientific) has to be operated in positive ionization mode. Perform the MS survey scan in the FT cell recording a window between 300 and 1,500 *m/z*.
   - Set the resolution to 60,000.
   - Set the automatic gain control (AGC) to 500,000 ions.
   - Put the *m/z* values triggering MS/MS on an exclusion list for 240 s.
   - The minimum MS signal for triggering MS/MS needs to be 500.
4. Apply a maximum of four MS/MS experiments for the four most intense signals exceeding a minimum signal of 500 in the survey scans.
   - Set the resolution for MS/MS to 7,500 for HCD; CID is detected in LTQ.
   - Set the first mass of HCD fixed at 100 *m/z*.
   - Set the isolation window at 4 *m/z*.

- Make a target value of 300,000 ions, with maximum accumulation time of 1 s.
- Perform fragmentation with normalized collision energy of 35% and an activation time of 40 ms for the CID, and a collision energy of 55% and an activation time of 40 ms for the HCD.

3.8. Database Searches and Quantitation

3.8.1. Protein Identification for 2-DE DIGE Method

1. Generated spectra have to be compared to a database in order to identify proteins; we used the Sequest search engine (BioWorks 3.2 software, ThermoElectron Inc.). In the TurboSEQUEST search parameters for Dta Generation, use the following settings:
   - Threshold of 50,000.
   - Precursor mass tolerance of 1.4 AMU.
   - Group scan of 1.
   - Minimum group count of 1.
   - Minimum ion count of 15.
2. In the TurboSEQUEST search parameters for a human database, use the following:
   - Peptide tolerance 2.0 AMU.
   - Fragment ions tolerance 1.0 AMU with charge state set on "Auto."
   - Allow carbamydomethylation and oxidated methionine in searches.

3.8.2. Protein Identification and Quantitation for iTRAQ® Method

Protein identification based on peptide profiling may be performed with several software using different algorithms. Among these software programs, ProteinPilot™ is dedicated to the identification of proteins and quantification of iTRAQ®-labeled peptides. This software reduces redundancy of identification by gathering proteins with high sequence overlap and evaluating the probability to identify one protein among a group. Then, ProteinPilot™ determines the relative expression of each peptide by comparison of the area under the curve of each iTRAQ® tag. Finally, this software allows the selection of the tag used as a reference after the identification and the determination of the ratio between tags.

1. Protein identification and quantitation with ProteinPilot™ are processed using Paragon Method. Set the identification options as follows:
   - Sample type: iTRAQ 4plex (peptide labeled).
   - Cys alkylation: Iodoacetamide.
   - Digestion: Trypsin.
   - Instrument: 4800.

- Special factors: none is selected.
- Species: *Homo sapiens*.
- Allow quantification by checking Quantitate Checkbox.
- All ID Focus are allowed (Biological modifications, Amino acid substitutions, and User-defined modifications).
- Database: use NCBInr (less than 6-month old).
- Search effort: select Thorough ID.
- Detection Protein Threshold [Unused ProtScore (conf)]: 0.10 (20%) (see Note 21).

3.9. Statistical Analysis

Usually, the first step in the validation of differentially expressed proteins is based on thorough statistical analysis. This part is performed by professional statisticians who apply various statistical and mathematical models (12, 13).

3.10. Validation

Last but not least, validation in proteomic profiling of CSF constitutes a challenging problem. It is necessary to validate biomarker candidates based on orthogonal, analytical, biological, and/or functional assays. Multiple reaction monitoring, ELISA, and quantitative Western blot are a few examples of such assays. It is also required that validation methods are standardized so that they are reproducible in independent laboratories by independent personnel (14, 15). Lack of such uniformly accepted procedures and protocols minimizes the impact of proteomic profiling studies of CSF. Considering the substantial variability of protein levels within any population of patients, validation assays have to be performed using relatively large cohorts of patients to verify which biomarkers are most promising and should be further investigated (10, 16).

4. Notes

1. Gel spot excision can be performed manually using razor blade or dedicated commercially available spot-cutter. Nevertheless, this may lead to human mistakes.
2. If a small blood vessel is punctured during spinal tap, CSF will be contaminated with plasma proteins and therefore will be disqualified from further studies. In some instances, contamination with blood is not visible, but if it is, hemoglobin will be detected during the profiling experiment.
3. Regardless of which profiling method is used, samples need to be cleaned up to remove contaminations. This is usually accomplished by protein precipitation. Generic methods such as acetone or methanol precipitation are frequently used, but

commercial kits are also available (e.g., Clean-up kit from GE Healthcare). There are two reasons for this step: (1) To remove salts and other nonproteinaceous contaminations that can interfere with further steps. (2) Proteins need to be reconstituted in a specific buffer (buffer exchange) for either enzymatic digestion or labeling.

4. IEF step is carried out at a constant temperature of 20°C with a total 45 kVh (500 V for 0.5 kVh, gradient to 1,000 V for 0.8 kVh, gradient to 8,000 V for 13.5 kVh, and 8,000 V for 30.2 kVh). For the preparative run after saturation labeling, 65 kVh is applied (500 V for 0.5 kVh, gradient to 1,000 V for 0.8 kVh, gradient to 8,000 V for 13.5 kVh, and 8,000 V for 50.2 kVh).
5. Second dimension is carried out with an Ettan Daltsix Electrophoresis System™ (GE Healthcare Piscataway, NJ) at 20°C. Power is held constant to 5 W per gel for the initial 30 min and 17 W per gel until bromophenol blue reached the bottom of the gel.
6. Signals are collected at excitation wavelength for Cy2-, Cy3-, and Cy5-labeled samples at 488, 520, and 620 nm, respectively, using Typhoon 9410 Variable Mode Imager (GE Healthcare Piscataway, NJ). Gels are scanned at 100 μm resolution and analyzed using DeCyder 2D 6.5™ software (GE Healthcare Piscataway, NJ). For visualization of protein spots used in identification after minimal labeling, gels are stained with Sypro Ruby and scanned at 400 nm on Typhoon 9410. Spots selected for protein identification after DeCyder analysis are picked from gels by automatic Ettan™ Spot Picker (GE Healthcare, Piscataway) with a 2-mm diameter tip.
7. iTRAQ® kits have been developed for different kinds of samples for the 4-plex version of the iTRAQ: there is regular iTRAQ® 4-plex for all samples and iTRAQ® 4-plex developed for plasma samples, which may be used for CSF samples. iTRAQ® 8-plex may be used for all samples without distinction between samples.
8. After a test of the iTRAQ® protocol with our sample, we determined that ethanol precipitation was more adapted to our sample. We also found that in order to ensure a better reproducibility, vortexing, spinning, and pipetting steps have to be as constant as possible and realized by only one operator to reduce variability induced by those steps.
9. iTRAQ®-labeled peptides generate a brown pellet with a viscous aspect.
10. To force liquids to pass through the resin in the cartridge, place a 2-mL rubber bulb on top of the cartridge. Make sure that the bottom of the bulb is carefully sealed to the cartridge and push

the liquid through the resin, but do not suck liquid back into the cartridge. Then hold the bulb squeezed and slowly slide the bulb off horizontally to remove it.

11. HPLC step gradient to desalt after SCX: load samples onto the C-18 HPLC column (Jupiter 4 u Proteo 90A, 50×4.6 mm, 4 μm, Phenomenex, Torrance, CA) in 95% mobile phase A (0.1% TFA) for 5 min. Elute peptides using 50% mobile phase B (0.1% TFA in acetonitrile).

12. 12-Well IPG strips may contain up to 100 μg of digested proteins according to the manufacturer. In order to fill each well with the optimal volume, 360 μL of sample has to be mixed with 1,440 μL of OFFGEL stock solution. Thus, make the calculation have the right concentration of peptides in order to make this solution.

13. Using tweezers remove the protective plastic film of the IPG OFFGEL strip and insert the strip in the OFFGEL tray with the gel facing up. The strip has to be placed all the way on the left (anode side). Place the frame against the mechanical stop on the left and clip it in the tray.

14. First place the fix electrode (anode side). Introduce tabs of the electrode in the slots of the tray and rotate down the electrode. Make sure that the electrode touches the electrode pads. Slowly insert the moveable electrode on the cathode side of the tray. The moveable electrode should touch the well frame. If the electrode pads are removed while the moveable electrode is placed, remove the electrode, replace the electrode pads, and insert the moveable electrode.

15. By convention, fraction 1 is positioned on the anode (+) and fraction 12 is on the cathode (−).

16. It is possible to optimize the peptide collection. After the first collection, pipette 200 μL of 50% methanol and 1% formic acid solution in each well. Incubate for 20 min at room temperature. Collect each fraction in a new tube, labeled "Rinsate." Speed vacuum dry the second collection. Transfer the first collection into the tube labeled "Rinsate" after drying.

17. Each tip will be in contact with different solutions and the samples involving cross-contamination. We recommend to prepare aliquots of each solution in Eppendorf tubes (1.5 mL) to limit the volume of contaminated solution.

18. Elution of peptides may be monitored with a UV cell at 214 nm absorbance.

19. In this chapter, we describe our method adapted for the LTQ-Orbitrap. Nevertheless, it is also possible to use the same strategy with a Q-Trap 4000 (AB Sciex).

20. Combining spectra for database searching is necessary only when using a search engine other than Proteome Discoverer (Thermo Scientific).

21. We use a low value of Detection Protein Threshold to evaluate the efficiency of the fragmentation and determine the loss of information due to a lack of "good" fragmentation.

References

1. Tsunoda, A., H. Mitsuoka, H. Bandai, T. Endo, H. Arai, and K. Sato. 2002. Intracranial cerebrospinal fluid measurement studies in suspected idiopathic normal pressure hydrocephalus, secondary normal pressure hydrocephalus, and brain atrophy. *J Neurol Neurosurg Psychiatry* **73**, 552–555.
2. Silverberg, G. D., M. Mayo, T. Saul, J. Carvalho, and D. McGuire. 2004. Novel ventriculo-peritoneal shunt in Alzheimer's disease cerebrospinal fluid biomarkers. *Expert Rev Neurother* **4**, 97–107.
3. Wittke, S., H. Mischak, M. Walden, W. Kolch, T. Radler, and K. Wiedemann. 2005. Discovery of biomarkers in human urine and cerebrospinal fluid by capillary electrophoresis coupled to mass spectrometry: towards new diagnostic and therapeutic approaches. *Electrophoresis* **26**, 1476–1487.
4. Ogata, Y., M. C. Charlesworth, and D. C. Muddiman. 2005. Evaluation of protein depletion methods for the analysis of total-, phospho- and glycoproteins in lumbar cerebrospinal fluid. *J Proteome Res* **4**, 837–845.
5. Johnston, I., and C. Teo. 2000. Disorders of CSF hydrodynamics. *Childs Nerv Syst* **16**, 776–799.
6. Maurer, M. H. 2010. Proteomics of brain extracellular fluid (ECF) and cerebrospinal fluid (CSF). *Mass Spectrom Rev.* **29**, 17–28.
7. Yuan, X., and D. M. Desiderio. 2003. Proteomics analysis of phosphotyrosyl-proteins in human lumbar cerebrospinal fluid. *J Proteome Res* **2**, 476–487.
8. Yuan, X., and D. M. Desiderio. 2005. Proteomics analysis of human cerebrospinal fluid. *J Chromatogr B Analyt Technol Biomed Life Sci* **815**, 179–189.
9. Rozek, W., M. Ricardo-Dukelow, S. Holloway, H. E. Gendelman, V. Wojna, L. Melendez, and P. Ciborowski. 2007. Cerebrospinal fluid proteomic profiling of HIV-1-infected patients with cognitive impairment. *J Proteome Res* **6**, 4189–99.
10. Ciborowski, P. 2009. Biomarkers of HIV-1-associated neurocognitive disorders: challenges of proteomic approaches. *Biomark Med* **3**, 771–785.
11. Lucchi, G., J. B. Hendra, D. Pecqueur, and P. Ducoroy. 2007. [Towards a standardization of the tools for the studies of clinical proteomics]. *Med Sci (Paris)* Spec No **1**, 19–22.
12. Oberg, A. L., D. W. Mahoney, J. E. Eckel-Passow, C. J. Malone, R. D. Wolfinger, E. G. Hill, L. T. Cooper, O. K. Onuma, C. Spiro, T. M. Therneau, and H. R. Bergen, 3 rd. 2008. Statistical analysis of relative labeled mass spectrometry data from complex samples using ANOVA. *J Proteome Res* 7, 225–233.
13. Storey, J. D., and R. Tibshirani. 2003. Statistical significance for genomewide studies. *Proc Natl Acad Sci U S A* **100**, 9440–9445.
14. Oh, J. H., S. Pan, J. Zhang, and J. Gao. 2010. MSQ: a tool for quantification of proteomics data generated by a liquid chromatography/matrix-assisted laser desorption/ionization time-of-flight tandem mass spectrometry based targeted quantitative proteomics platform. *Rapid Commun Mass Spectrom* **24**, 403–408.
15. Pan, S., J. Rush, E. R. Peskind, D. Galasko, K. Chung, J. Quinn, J. Jankovic, J. B. Leverenz, C. Zabetian, C. Pan, Y. Wang, J. H. Oh, J. Gao, J. Zhang, T. Montine, and J. Zhang. 2008. Application of targeted quantitative proteomics analysis in human cerebrospinal fluid using a liquid chromatography matrix-assisted laser desorption/ionization time-of-flight tandem mass spectrometer (LC MALDI TOF/TOF) platform. *J Proteome Res* 7, 720–730.
16. Silberring, J., and P. Ciborowski. Biomarker discovery and clinical proteomics. *Trends Analyt Chem* **29**, 128.

Chapter 14

New Nanotechnology Applications in Single Cell Analysis: Why and How?

Gradimir N. Misevic, Gerard BenAssayag, Bernard Rasser, Philippe Sales, Jovana Simic-Krstic, Nikola Misevic, and Octavian Popescu

Abstract

Cell heterogeneity is intrinsic to both genetically programmed differentiation and stochastic/epigenetic variation. The scientific and technological challenge is to quantitatively study the nature and extent of the heterogeneity of populations of cells. In order to reach this goal, scientists need to measure the complete molecular content "omes" of single cells. This is not achievable by the classical approach implementing chromatography/electrophoresis microsystem separation and analysis by mass spectroscopy and nuclear magnetic resonance, due to their lack of high throughput technology and their lack of sufficiently high detection sensitivity. Here we propose that single cell "omic" measurements can be realized with the new interdisciplinary nanotechnology combining physics and chemistry with biology. Our nanoscience approach is based on the implementation of novel Nano in Micro Array (NiMA) biosensor chip platform that can analyze the complete proteome and glycome by means of accommodating up to 2,500 different cell samplings (positioned in microwells) and 250,000 probe markers (positioned in nanowells) per chip. Using a combination of chemical, mechanical, optical, and electrical detection with Secondary Ion Mass Spectrometry (SIMS) and by Scanning Probe Microscopy (SPM), we can quantify all biomolecules approaching detection of a single protein molecule. The gained knowledge about molecular heterogeneity quantified at the single molecular level within each individual cell in the form of "omes" (proteome, glycome, transcriptome, and metabolome) is fundamental to our understanding of causative relationships and formulations of natural laws. This will be a large step toward comprehension and prediction of processes associated with complex living systems like the evolution of life, embryogenesis and morphogenesis, immunity, adaptivity, self–nonself recognition, neural plasticity, and learning, as well as driving forces leading to diseased states of living organisms such as cancer, bacterial, and viral disease, neurodegenerative disorders, and autoimmunity.

Key words: Nanosciences and technologies, Single cell analysis, High throughput analysis, Proteome, Multicellularity, Scanning Probe Microscopy, Secondary Ion Mass Spectrometry, Focus Ion Beam, Scanning Electron Microscopy, Nano-in-Micro Arrays (NiMA)

Yannis Karamanos (ed.), *Expression Profiling in Neuroscience*, Neuromethods, vol. 64,
DOI 10.1007/978-1-61779-448-3_14, © Springer Science+Business Media, LLC 2012

1. Introduction

The essential goal of fundamental research is to find causative relationships between and within matter/energy composites of any existing or imaginable system. The emerging knowledge should establish the laws which would be the basis for predicting future events occurring in such a natural and/or artificial system independent of its complexity. Physics and chemistry together with mathematics have reached considerably high levels of knowledge substantiating many natural laws and thus can be described as well-developed theoretical and experimental branches of natural sciences. Contrary, biology was slowly emerging as a quantitative science. Only during the second half of the past century has some progress been made with the interdisciplinary molecular biology approach integrating chemistry and physics with biology. In the past two decades, a new extended multidisciplinary opportunity unified into nano- and microscience and their derivative technologies started to appear on the horizon. The steadily increasing number of research publications was noticeable and was often followed by oversimplified and incorrect and sensational press reports. As usual for the emerging science disciplines, also noticeable is the large number of overinterpreted, and/or obviously wrong results with unprofessional approaches that have been published in a variety of scientific journals. Such embarrassing pollutions of this young field seem to be a general common nominator in the modern sciences because it is in part a consequence of the growing restrictive funding policy of the fundamental and novel research with invitations for "quick and dirty" results. In spite of the increasing negative selection in funding policy for fundamental research also a large number of excellent and original researches has been conducted and published in all fields including the nano-bio-sciences and technologies. This indeed supports the historical facts that innovations are unstoppable because the driving power of the curious human mind is constantly searching for causalities and predictions of the events while asking: Why? How? What? When? Where?

Nano-bio-science and associated technologies are rapidly developing interdisciplinary field operating as the connecting catalytic interface between physics and chemistry on one side and biology on the other side. The goal of the nano-research is to discover new laws by implementing classical and modern natural sciences physics and chemistry into a study of complex biological system at the level of the single biomolecular measurements. Such a nanoscience approach shall ideally enable high throughput and high sensitivity detection down to a single molecule, without using any labeling and modification, thus providing complete "omic" quantification with structure to function-related studies for all molecular components present in each single cell. The generated output will be the complete information about phenotype and genotype at the single cell level for all cells within each organism and will provide a solid basis for formulation of causative relationships and prediction of events in living complex systems.

2. Why Single Cell Analysis?

Although the heterogeneity of populations of cells is believed to play a major role in the resistance of cancer cells to treatment and of bacteria to antibiotics, the true extent, nature, and causes of heterogeneity are largely unknown. Genetically programmed differentiation and stochastic/epigenetic variation are the molecular basis for cell heterogeneity which is fundamental to all life processes. They can be clearly associated with (1) evolution of complex multicellular organisms, (2) fundamental life processes of cell recognition and adhesion, (3) fertilization, (4) embryogenesis and morphogenesis, (5) immunity and self–nonself recognition, (6) adaptivity, learning and neural plasticity, and (7) many pathological processes such as in cancer, bacterial and viral disease, neurodegenerative disorders and autoimmunity.

To establish the causative relationship between gene expression and epigenetic factors, biologists must study the "omes" (proteome, glycome, transcriptome, genome, and metabolome) of the individual cell if they are ever to fully understand cells and hence develop new therapies to diseases. Only limited knowledge about such molecular compositions, "omes" of individual cells is available because technological limitations are intrinsic to macro- and microscale approaches commonly used to study "omes." The choice is either to study population of cells, thus leading to averaging of the cellular compositions, or study only a few abundant proteins in one cell. Such averaging results when cell population is studied in tissue biopsies or body fluids may give wrong picture dramatically effecting diagnosis of cancer in the early stages. Imagine the case of high expression of specific cancer marker in only few precancerous cells in microbiopsy of 1 mm^3 or a 1 μl drop of body fluid taken from the patients. This sample contains over one million cells. Using classical proteome analysis either based on array approach or separation techniques followed by mass spectrometry, the highly abundant cancer marker present only in a few precancerous cells out of million cells will be invisible due to labeling and/or detection and/or dilutions in extracts from such cell population during proteome analysis. A similar problem arises in clinicians trying to diagnose early infection stages by bacteria resistant to antibiotics. Also early diagnosis of neurodegenerative diseases is very problematic due to lack of appropriate technologies capable of performing quantitative single cell high throughput "omic" analysis. Therefore, scientists need multidisciplinary nanoscale quantitative technology to gain information about single cell "omics." To reach this goal, a breakthrough in high throughput nanotechnology using novel biosensor chip design and analytical detection of single molecules without labeling and modification is needed.

Many bacterial species are confronted with the problem of a changing environment in which different and sometimes incompatible strategies are required for survival and for growth. This is resolved at the population level by the generation of both phenotypic diversity (1–3) and genetic diversity (4, 5). In generating phenotypic diversity, transcription factors are clearly important and, since they are often present in small numbers, there is a role to be played by stochastic noise (6–8); another role, it has been argued, is played by the cell cycle (9) which leads to the presence of two or more chemically identical chromosomes within the same cytoplasm that spontaneously adopt complementary patterns of expression to equip the future daughter cells for life in different environments (10). In generating genetic diversity, there is an interesting phenomenon whereby certain individuals in a stressed population undergo mutations in proofreading genes that lead to a high level of mutations; when these unhealthy individuals lyse fragments of their DNA might be taken up and used by other individuals which may thus acquire a beneficial mutation (5, 11). Due to the lack of appropriate nanoseparation and detection technology, most of the single cell studies were limited to transcriptome rather than to proteome (12). A holy grail for many biologists would be to study heterogeneity by studying the complete phenotypes of individual cells (13, 14). This entails obtaining the entire proteomes of individual cells, and this is the technical challenge that must eventually be met.

3. How to Perform Single Cell Analysis Using New Nanotechnology?

The state-of-the-art of how "omic" analysis is performed can be considered from either the point of view of the tool developer/manufacturer or from that of the clinical user aspect, although of course they should ideally be very closely aligned. Two different approaches for proteome analysis are currently employed. The first one is based on array technology where the proteome is screened via detection of selective antigen target interaction with antibody probes. The second one is based on the electrophoresis and/or chromatography separation of proteins and/or their fragmentation using differences in their physicochemical properties followed by Electrospray Ionization Mass Spectrometry (ESI-MS) and/or Matrix Assisted Laser Desorption/Ionization Time of Flight Mass Spectrometry (MALDI-TOF MS). Parallel sampling capabilities of protein array technology has superior throughput to more serial processing of samples by separation techniques. However, the advantage of separation approaches is that they offer direct identification of proteins, rather than the indirect protein target binding to antibody probe. Classification of different types of proteomic approaches can be also made according to different principles (1) top–bottom or bottom–up, (2) parallel or serial, (3) biochemical bases of protein identification, (4) methodological sequence of processes, (5) detection methods, (6) historical developments, (7) usage of different marketing brands equipment. However, for the developer the main question remains on how to design and manufacture the appropriate tools keeping advantages and overcoming limitations of each of the proteome approaches and considering clinical user requests for highly sensitive, quantitative, and complete proteome analyzes in single cells with high throughput capability measurements of many samples screened for many markers on a single chip.

Both classical "omic" approaches now also use microsystems array and microseparation (chromatography and electrophoresis) and analysis by mass spectroscopy and nuclear magnetic resonance. Unfortunately, these technologies still lack the capability to (1) engage quantitative and complete single cell "omic" analysis, (2) high throughput capacity, and (3) sufficiently high detection sensitivity. To increase sensitivity, commonly labeling and modification of cellular components are performed. Therefore, analysis of only cell populations, and not single cells, can be performed. At least one million cells are needed and only detections of more abundant molecules can be achieved. Due to this physical limitation, the expectation of classical technological approach for the complete and quantitative single cell "omic" analysis is a science fiction (Table 1).

So far, more studies of heterogeneity have been performed on bacteria than on eukaryotic cells using chemical fluorescent probe labeling and modifications of complex and unknown mixtures of cellular protein extracts followed by their separation in capillaries and optical detections (15). Alternative is to genetically modify proteins with green fluorescent protein. The main problem of these approaches is the obviously nonuniform labeling and/or not labeling of different species of proteins which prevents quantitative evaluations and comparisons between different samples. Such studies unfortunately generated inconclusive results making more damage than use for our knowledge. The other approach for single cell analysis was to use capillary electrophoresis or microfluidics with even less sensitive mass spectrometric and electrochemical detection (16, 17).

To try to evaluate the heterogeneity in populations of cells, biologists sometimes amplify the mRNA extracted from a few cells but this is fraught with artifacts and the principal interpretation made of mRNA is indirect insofar as it is in terms of the proteins encoded—and there is no way to amplify proteins. While valuable

Table 1
Array proteomic technologies Nano-in-Micro Arrays versus Micro Arrays

	Nano-in-Micro Array	APiX™ Protein Micro Array
Detection principal	Direct SIMS	Indirect Fluorescence
Minimal detectible concentration	aM	pM
Minimal detectible amount	70 zg 1 protein molecule	100 fg 1,500,000 protein molecules
Minimal usable volume	Picoliters	Microliters
Number of samples/chip	2,500 50×50 array	8
Number of markers/sample	2.5×10^5 500×500 array	46 64×64 array
Minimal spot size of probe	100 nm	300 μm

Calculation was made assuming that protein has 40 kDa molecular weight

techniques exist to allow the study of the heterogeneity of a single species of mRNA or protein in a population of cells, the state of the remaining thousands of mRNAs or proteins is unknown: an integrative picture of the phenotype of the individual cells in the population can only be guessed at. This obliges biologists to work with mRNA or proteins extracted from large numbers of cells (and, hence, to perform global analyses on the "average cell," which may not even exist) or to conduct investigations at the level of expression of a single gene or group of genes or the generation of a particular phenotype. Leading specialists are manipulating single cells, extracting their contents, and are attempting to separate them. Development of new miniaturization and detection nanotechnology is, therefore, timely.

Classical fluorescent optical microscopic analysis and flow cytometry analysis, using fluorescent-labeled or color-marked antibodies against cellular markers reveals very useful information about the presence and/or spatial position of specific antigens in the single cells with the possibility of relatively rapidly examining many cells. Unfortunately, these very useful and excellent techniques provide semiquantitative information about the presence of only a few glycoproteins and not the whole proteome and glycome.

Innovative combination of knowledge and technologies of nano- and microsciences shall provide realistic possibilities to develop novel approaches for implementation of complete and quantitative single cell high throughput "omics" analysis. Obtained knowledge will help the scientist to understand and formulate quantitative natural laws about the living biological system.

This will enable prediction of the future events in living organisms which are particularly relevant to medicinal diagnostics and prognosis in diseases, such as cancer, infection and autoimmunity, and neurodegeneration. Here, we describe how nanoscience and technologies can be implemented in single cell analysis.

We have developed a concept of 1CellOmes project which is a high-throughput single cell "omic" analysis by chemical measurements using Secondary Ion Mass Spectrometry (SIMS) and mechanical, optical, and electrical Scanning Probe Microscopy (SPM) measurements on novel Nano in Micro Arrays (NiMA) biosensor chips (18–21). Therefore, the characteristic of 1CellOmes concept is to possess two units (a) the biosensor part based on antibody and phage display library probes functionalized on NiMA chips and (b) analytic readout of NiMA by improved SMP nanoprecision and molecular sensitivity without any labeling. Objectives of 1CellOmes is to develop NiMA SPM high-throughput platform tool to study the proteome and glycome of a single cells in order to (a) fully understand genotype and dynamics of individual cell phenotype, (b) improve diagnostic, prognostic, and therapy monitoring to diseases, (c) advance in methods and instrumentation for the full omic fingerprint analyses of single cells, (d) improve data and specimen (biobanks) standardization, acquisition, and analysis for phenotyping applied to bioinformatics and systems biology "omics." 1CellOmes is a high-throughput biosensor chip that can analyze the complete proteome and glycome by means of accommodating up to 2,500 different cell samplings (positioned in microwells) and 250,000 probe markers (positioned in nanowells) per chip with SIMS and SPM mechanical, optical, and electrical sensitivity approaching single protein molecule detection (Figs. 1 and 2). NiMA biosensor is prepared by Focus Ion Beam (FIB) engraving and Scanning Electron Microscopy (SEM) monitoring or by e-beam lithography (Fig. 1).

SIMS is 3D chemical microscopy (22). It is one of the emerging applications in biology. Tissues, single cells, extracellular matrices, polymer films, and individual biomolecules can be imaged at present with a lateral resolution of tens of nm and depth resolution of 1 nm using the latest generation of instrumentation. SIMS imaging is based on an ion source emitting primary ions in the energy range from few to tens keV. The primary ion beam (which determines lateral resolution of the SIMS chemical microscope) can be at present focused down to a diameter of 50 nm (in the case of CAMECA NanoSIMS 50 equipped with Cs source; http://www.cameca.com) on the surface of a solid sample in ultrahigh vacuum. The sample surface is scanned with the primary beam at the desired speed. The primary ions at the impact point sputter the most superficial molecular layer by the complex and not yet completely understood process of atomization and ionization. The degree and mechanism of sputtering depend on the chemical

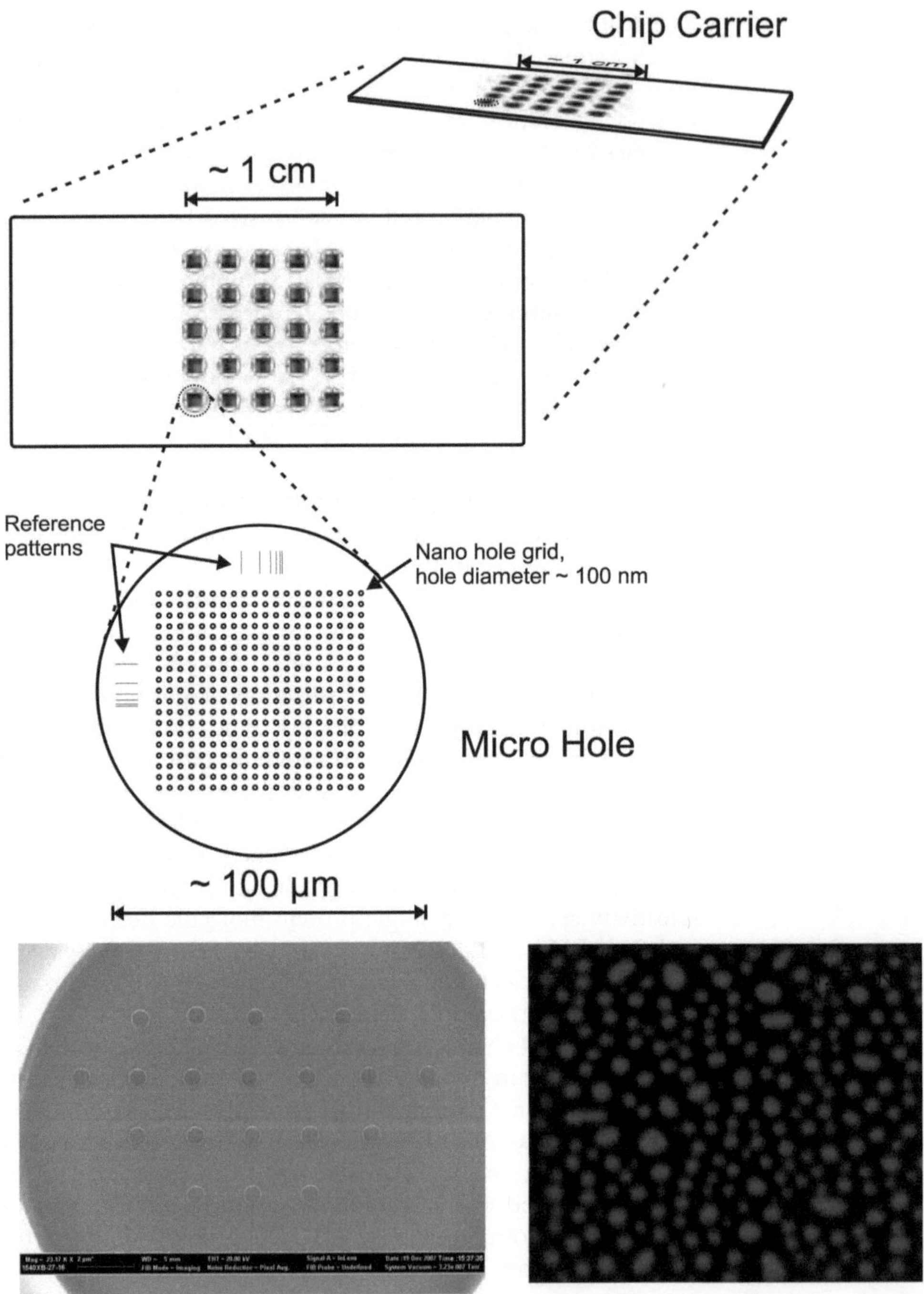

Fig. 1. Design of Nano-in-Micro Array (NiMA). (a) *top figure*: schematic presentation, as an example, only few micro-wells are shown which size is drawn in great disproportion to the chip size, (b) *bottom left*: scanning electron microscope image of preliminary prototype of NiMA manufactured by FIB (micro-well with 20 μm diameter with array of nano-wells with 600 nm diameter and pitch of 2.4 μm), (c) *bottom right*: nanoSIMS image of protein Bovine Serum Albumin small clusters (50 nm) in polyvinylalcohol are (total: 30,000 counts of CN-, total area: $4 \times 4\ \mu m^2$).

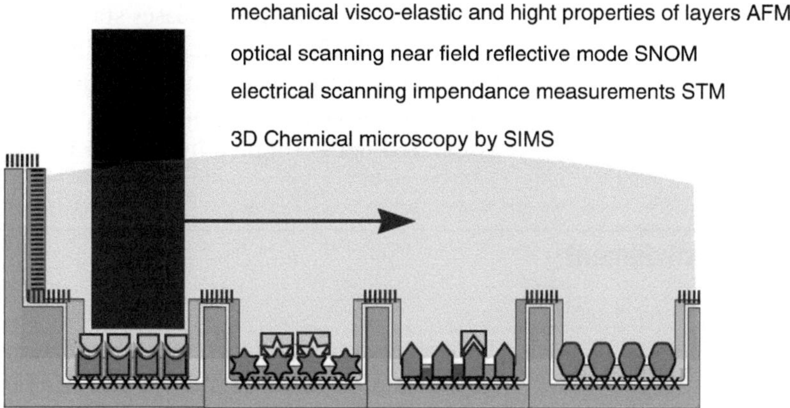

Fig. 2. Schematic presentation of NiMA bio-sensor chip analysis by SPM.

nature of the primary ion source, their energy, and chemical nature of the analyzed sample. Secondary ions are collected and separated in a high performance magnetic sector mass spectrometer and detected with a photomultiplier (Fig. 2). The value of counts collected for each selected secondary ion mass at each scanned point provides information for the construction of a series of images, each representing the selected ion mass of a constituent chemical element (Fig. 1). The focus diameter and the precision of the primary ion beam at each scanned point determine the lateral resolution of the chemical image. Since the scanning process can be repeated several times, three dimensional, but destructive, analyses of the samples can be achieved with a depth resolution of 1 nm. Typically, SIMS allows the detection of a few ppm of any isotope of most elements at a mass resolution up to four decimal places. Our preliminary results have shown that using combined SIMS and SPM chemical, electrical, optical, and mechanical nano-sensors detection and identification of single protein molecules can be achieved (Figs. 1 and 2; Table 1).

Nano in Micro Array will enable to quantify all proteins, glycans, and glycoconjugates by Scanning Probe Microscopy and SIMS in a single cell. These will provide high throughput technology platform for single cell "omes" analysis relevant to dynamics of single cell phenotype/genotype. It will also have a large impact on systems biology because it will catalyze progress for further developing research tools for system biology that will enhance data generation and improve data and specimen of "omes" biobanks with implementation for real standardization, acquisition, and analysis in bioinformatics. Furthermore, NiMA single cell "omics" will have significant impact in medicine and in the biotechnology by improved diagnostics such as cancer, infection, and neurodegenerative diseases, monitoring of the disease progression and therapeutic effects, as well as the discovery of new cellular markers associated with different cell differentiation and diseases states.

In conclusion, our nanoscience approach with high throughput quantitative single cell "omic" single analysis shall provide a significant step toward better comprehension and prediction of processes associated with complex living systems.

Acknowledgment

This work was mainly financed by private GNM funds supporting the realization of original ideas and fundamental sciences and technologies and thus contributing to the prevention of the cancerous spread of mediocrity and dogmatism.

References

1. Balaban NQ, Merrin J, Chait R, Kowalik L, Leibler S (2004) Bacterial persistence as a phenotypic switch. *Science* 305:1622–1625
2. Booth IR (2002). Stress and the single cell: intrapopulation diversity is a mechanism to ensure survival upon exposure to stress. *International J Food Microbiol*, 78:19–30
3. Tolker-Nielsen T, Holmstrom K, Boe L, Molin S (1998). Non-genetic population heterogeneity studied by in situ polymerase chain reaction. *Mol Microbiol*, 27:1099–1105
4. Gabrovsky V, Yamamoto ML, Miller JH (2005). Mutator effects in Escherichia coli caused by the expression of specific foreign genes. *J Bacteriol* 187:5044–5048
5. Matic I, Taddei F, Radman M (2004). Survival versus maintenance of genetic stability: a conflict of priorities during stress. *Res Microbiol* 155:337–341
6. Cai L, Friedman N, Xie XS (2006). Stochastic protein expression in individual cells at the single molecule level. *Nature* 440(7082):358–62
7. Elowitz MB, Levine AJ, Siggia ED, Swain PS (2002). Stochastic gene expression in a single cell. *Science* 297:1183–1186
8. Thattai M, van Oudenaarden A (2004). Stochastic gene expression in fluctuating environments. *Genetics* 167:523–530
9. Segre D, Ben-Eli D, Lancet D (2000). Compositional genomes: prebiotic information transfer in mutually catalytic noncovalent assemblies. *Proc Nat Acad Sci* USA 97:4112–4117
10. Minsky A, Shimoni E, Frenkiel-Krispin D (2002). Stress, order and survival. *Nat Rev Mol Cell Biol* 3:50–60
11. Miller JH (1996) Spontaneous mutators in bacteria: insights into pathways of mutagenesis and repair. *Ann Rev Microbiol* 50:625–643
12. Levsky MJ, Singer RH (2003) Gene expression and the myth of average cell. *Trends Cell Biol.* 13:4–6
13. Hu S, Michels D, Fazal MA, Ratisoontorn C, Cunningham ML, Dovichi NJ (2004) Capillary sieving electrophoresis/micellar electrokinetic capillary chromatography for two-dimensional protein fingerprinting of single mammalian cells. *Anal Chem* 76 1:4044–4049
14. Gutstein HB, Morri JS, Annangudi SP, , Sweedler JV (2008). Microproteomics: Analysis of protein diversity in small samples. *Mass Spectrom Rev* 27:316–330
15. Dovichi NJ (2003) Cell cycle-dependent protein fingerprint from a single cancer cell: image cytometry coupled with single-cell capillary sieving electrophoresis. *Anal Chem*, 75:3495–3501

16. Woods LA, Roddy TP, Ewing AG. (2004). Capillary electrophoresis of single mammalian cells. *Electrophoresis*, 25:1181–1187
17. Vyawahare S, Griffiths AD, Merten CA (2010). Miniaturization and Parallelization of Biological and Chemical Assays in Microfluidic Devices. *Chem Biol*, 17:1052–1065
18. Dammer U, Popescu O, Wagner P, Anselmetti D, Güntherodt HJ, Misevic GN (1995). Binding strength between cell adhesion proteoglycans measured by atomic force microscopy. *Science* 267:1173–1175
19. Misevic G., Ripoll C, Norris V (2008). US Patent Application
20. Misevic G., Rasser B, Norris V, Dérue C, Gibouin D, Lefebvre F, Verdus MC, Delaune A, Legent G., Ripoll C. Chemical microscopy of biological samples by dynamic mode secondary ion mass spectrometry (2009). *Methods Mol Biol* 522:163–173
21. Misevic G., Karamanos Y, Misevic N, (2009). Atomic force microscopy measurements of intermolecular binding forces (2009). *Methods Mol Biol*, 522:143–150
22. Slodzian, G., Hillionb, F., Stadermannc, F.J., Zinner, E. (2004). QSA influences on isotopic ratio measurements. *Applied Surface Science*, 874:231–232

Index

A

Acetylation 38, 41–44, 47
Alzheimer's disease 101, 140, 154, 162, 210, 213, 214
Amplification 3, 5–6, 10–13, 16, 21–31, 39, 51–53, 58, 73, 83, 84, 275
Amyotrophic lateral sclerosis (ALS) 101, 154
Arterioles 63, 65, 70, 72

B

Bioinformatics 89, 90, 103, 120, 122, 129–133, 194–195, 277, 279
Biomarker 47, 139, 201, 214, 247, 267
BLAST search 239
Blood–brain barrier (BBB) 63–74, 154, 161–178, 225, 245
Blotting 43–45, 137, 151, 152, 188
Brain 1–19, 22, 35–39, 41–44, 49–51, 53, 64, 65, 67–69, 71–74, 92, 101, 102, 108, 139–157, 161–178, 185, 188, 189, 191–193, 195, 198, 200, 202, 203, 207, 209–214, 245, 247
Brain disorders 35–60

C

Capillaries 15, 22, 39, 63, 66, 70, 71, 74, 102, 124, 161–178, 190, 275
Captured microvessels 70, 71
Cell isolation 103–104, 108–109
Central nervous system (CNS) 63, 64, 101, 140, 155, 161, 162, 204–207, 209–213, 247
Cerebellum 1
Cerebral endothelial cells 161
Cerebrospinal fluid (CSF) 140, 151, 225–243, 245–270
Cryosections 50–51
Cytoskeleton 168, 170, 208

D

Databases 16, 47, 90, 93, 95, 113, 114, 121, 122, 129–131, 137, 150, 165, 173–176, 201, 229, 237, 239–240, 243, 246, 266–267, 270
Data mining 114
Differential expression 43, 49, 84, 85, 91, 92, 94, 95, 97–99, 102, 136
DNA microarrays 1–19, 39–40, 48, 49, 214
DNA modifications 38, 41–43
Dopaminergic 36, 213, 214

E

Electroendosmosis 169
Endothelial cells 63–74, 154, 161–178
Enzyme-linked immunosorbent assay (ELISA) 43, 58, 60, 267
Epifluorescence 64, 70, 71
Epigenetics 41, 42, 44, 273
Epilepsy 92, 97
Expression profiling 2–3, 21–31, 34–60, 77–86, 89–99, 102, 114
Expression proteomics 122

F

Focus ion beam (FIB) 277, 278
Functional genomics 87

H

High throughput analysis 77, 120
Histofluorescence 22
Hybridization 2, 3, 5–6, 10–13, 16, 17, 21, 26, 29–31, 39, 40, 42, 43, 51–52, 78

I

Immunoblotting 44
Immunofluorescence 64, 65
Internet 30, 92, 113–114
In vitro transcription (IVT) 10, 23–26, 29, 39, 40, 51
Iodoacetamide 105, 164, 165, 169, 171, 172, 177, 228, 235, 249, 250, 256, 257, 266
*I*sobaric *t*ags for *r*elative and *a*bsolute *q*uantification (iTRAQ) 119–137, 148, 155, 248–250, 257–258, 262–268

L

Laser capture microdissection (LCM) 25–27, 63–74, 140
Lou Gehrig's disease .... 101 *See also* ALS

Yannis Karamanos (ed.), *Expression Profiling in Neuroscience*, Neuromethods, vol. 64,
DOI 10.1007/978-1-61779-448-3,

M

MALDI-MSI 182–215
Mammalian target of rapamycin (mTOR) 121, 122
Mass spectrometry........ 46, 102, 103, 111, 120–122, 124, 128–129, 131, 140, 142, 162, 165, 169, 171–176, 181, 182, 191–193, 196, 202, 228–229, 234–242, 246, 248–250, 255, 257, 258, 261–264, 273, 274
Medulloblastoma........ 1–2
Microarray........ 1–19, 22, 23, 30, 31, 39–42, 47–49, 51, 52, 59–60, 78, 79, 84–85, 102, 114, 119, 214
Multicellularity........ 273
Myelin........ 114, 121

N

Nanochips 4
Nano-in-micro arrays (NiMA)........ 276–279
Nano-sciences and technologies........ 271–280
Neural progenitor cells 101
Neural proliferation........ 1, 2
Neural stem cells 101–116
Neurodegenerative diseases 101, 139, 201, 212, 214, 273, 279
Neuroproteomics........ 120, 121, 139, 140, 142
Neuroscience 21–31, 91, 139, 154, 181–216

O

Oligodendrocyte progenitor cell (OPC)........ 121, 122, 131, 133, 137

P

Parallel sequencing 78
Parkinson's disease (PD)........ 101, 140, 154, 213, 214
PCR 23, 25, 26, 28, 30, 39–45, 52, 53, 56–59, 63–74, 78, 79, 236
Polyadenylylation........ 81
Polyvinylidene difluoride (PVDF)........ 43, 44, 145, 151
Post-mortem human brain 37, 43, 140
Proteome 37, 43–49, 120, 124, 131, 144, 161–178, 213, 227, 239, 240, 247, 248, 252, 270, 273–277
Proteomics........ 45, 47, 102, 113, 114, 122, 131, 136, 139–157, 162, 182, 195, 196, 201, 214, 237
Proteomics profiling........ 119–137
Pulsed SILAC (pSILAC)........ 141, 154
Purkinje neurons........ 1

R

Rapamycin........ 121, 122, 130–133, 137
Reference gel 112, 115, 116, 171, 177
Repositories........ 22, 89, 90
RNA amplification........ 5, 21–31, 39, 51
RNA integrity number (RIN) 15, 16, 22, 39
RNA isolation 3–9, 14, 65

S

Scanning electron microscopy (SEM) 277
Scanning probe microscopy (SPM)........ 277, 279
Secondary ion mass spectrometry (SIMS)........ 182, 196, 202, 276–279
Shotgun proteomics........ 47, 131, 136
Single cell analysis 271–280
Smooth muscle actin 65, 67
Sonic hedgehog (Shh) 1–2
Stable isotope labeling of amino acids in cell (SILAC) 120, 141–145, 147–148, 150, 154, 156
Streptavidin 43, 57, 59
Syncytium........ 64

T

3'-Tag digital gene expression profiling 77–86
Terminal continuation (TC)........ 21–31
Thermocycler 5, 10, 16
Tissue extraction........ 142, 143, 145, 146
Transcriptome 36–43, 77, 78, 86, 273, 274
Transcriptome profiling........ 77, 86
Tumorigenesis 1
Two-dimensional gel electrophoresis (2DGE)........ 45, 102–103, 105–107, 109–111, 162, 234, 248, 249

V

Validation 145, 150–154, 156, 173–176, 178, 198, 201, 246, 267
Venules 63, 70, 71, 74

W

Western blot........ 43–45, 48, 137, 142, 150, 151, 153, 154, 267
World Wide Web........ 89

MIX
Papier aus verantwortungsvollen Quellen
Paper from responsible sources
FSC® C105338

If you have any concerns about our products,
you can contact us on
ProductSafety@springernature.com

In case Publisher is established outside the EU,
the EU authorized representative is:
Springer Nature Customer Service Center GmbH
Europaplatz 3, 69115 Heidelberg, Germany

Printed by Libri Plureos GmbH
in Hamburg, Germany